32nd Winter Workshop on Nuclear Dynamics (WWND 2016)

AA005652

Journal of Physics: Conference Series Volume 736

Guadeloupe
28 February – 5 March 2016

Editors:

Joerg Aichelin
Rene Bellwied
Elena Bratkovskaya
Carsten Greiner
Christina Markert

ISBN: 978-1-5108-2922-0
ISSN: 1742-6588

Printed from e-media with permission by:

Curran Associates, Inc.
57 Morehouse Lane
Red Hook, NY 12571

Some format issues inherent in the e-media version may also appear in this print version.

Printed by Curran Associates, Inc. (2016)

For permission requests, please contact the Institute of Physics
at the address below.

Institute of Physics
Dirac House, Temple Back
Bristol BS1 6BE UK

Phone: 44 1 17 929 7481
Fax: 44 1 17 920 0979

techtracking@iop.org

Additional copies of this publication are available from:

Curran Associates, Inc.
57 Morehouse Lane
Red Hook, NY 12571 USA
Phone: 845-758-0400
Fax: 845-758-2633
Email: curran@proceedings.com
Web: www.proceedings.com

Table of contents

Volume 736

32nd Winter Workshop on Nuclear Dynamics
28 February to 5 March 2016, Guadeloupe

Accepted papers received: 26 July 2016
Published online: 24 August 2016

Preface

Papers

Journal of Physics: Conference Series **736** (2016) 011001 doi:10.1088/1742-6596/736/1/011001

PREFACE

These are the proceedings of the 32nd Winter Workshop on Nuclear Dynamics, which was held in Guadeloupe, Caribbean, in February/March 2016. As in previous years, the unique character of this conference series has allowed us to bring together nuclear scientists with very different interests to discuss recent progress and scientific achievements.

Out of the 44 contributions at WWND 2016 we have selected these 27 representative manuscripts. The topics capture the range of theoretical and experimental advances in our field, from very exciting results from the RHIC beam energy scan program to the small systems measurements at the LHC. On the theory side this system size dependence led to a detailed evaluation of the initial conditions and subsequent hydrodynamical or transport approaches, as well as a profound discussion on how small a QGP system can be. These results were complemented by the most recent continuum extrapolated data from lattice in order to model the complete evolution of the relativistic heavy ion system.

These proceedings of the 32nd Winter Workshop on Nuclear Dynamics again provide a snapshot of the status of the field. The articles, many of which were written by some of the most promising young scientists in the field, are documenting the excitement and achievements that are characteristic for modern day nuclear science.

Joerg Aichelin, SUBATECH, Ecole des Mines, France

Rene Bellwied, University of Houston, USA

Elena Bratkovskaya, FIAS, Johann Wolfgang Goethe-Universitaet, Germany

Carsten Greiner, Johann Wolfgang Goethe-Universitaet, Germany

Christina Markert, University of Texas at Austin, USA

Journal of Physics: Conference Series **736** (2016) 011002 doi:10.1088/1742-6596/736/1/011002

Peer review statement

All papers published in this volume of *Journal of Physics: Conference Series* have been peer reviewed through processes administered by the proceedings Editors. Reviews were conducted by expert referees to the professional and scientific standards expected of a proceedings journal published by IOP Publishing.

Journal of Physics: Conference Series **736** (2016) 012001 doi:10.1088/1742-6596/736/1/012001

Recent results on lattice QCD thermodynamics

Claudia Ratti

Department of Physics, University of Houston, Houston, TX 77204, USA

E-mail: cratti@uh.edu

Abstract. I review recent results on QCD thermodynamics from lattice simulations. In particular, I will focus on the QCD equation of state at zero and finite chemical potential, the curvature of the phase diagram and fluctuations of conserved charges. The latter are compared to experimental data, to the purpose of extracting the chemical freeze-out temperature and chemical potential from first principles.

1. Introduction

The deconfined phase of Quantum Chromodynamics (QCD), the Quark-Gluon Plasma (QGP) can be created in the laboratory in heavy ion collision experiments currently running at RHIC (Brookhaven National Laboratory) and at the LHC (CERN). These experiments allow to explore the QCD phase diagram and to extract the properties of this new phase of matter. The QGP turns out to be the most ideal fluid ever observed, which led to the idea that the system created in the collisions is strongly interacting and cannot be studied by means of perturbative techniques.

Simulations on a discretized grid in space-time (lattice) represent the most reliable tool to solve the theory of strong interactions in its non-perturbative regime. The precision achieved by recent lattice QCD simulations allows a quantitative comparison between theory and experiment for the first time in our field. This is due to a steady improvement in both numerical algorithm and computational resources, as well as in our physical understanding which manifests itself in physical techniques (e.g. the Wilson-flow scale setting introduced in Ref. [1]).

The low temperature phase of strongly interacting matter can be understood in terms of a non-interacting gas of hadrons and resonances: the Hadron Resonance Gas (HRG) model provides a very good description of most thermodynamic observables. In the infinite temperature limit, matter behaves again as a non-interacting gas, but in this case the degrees of freedom are quarks and gluons. Reducing the temperature, it is possible to treat the theory as weakly coupled and to calculate thermodynamic quantities perturbatively. Resummation techniques improve the convergence of the perturbative series and bring the agreement with lattice QCD results down to $T \simeq 2.5T_c$. The temperature range between these two limiting situations corresponds to the non-perturbative regime of strongly interacting matter: the coupling constant becomes large and it is not possible to treat the system perturbatively. This is the region which is accessible experimentally, and in which QCD can only be solved numerically by means of lattice simulations.

2. QCD equation of state

The QCD equation of state at zero chemical potential is now known with high accuracy: continuum extrapolated results are available for the pressure, energy density, entropy density, interaction measure and speed of sound of a system of 2+1 quark flavors with physical quark masses [2, 3, 4]. A selection of these results is shown in Fig. 1, in which curves from the WB collaboration obtained with the 2stout action (gray) are compared to those from the HotQCD collaboration obtained with the HISQ action (colored). The agreement between the two sets of curves is a fundamental test of the validity of the lattice approach to solve QCD.

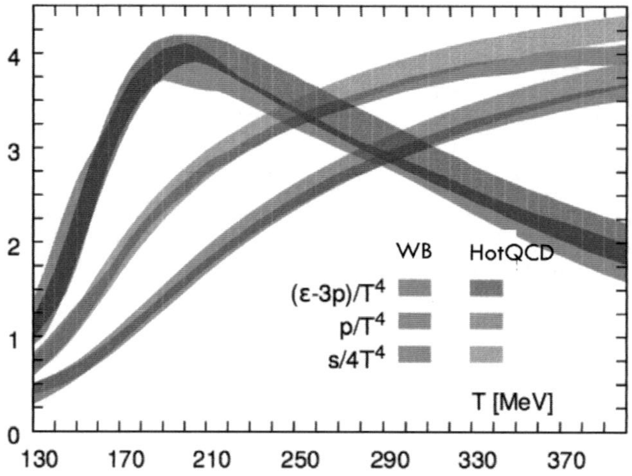

Figure 1. Scaled pressure, entropy density and interaction measure as functions of the temperature. The gray curves have been obtained by the WB collaboration with the 2stout action [2, 3], the colored ones by the HotQCD collaboration with the HISQ action [4].

Simulations on the lattice at finite density are not possible at the moment, due to the sign problem. Nevertheless, it is possible to explore a low-density region of the phase diagram by means of alternative methods. Here I will focus on the Taylor expansion of thermodynamic observables around $\mu_B = 0$ [5, 6] (which can be considered as a truncated version of the multiparameter reweighting technique [7]) and on the analytical continuation from imaginary chemical potentials [8, 9, 10].

Thermodynamic quantities can be expanded in Taylor series around $\mu_B = 0$; for example for the pressure such an expansion reads:

$$\frac{p(\mu_B)}{T^4} = c_0(T) + c_2(T)\left(\frac{\mu_B}{T}\right)^2 + c_4(T)\left(\frac{\mu_B}{T}\right)^4 + c_6(T)\left(\frac{\mu_B}{T}\right)^6 + \mathcal{O}(\mu_B^8). \tag{1}$$

The coefficients $c_i(T)$ in the above expansion can be calculated on the lattice. Several results exist in the literature: $c_2 ... c_6$ have been calculated long ago on coarse lattices and with heavier than physical quark masses [11]; continuum extrapolated results for c_2 at the physical quark mass were published for the first time in Ref. [12]; results for c_4 at finite lattice spacing were shown in Ref. [13]. In Fig. 2 I show preliminary results for c_2, c_4 and c_6 in the continuum limit and for physical quark masses. The systematic error is not included in the plot. The results have been obtained with the method of analytical continuation from imaginary chemical potential [14, 15]. It is important to notice that the chemical potentials μ_B, μ_S and μ_Q are not independent of each other: μ_S and μ_Q are both functions of T and μ_B, such that the following experimental conditions are satisfied:

$$\rho_S = 0, \qquad \rho_Q = 0.4\rho_B, \tag{2}$$

where ρ_S, ρ_Q and ρ_B are the densities of strangeness, charge and baryonic number, respectively. The coefficients $c_2 ... c_6$ in Eq. (1) are full derivatives with respect to μ_B/T, namely they take into account the dependence of μ_S and μ_Q on μ_B and T.

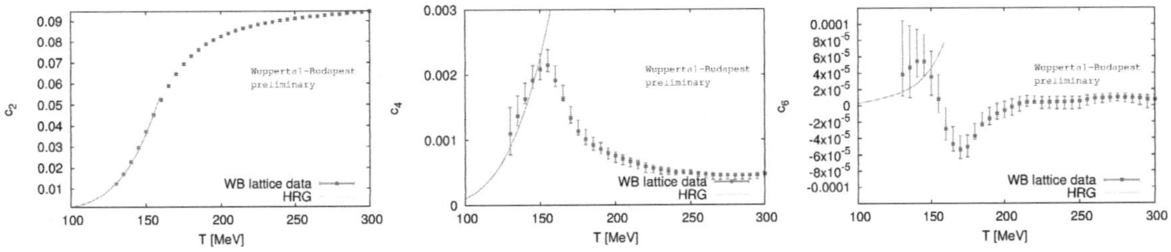

Figure 2. Preliminary results for the Taylor coefficients c_2, c_4 and c_6 as functions of the temperature from the WB collaboration, obtained from imaginary μ_B simulations. The data are continuum extrapolated; the error-bars are only statistical: the systematics of the μ_B fitting are not included [14, 15].

3. The QCD phase diagram

Due to the sign problem, only a small portion of the QCD phase diagram can be investigated by means of lattice QCD calculations at the moment. It was unambiguously shown [16] that at $\mu_B = 0$ the phase transition is an analytical crossover, which takes place over a broad range of temperatures around $T_c \simeq 155$ MeV [17, 18, 19, 20]. The transition temperature can be defined by locating the inflection point of relevant observables. By following the change in the position of this inflection point as the chemical potential is increased, it is possible to express the μ_B−dependence of T_c in the following way:

$$\frac{T_c(\mu_B)}{T_c(\mu_B = 0)} = 1 - \kappa \left(\frac{\mu_B}{T_c(\mu_B)}\right)^2 + \lambda \left(\frac{\mu_B}{T_c(\mu_B)}\right)^4 + \dots . \tag{3}$$

κ is the curvature of the phase diagram, and it was recently investigated by several groups. By looking at three different observables (chiral condensate, chiral susceptibility and strange quark susceptibility) the WB collaboration recently published a value of $\kappa = 0.0149 \pm 0.0021$ [21]; this has been obtained by fixing the strange quark chemical potential to impose strangeness neutrality. The phase diagram corresponding to this value of κ is showed in the left panel of Fig. 3, together with a compilation of freeze-out parameters obtained with different methods. The shaded band indicates the broadness of the QCD transition, while the dark blue band shows that it is possible to find a value for T_c with great accuracy when looking at a single observable (in this case the chiral condensate). Similar results have been obtained recently by two other groups: P. Cea *et al.* obtain a value of $\kappa = 0.020(4)$ by fixing $\mu_s = \mu_l$ [22], while Bonati *et al.* find $\kappa = 0.0135(20)$ both with $\mu_s = 0$ and $\mu_s = \mu_l$ [23]. In the right panel of Fig. 3, the phase diagram with the curvature from Ref. [22] is shown.

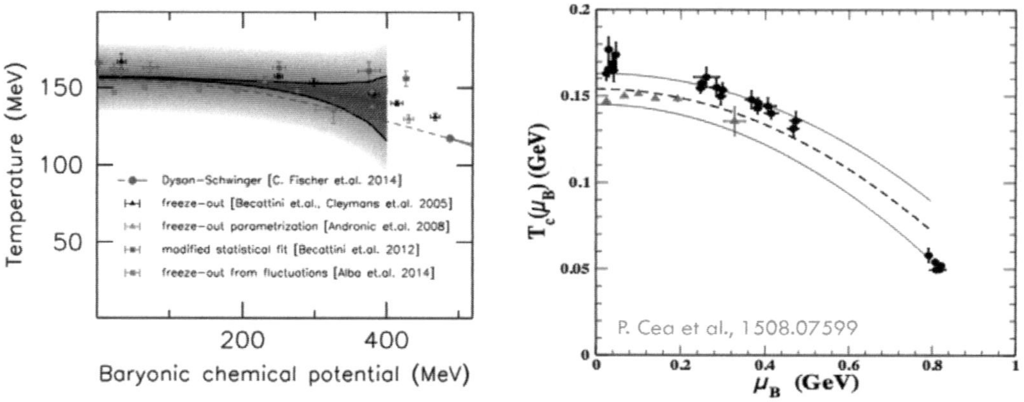

Figure 3. Left: The phase diagram based on the μ_B-dependent T_c from the chiral condensate, analytically continued from imaginary chemical potential [21]. The blue band indicates the width of the transition. The shaded black region shows the transition line obtained from the chiral condensate. The widening around 300 MeV is coming from the uncertainty of the curvature and from the contribution of higher order terms, thus the application range of the results is restricted to smaller values. We also show some selected non-lattice results: the Dyson-Schwinger result [24], and the freeze-out data of Refs. [25]-[31]. Right: analogous plot from Ref. [22].

4. Fluctuations of conserved charges and chemical freeze-out

Fluctuations of conserved charges have received increasing interest in recent years. The reason is that they can be measured in heavy ion collision experiments, and they can be simulated on the lattice. Thus, by comparing the experimental observables with the theoretical predictions, it is possible to gain information on the evolution of a heavy ion collision from first principles. Fluctuations have been proposed long ago as a signature for the QCD critical point [32, 6]. Besides, since they are fixed at the chemical freeze-out, they can be used to extract the freeze-out parameters [33, 35]. They are defined as derivatives of the pressure with respect to the chemical potential of the corresponding conserved charge:

$$\chi_{lmn}^{BQS}(T,\mu_B) = \frac{\partial^{l+m+n} p/T^4}{\partial(\mu_B/T)^l \partial(\mu_Q/T)^m \partial(\mu_S/T)^n} \tag{4}$$

and they can be related to the cumulants of the event by event distribution of the corresponding net conserved charge measured in experiments. In particular, it is possible to define volume-independent ratios which only depend on T and μ_B:

$$\begin{aligned} M/\sigma^2 &= \chi_1/\chi_2 \qquad & S\sigma^3/M = \chi_3/\chi_1 \\ S\sigma &= \chi_3/\chi_2 \qquad & \kappa\sigma^2 = \chi_4/\chi_2 \end{aligned} \tag{5}$$

where M is the mean of the experimental distribution, σ the variance, S the skewness and κ the kurtosis. The lattice QCD results for the above ratios will depend on T and μ_B: by comparing them to the experimental value it is therefore possible to extract the freeze-out temperature and chemical potential.

One has to keep in mind, that the outcome of lattice QCD simulations are thermal fluctuations. In order to compare them to experiment, one has to make sure that any source of non-thermal fluctuations which might affect the experimental data is under control. The system created in a heavy-ion collision is a canonical ensemble, in which net-charges such as electric charge, baryon number and strangeness are strictly conserved. However, due to experimental limitations in the detector acceptance, the system can be treated as a Grand Canonical Ensemble, with net-charges fluctuating event-by-event [36]. Sources of non-thermal fluctuations may include fluctuations in the initial volume, protons coming from the interaction with the beam pipe, effects due to cuts in rapidity and transverse momentum, and so on. A variety of effects has been identified and studied in the literature [37]-[43]. In 2014 the WB collaboration found an upper limit for the chemical freeze-out temperature of $T_{ch} \leq 151 \pm 4$ MeV. It was also pointed out that, by analyzing the fluctuations of electric charge and baryon number independently, consistency is found between the freeze-out chemical potentials corresponding to the highest RHIC energies [44, 45, 46]. Recently, the authors of Ref. [47] were able to obtain both the freeze-out temperature and the curvature of the freeze-out line. The value of the freeze-out temperature ($T_f = (147 \pm 2)$ MeV) is in agreement with the one obtained in Ref. [44]. The left panel of Fig. 4 shows the ratio of ratios of χ_1/χ_2 for electric charge and proton number used for this fit. Since consistently between the freeze-out lines for electric charge and baryon number was found, the WB collaboration performed a combined fit of χ_1/χ_2 for electric charge and proton number and found the freeze-out temperature and chemical potential for the highest RHIC energies. These preliminary results are shown in the right panel of Fig. 4, together with the isentropic lines which match the freeze-out data, the contours for constant mean/variance of net-electric charge from the lattice, and the results of a previous analysis based on the HRG model [31]. The WB results agree with the ones of Ref. [47] and with the HRG model ones. Unfortunately, fluctuation data are not yet available at the LHC. However the authors of Ref. [50], assuming that the lower moments follow a Skellam distribution, expressed the second moments in terms of the particle yields and compared the lattice results to the ALICE experimental data, finding a slightly higher freeze-out temperature than the ones obtained at RHIC. Recently, the study of fluctuations has been extended to very large temperatures [51, 52] to extract the onset of the HTL perturbative expansion [53, 54], which is found to be $T \simeq 250$ MeV. Fluctuations of conserved charges can also be used to identify the relevant

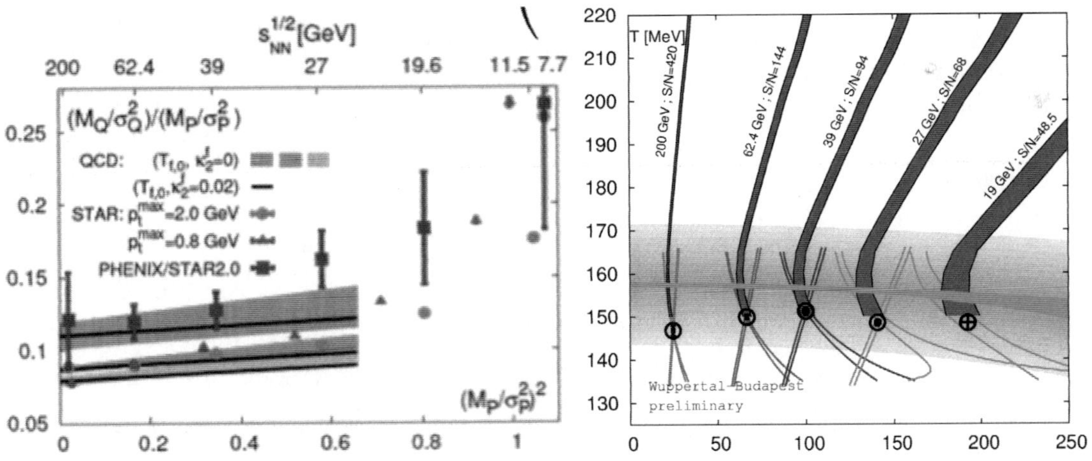

Figure 4. Left: From Ref. [47]: the ratio of ratios of χ_1/χ_2 for net electric charge and net-proton fluctuations measured by the STAR and PHENIX Collaborations [45, 46, 48, 49]. Right: Preliminary results of the WB collaboration. The colored full and dashed lines are the contours at constant mean/variance ratios of the net electric charge from lattice simulations. The contours that correspond to STAR data intersect in the freeze-out points of Ref. [31]. The red band is the QCD phase diagram shown in Fig. 3. Also shown are the isentropic contours that match the chemical freeze-out data [14, 15].

degrees of freedom in the vicinity of the QCD phase transition: by using linear combinations of charm fluctuations and correlations, in Ref. [55] it was pointed out that even if the onset of deconfinement for the charm quark takes place around $T \simeq 165$ MeV, it becomes the dominant degree of freedom in the thermodynamics of the charm sector only at $T \simeq 200$ MeV, while between these two temperatures the dominant contribution to the charmed pressure is given by open charm meson- and baryon-like excitations with integral baryonic charge.

In conclusion, the progress and the precision achieved by lattice QCD simulation in the last few years is really impressive. Precise results are available for QCD thermodynamics at zero and small chemical potentials, which allow a quantitative comparison with experimental results for the first time. This will enable us to achieve a comprehensive understanding of bulk properties of QCD matter from first principles.

Acknowledgments

This material is based upon work supported by the National Science Foundation under grant no. PHY-1513864, and by the U.S. Department of Energy, Office of Science, Office of Nuclear Physics, within the framework of the Beam Energy Scan Theory (BEST) Topical Collaboration. An award of computer time was provided by the INCITE program. This research used resources of the Argonne Leadership Computing Facility, which is a DOE Office of Science User Facility supported under Contract DE-AC02-06CH11357.

References

[1] S. Borsanyi *et al.*, JHEP **1209**, 010 (2012) doi:10.1007/JHEP09(2012)010 [arXiv:1203.4469 [hep-lat]].
[2] S. Borsanyi, G. Endrodi, Z. Fodor, A. Jakovac, S. D. Katz, S. Krieg, C. Ratti and K. K. Szabo, JHEP **1011** (2010) 077 doi:10.1007/JHEP11(2010)077 [arXiv:1007.2580 [hep-lat]].
[3] S. Borsanyi, Z. Fodor, C. Hoelbling, S. D. Katz, S. Krieg and K. K. Szabo, Phys. Lett. B **730** (2014) 99 doi:10.1016/j.physletb.2014.01.007 [arXiv:1309.5258 [hep-lat]].
[4] A. Bazavov *et al.* [HotQCD Collaboration], Phys. Rev. D **90** (2014) 094503 doi:10.1103/PhysRevD.90.094503 [arXiv:1407.6387 [hep-lat]].
[5] C. R. Allton, S. Ejiri, S. J. Hands, O. Kaczmarek, F. Karsch, E. Laermann, C. Schmidt and L. Scorzato, Phys. Rev. D **66** (2002) 074507 doi:10.1103/PhysRevD.66.074507 [hep-lat/0204010].
[6] R. V. Gavai and S. Gupta, Phys. Rev. D **78** (2008) 114503 doi:10.1103/PhysRevD.78.114503 [arXiv:0806.2233 [hep-lat]].
[7] Z. Fodor and S. D. Katz, Phys. Lett. B **534** (2002) 87 doi:10.1016/S0370-2693(02)01583-6 [hep-lat/0104001].
[8] P. de Forcrand and O. Philipsen, Nucl. Phys. B **642** (2002) 290 doi:10.1016/S0550-3213(02)00626-0 [hep-lat/0205016].
[9] M. D'Elia and M. P. Lombardo, Phys. Rev. D **67** (2003) 014505 doi:10.1103/PhysRevD.67.014505 [hep-lat/0209146].
[10] L. K. Wu, X. Q. Luo and H. S. Chen, Phys. Rev. D **76** (2007) 034505 doi:10.1103/PhysRevD.76.034505 [hep-lat/0611035].
[11] C. R. Allton, M. Doring, S. Ejiri, S. J. Hands, O. Kaczmarek, F. Karsch, E. Laermann and K. Redlich, Phys. Rev. D **71** (2005) 054508 doi:10.1103/PhysRevD.71.054508 [hep-lat/0501030].
[12] S. Borsanyi, G. Endrodi, Z. Fodor, S. D. Katz, S. Krieg, C. Ratti and K. K. Szabo, JHEP **1208** (2012) 053 doi:10.1007/JHEP08(2012)053 [arXiv:1204.6710 [hep-lat]].
[13] P. Hegde [BNL?Bielefeld?CCNU Collaboration], Nucl. Phys. A **931**, 851 (2014) doi:10.1016/j.nuclphysa.2014.08.089 [arXiv:1408.6305 [hep-lat]].
[14] R. Bellwied, S. Borsanyi, Z. Fodor, J. Gunther, S. D. Katz, A. Pasztor, C. Ratti and K. K. Szabo, arXiv:1601.00466 [nucl-th].
[15] C. Ratti, arXiv:1601.02367 [hep-lat].
[16] Y. Aoki, G. Endrodi, Z. Fodor, S. D. Katz and K. K. Szabo, Nature **443** (2006) 675 doi:10.1038/nature05120 [hep-lat/0611014].
[17] Y. Aoki, Z. Fodor, S. D. Katz and K. K. Szabo, Phys. Lett. B **643** (2006) 46 doi:10.1016/j.physletb.2006.10.021 [hep-lat/0609068].
[18] Y. Aoki, S. Borsanyi, S. Durr, Z. Fodor, S. D. Katz, S. Krieg and K. K. Szabo, JHEP **0906** (2009) 088 doi:10.1088/1126-6708/2009/06/088 [arXiv:0903.4155 [hep-lat]].

[19] S. Borsanyi *et al.* [Wuppertal-Budapest Collaboration], JHEP **1009** (2010) 073 doi:10.1007/JHEP09(2010)073 [arXiv:1005.3508 [hep-lat]].
[20] A. Bazavov *et al.*, Phys. Rev. D **85** (2012) 054503
[21] R. Bellwied, S. Borsanyi, Z. Fodor, J. Gnther, S. D. Katz, C. Ratti and K. K. Szabo, Phys. Lett. B **751** (2015) 559
[22] P. Cea, L. Cosmai and A. Papa, arXiv:1508.07599 [hep-lat].
[23] C. Bonati, M. D?Elia, M. Mariti, M. Mesiti, F. Negro and F. Sanfilippo, Phys. Rev. D **92** (2015) 5, 054503
[24] C. S. Fischer, J. Luecker and C. A. Welzbacher, Phys. Rev. D **90** (2014) 3, 034022
[25] J. Cleymans, B. Kampfer, M. Kaneta, S. Wheaton and N. Xu, Phys. Rev. C **71** (2005) 054901
[26] F. Becattini, J. Manninen and M. Gazdzicki, Phys. Rev. C **73** (2006) 044905
[27] A. Andronic, P. Braun-Munzinger and J. Stachel, Phys. Lett. B **673** (2009) 142 [Phys. Lett. B **678** (2009) 516]
[28] F. Becattini, M. Bleicher, T. Kollegger, T. Schuster, J. Steinheimer and R. Stock, Phys. Rev. Lett. **111** (2013) 082302
[29] J. Stachel, A. Andronic, P. Braun-Munzinger and K. Redlich, J. Phys. Conf. Ser. **509** (2014) 012019
[30] A. Andronic, Int. J. Mod. Phys. A **29** (2014) 1430047
[31] P. Alba, W. Alberico, R. Bellwied, M. Bluhm, V. Mantovani Sarti, M. Nahrgang and C. Ratti, Phys. Lett. B **738** (2014) 305
[32] M. A. Stephanov, K. Rajagopal and E. V. Shuryak, Phys. Rev. D **60** (1999) 114028
[33] F. Karsch, Central Eur. J. Phys. **10** (2012) 1234
[34] A. Bazavov *et al.*, Phys. Rev. Lett. **109** (2012) 192302
[35] S. Borsanyi, Z. Fodor, S. D. Katz, S. Krieg, C. Ratti and K. K. Szabo, Phys. Rev. Lett. **111** (2013) 062005
[36] V. Koch, Chapter of the book "Relativistic Heavy Ion Physics", R. Stock (Ed.), Springer, Heidelberg, 2010, p. 626-652. (Landolt-Boernstein New Series I, v. 23). (ISBN: 978-3-642-01538-0, 978-3-642-01539-7 (eBook)) [arXiv:0810.2520 [nucl-th]].
[37] V. Skokov, B. Friman and K. Redlich, Phys. Rev. C **88** (2013) 034911
[38] A. Bzdak and V. Koch, Phys. Rev. C **86** (2012) 044904
[39] S. Jeon and V. Koch, Phys. Rev. Lett. **85** (2000) 2076
[40] M. Kitazawa and M. Asakawa, Phys. Rev. C **85** (2012) 021901
[41] M. Nahrgang, M. Bluhm, P. Alba, R. Bellwied and C. Ratti, Eur. Phys. J. C **75** (2015) 12, 573
[42] P. Alba, R. Bellwied, M. Bluhm, V. M. Sarti, M. Nahrgang and C. Ratti, Phys. Rev. C **92** (2015) 6, 064910
[43] F. Becattini, M. Bleicher, T. Kollegger, T. Schuster, J. Steinheimer and R. Stock, Phys. Rev. Lett. **111** (2013) 082302
[44] S. Borsanyi, Z. Fodor, S. D. Katz, S. Krieg, C. Ratti and K. K. Szabo, Phys. Rev. Lett. **113** (2014) 052301
[45] L. Adamczyk *et al.* [STAR Collaboration], Phys. Rev. Lett. **112** (2014) 032302
[46] L. Adamczyk *et al.* [STAR Collaboration], Phys. Rev. Lett. **113** (2014) 092301
[47] A. Bazavov *et al.*, arXiv:1509.05786 [hep-lat].
[48] X. Luo [STAR Collaboration], PoS CPOD 2014, 019 (2015).
[49] A. Adare *et al.* [PHENIX Collaboration], arXiv:1506.07834 [nucl-ex].
[50] P. Braun-Munzinger, A. Kalweit, K. Redlich and J. Stachel, Phys. Lett. B **747** (2015) 292
[51] R. Bellwied, S. Borsanyi, Z. Fodor, S. D. Katz, A. Pasztor, C. Ratti and K. K. Szabo, Phys. Rev. D **92** (2015) 11, 114505
[52] H.-T. Ding, S. Mukherjee, H. Ohno, P. Petreczky and H.-P. Schadler, Phys. Rev. D **92** (2015) 7, 074043
[53] N. Haque, A. Bandyopadhyay, J. O. Andersen, M. G. Mustafa, M. Strickland and N. Su, JHEP **1405** (2014) 027
[54] S. Mogliacci, J. O. Andersen, M. Strickland, N. Su and A. Vuorinen, JHEP **1312** (2013) 055
[55] S. Mukherjee, P. Petreczky and S. Sharma, arXiv:1509.08887 [hep-lat].

Journal of Physics: Conference Series **736** (2016) 012002 doi:10.1088/1742-6596/736/1/012002

Energy Dependence of Moments of Net-Proton, Net-Kaon, and Net-Charge Multiplicity Distributions at STAR

Ji Xu (for the STAR Collaboration)

Key Laboratory of Quark & Lepton Physics (MOE) and Institute of Particle Physics, Central China Normal University, Wuhan 430079, China

Lawrence Berkeley National Laboratory, Berkeley, CA 94720, US

E-mail: `jixu@lbl.gov`

Abstract. One of the main goals of the RHIC Beam Energy Scan (BES) program is to study the QCD phase structure, which includes the search for the QCD critical point, over a wide range of chemical potential (μ_B). Theoretical calculations predict that fluctuations of conserved quantities, such as baryon number (B), charge (Q), and strangeness (S), are sensitive to the correlation length of the dynamical system. Experimentally, higher moments of multiplicity distributions have been utilized to search for the QCD critical point in heavy-ion collisions. In this paper, we report recent efficiency-corrected cumulants and cumulants ratios of the net-proton, net-kaon, and net-charge multiplicity distributions in Au+Au collisions at $\sqrt{s_{NN}}$ = 7.7, 11.5, 14.5, 19.6, 27, 39, 62.4, and 200 GeV collected in the years 2010, 2011, and 2014 with STAR at RHIC. The centrality and energy dependence of the cumulants up to the fourth order, as well as their ratios, are presented. Furthermore, the comparisons with baseline calculations (Poisson) and non-critical-point models (UrQMD) will also be discussed.

1. Introduction

Fluctuations of conserved quantities, such as net-baryon (ΔN_B), net-charge (ΔN_Q) and net-strangeness (ΔN_S), have long been predicted to be sensitive to the QCD phase transition and the QCD critical point [1∼3]. Experimentally, one can measure various order moments (variance (σ^2), skewness (S), and kurtosis (κ)) of the conserved quantity distributions in heavy-ion collisions. These moments are sensitive to the correlation length (ξ) of the hot dense matter created in the initial collision and are also connected to the thermodynamic susceptibilities computed with Lattice QCD [4∼6] and in the Hadron Resonance Gas (HRG) model [7∼10]. It has been proposed that variance, skewness, and kurtosis is related to different powers of the correlation length as ξ^2, $\xi^{4.5}$, and ξ^7 [1], respectively. Furthermore, the nth order susceptibility $\chi^{(n)}$ are related to cumulant as $\chi^{(n)} = C_n/VT^3$ [11], where V, T are the volume and temperature of the system, and C_n is the nth order cumulant of a multiplicity distribution. In order to compare with theoretical calculations, cumulant ratios ($S\sigma = C_3/C_2$, $\kappa\sigma^2 = C_4/C_2$) are constructed to cancel the volume (VT^3). Thus, those cumulant ratios are also directly related to the ratios of various order susceptibilities as $C_4/C_2 = \chi_i^{(4)}/\chi_i^{(2)}$ and $C_3/C_2 = \chi_i^{(3)}/\chi_i^{(2)}$, where i indicates the conserved quantity.

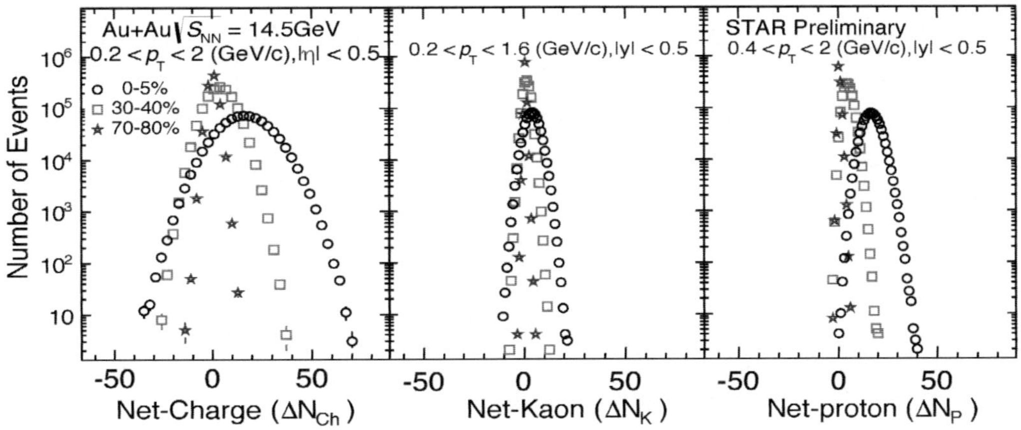

Figure 1. Uncorrected raw event-by-event net-particle multiplicity distributions for Au+Au collisions at $\sqrt{s_{NN}}$ = 14.5 GeV for ΔN_Q (left panel), ΔN_K (middle panel), and ΔN_P (right panel) for 0-5% top central (black circles), 30-40% central (red squares), and 70-80% peripheral collisions (blue stars).

Experimentally, it is very hard to measure the net-baryon (ΔN_B) and the net-strangeness (ΔN_S) distributions, so we use net-proton (ΔN_P) and net-kaon (ΔN_K) as proxies respectively. These have been widely studied experimentally and theoretically [1~3, 7, 11~14]. In this work, we report recent efficiency-corrected cumulants and cumulant ratios of the net-proton (ΔN_P), net-kaon (ΔN_K) and net-charge (ΔN_Q) multiplicity distributions in Au+Au collisions at $\sqrt{s_{NN}}$ = 7.7, 11.5, 14.5, 19.6, 27, 39, 62.4, and 200 GeV collected in 2010, 2011, and 2014 by STAR at RHIC.

2. Analysis Details

The STAR (Solenoidal Tracker At RHIC) detector at BNL has a large uniform acceptance at mid-rapidity and excellent particle identification capabilities. Energy loss (dE/dx) in the Time Projection Chamber and mass-squared (m^2) from the Time-Of-Flight detector are used to identify protons and kaons. The protons and anti-protons are obtained at mid-rapidity ($|y|$<0.5) and within the transverse momentum range 0.4<p_T<2.0 GeV/c. The K^+ (K^-) for net-kaon are measured at mid-rapidity ($|y|$<0.5) within the transverse momentum 0.2<p_T<1.6 GeV/c. The charged particles in the net-charge fluctuations are measured at pseudo-rapidity range $|\eta|$<0.5 and within the transverse momentum 0.2<p_T<2.0 GeV/c.

For the centrality selection, the uncorrected multiplicity distribution of primary charged particles within 0.5<$|\eta|$<1.0 has been used to determine the centrality for the ΔN_Q measurement, while the uncorrected multiplicity distribution of primary charged particles, excluding kaons/anti-kaons (protons/anti-protons) within $|\eta|$<1.0 has been used for ΔN_K (ΔN_P) to avoid auto-correlations. The results are corrected for a finite centrality bin width [15].

Figure 1 shows the uncorrected event-by-event net-particle multiplicity distributions for Au+Au collisions at $\sqrt{s_{NN}}$ = 14.5 GeV for ΔN_Q, ΔN_K and ΔN_P in three centrality intervals. We observed that the net-charge multiplicity distributions have the largest standard deviation, σ, compared with the net-proton and net-kaon results. This can lead to a large statistical error for the net-charge measurement. Detailed discussions about the efficiency correction and error estimation can be found in [15, 16].

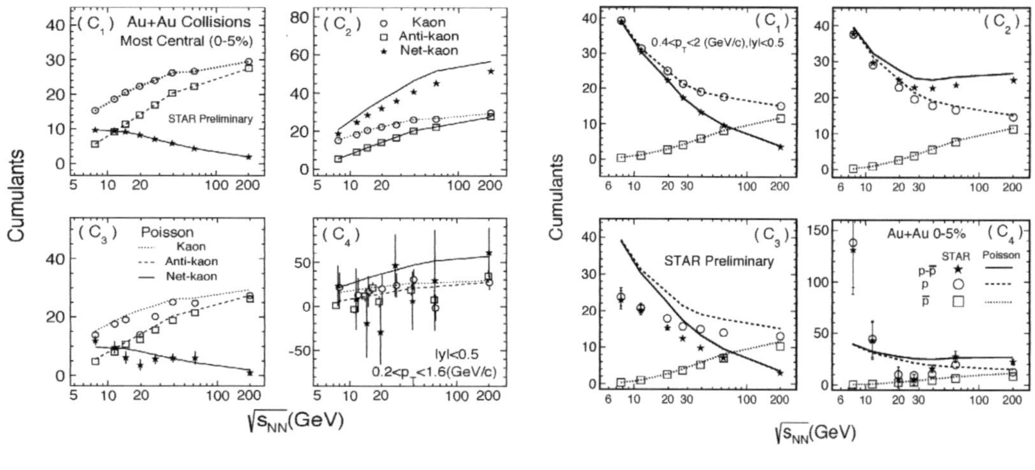

Figure 2. Left panel: Energy dependence of cumulants (C_1, C_2, C_3, and C_4) for kaon, anti-kaon, and net-kaon multiplicity distributions. Right panel: Energy dependence of cumulants (C_1, C_2, C_3, and C_4,) for proton, anti-proton, and net-proton multiplicity distributions.

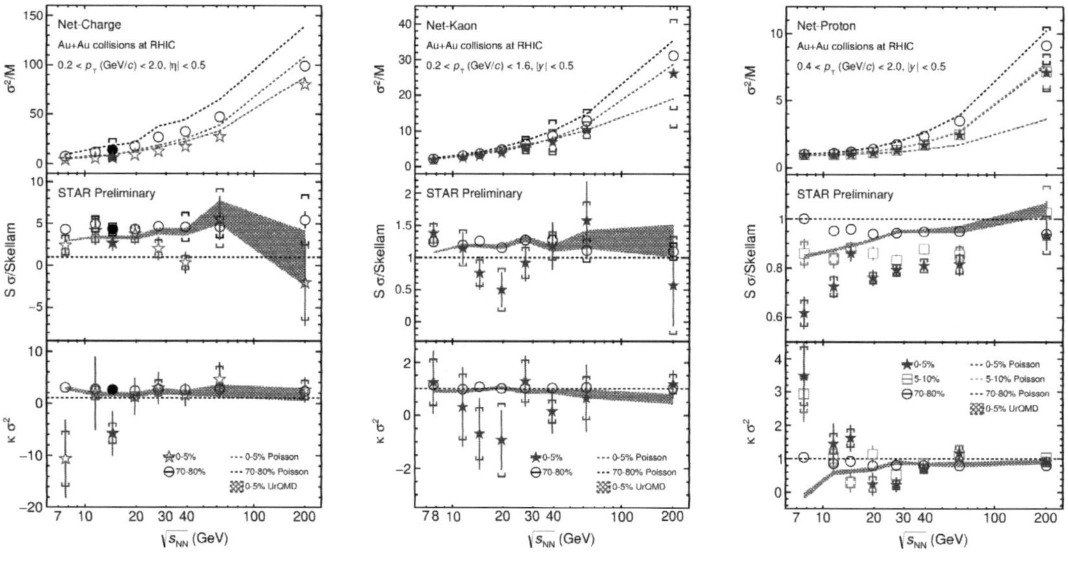

Figure 3. Energy dependence of cumulant ratios (σ^2/M, $S\sigma$/Skellam, $\kappa\sigma^2$) of net-charge, net-kaon, and net-proton multiplicity distributions for top 0-5% central (red stars), 5-10% central (green squares), and 70-80% peripheral (black circles) collisions. The Poisson expectations are denoted as dotted lines and UrQMD calculations are shown as blue bands.

3. Results and Discussion

All results have been corrected for finite detector reconstruction efficiencies, with the statistical uncertainties calculated based on the Delta Theorem [16]. The left panel of Fig. 2 shows the energy dependence of cumulants (C_1, C_2, C_3, and C_4) for kaon, anti-kaon, and net-kaon multiplicity distributions. Dashed lines indicate the Poisson expectations. In general, the cumulants for net-kaon, kaon and anti-kaon are consistent with the Poisson baseline within uncertainties. The right panel of Fig. 2 shows the energy dependence of cumulants (C_1, C_2, C_3, and C_4) for proton, anti-proton, and net-proton multiplicity distributions. One can see that the higher the order of the cumulant, the larger the deviations from the Poisson expectation for the net-protons and protons.

The energy dependence of the volume independent cumulant ratios (σ^2/M, $S\sigma$/Skellam, $\kappa\sigma^2$) for net-proton, net-kaon, and net-charge multiplicity distributions in Au+Au collisions are presented in Figure 3. The solid circles on the left figure represent the results of $\sqrt{s_{NN}}$ = 14.5 GeV obtained from Run 14 at RHIC, added to the published net-charge results [17] (open symbols). The blue bands give the predictions from the UrQMD calculations which do not include critical physics processes. No significant deviation from the Poisson expectation (dashed lines in the figure) is observed within the uncertainties for net-charge and net-kaon cumulant ratios. Both the data and the UrQMD simulation show no energy dependence for $S\sigma$/Skellam and $\kappa\sigma^2$ in net-charge and net-kaon measurement.

However, a deviation from the Poisson expectation, as well as from the UrQMD calculations is observed for $S\sigma/Skellam$ and $\kappa\sigma^2$ of the net-proton distributions. A non-monotonic behavior of the net-proton $\kappa\sigma^2$ can be seen in top 0-5% central collisions; the value of $\kappa\sigma^2$ go down below unity and then rise up from high to low collision energies. UrQMD calculations show suppression at lower energies due to baryon number conservation.

The upcoming RHIC BES II in 2019-2020, will include an upgraded STAR detector. An i(nner)TPC and Endcap TOF upgrade will enlarge the phase-space up to $|\eta|<1.5$ and down to p_T = 60 MeV/c. The Event-Plane Detector at forward rapidities will allow for a better centrality estimation, suppressing auto-correlations.

4. Summary

In this paper, we present the energy dependence on cumulant ratios from $\sqrt{s_{NN}}$ = 7.7 to 200 GeV Au+Au collisions for net-kaon, net-charge and net-proton multiplicity distributions. Within uncertainties (statistical and systematic), net-charge and net-kaon results follow the Poisson expectation, while a non-monotonic behavior is observed in the energy dependence for the net-proton $\kappa\sigma^2$. The energy dependence of cumulants (C_1 , C_2, C_3, and C_4) for kaon and proton multiplicity distributions have also been shown, and in general, the cumulants are consistent with the Poisson baseline within uncertainties for kaon, anti-kaon, and net-kaon. We can see that for net-proton and proton distributions, the deviations from Poisson expectations are larger for the higher order cumulants. The RHIC BES II will bring a larger event sample and a wider phase-space to enhance the search for the QCD critical point.

Acknowledgments

The work was supported in part by the MoST of China 973-Project No.2015CB856901, NSFC under grant No. 11575069, 11221504.

References

[1] M. A. Stephanov, Phys. Rev. Lett. 102, 032301 (2009).
[2] M. Asakawa *et al.*, Phys. Rev. Lett. 103 262301 (2009).
[3] M. A. Stephanov, Phys. Rev. Lett. 107 052301 (2011).
[4] R. V. Gavai and S. Gupta, Phys. Rev. D 78, 114503 (2008).
[5] M. Cheng *et al.*, Phys. Rev. D 79, 074505 (2009); A. Bazavov *et al.*, arXiv:1203.0784.
[6] S. Borsanyi *et al.*, JHEP01, 138 (2012).
[7] F. Karsch and K. Redlich, Phys. Lett. B 695, 136 (2011).
[8] P. Garg *et al.*, Phys. Lett. B 726, 691 (2013).
[9] Marlene Nahrgang *et al.*, arXiv: 1402.1238.
[10] J. Fu, Phys. Lett. B 722, 144 (2013).

Journal of Physics: Conference Series **736** (2016) 012002 doi:10.1088/1742-6596/736/1/012002

[11] S. Gupta, X. Luo, B. Mohanty, H. G. Ritter and N. Xu, Science 332, 1525 (2011).
[12] Aggarwal M M *et al.* (STAR Collaboration) 2010 Phys. Rev. Lett. 105 022302
[13] X. Luo (STAR Collaboration), Nucl. Phys. A 904 911c (2013).
[14] X. Luo, arXiv:1512.09215
[15] X. Luo, J. Phys. G: Nucl. Part. Phys. 39 025008 (2012).
[16] X. Luo, Phys. Rev. C 91, 034907 (2015).
[17] L. Adamczyk *et al.*, (STAR Collaboration), Phys. Rev. Lett.113, 092301 (2014).

Journal of Physics: Conference Series **736** (2016) 012003 doi:10.1088/1742-6596/736/1/012003

Hyperons at CBM-FAIR

Hans Rudolf Schmidt for the CBM Collaboration

University of Tuebingen, Auf der Morgenstelle14, Tuebingen, Germany

Hans-Rudolf.Schmidt@Uni-Tuebingen.de

Abstract. The CBM experiment is one of the four scientific pillars of the Facility for Antiprotons and Ion Research (FAIR) in Darmstadt, Germany. Its discovery potential – complementary to heavy-ion experiments at colliders – is based on high-luminosity ion beams. This enables access to extremely rare probes such as charmed particles, vector mesons or multi-strange hyperons with high statistics. However, 3rd generations readout systems and detectors are required to handle the large interaction rates (up to 10 MHz for Au+Au) with sufficient precision and bandwidth. In this contribution we will outline the unique CBM physics program focusing onto hyperons .

1. Introduction

The observation of a mass $(1.97 \pm 0.04)\ M_{\odot}$ neutron star [1] has spurred interest into models of nuclear matter capable to stabilize nuclear matter at high densities [2,3,4], i.e., to prevent the collapse of heavy neutron stars into black holes. The relevant degrees of freedom in nuclear matter at high densities are nucleons and hyperons; as well as quarks in the case of the onset of a deconfinement transition.

The future experiments at FAIR (Darmstadt) or at NICA (Dubna) will employ high intensity heavy-ion beams in the range 2-35 AGeV ($\sqrt{s_{NN}} = 2.7 - 8.3\ GeV$) and $\sqrt{s_{NN}} = 4 - 11\ GeV$), respectively. These experiments will allow to investigate rather exotic matter at baryon densities up to $7 \times \rho_0$ with good statistics for the first time. This will enable to explore yet uncharted territory with rare probes, such as charmed particles, di-leptons, double hypernuclei or multi-strange hyperons [5,6].

2. Rare Probes

Rare probes such as low mass di-leptons, charmonium and open charm will allow to characterize the medium at high net baryon densities much more detailed then previously possible. While bulk observables (flow, spectra of proton and pions) as well as correlations and fluctuations are accessible to all experiments (STAR/PHENIX@RHIC, NA61@SPS, MPD@NICA) at the relevant low beam energies ($\sqrt{s_{NN}} \sim 8\ GeV$) , the above mentioned rare probes can only be measured at CBM@FAIR.

In this paper, however, we will not further focus on leptons and charm. Instead, we will discuss (multi-)strangeness as probes of dense matter: flow and excitation functions, subthreshold production of multi-strange hyperons, double-strange hypernuclei and strange di-baryons, the latter also called MEMOs (metastable exotic multi-hypernuclear objects).

2.1. Flow

The experimental investigation of flow has basically ceased in the last century with the decommissioning of the FOPI experiment at the GSI SIS18 and the E877/895 experiments at the BNL AGS. One of the central topics of flow physics is the quest for the nuclear equation of state. Besides being interesting in itself as describing the property of nuclear matter the knowledge of the nuclear EoS is of outmost importance to understand the stability of heavy neutron stars.

As can be seen from **Figure 1** the description of flow data by simulations is unsatisfactory: no single EoS describes all types of anisotropies at all energies. At the same time the experimental errors allow for an uncertainty in the incompressibility parameter K of up to 150 MeV. However, protons represent only a small part of the flowing species. The CBM experiment will dramatically improve the data situation by measuring the flow of identified particles in the relevant energy range, including multi-strange hyperons and dileptons. Of particular interest is the flow of particles not suffering from rescattering like Ω hyperons or ϕ mesons, for which no experimental data exist. These measurements will significantly contribute to our understanding of the QCD-matter equation-of-state at neutron star core densities.

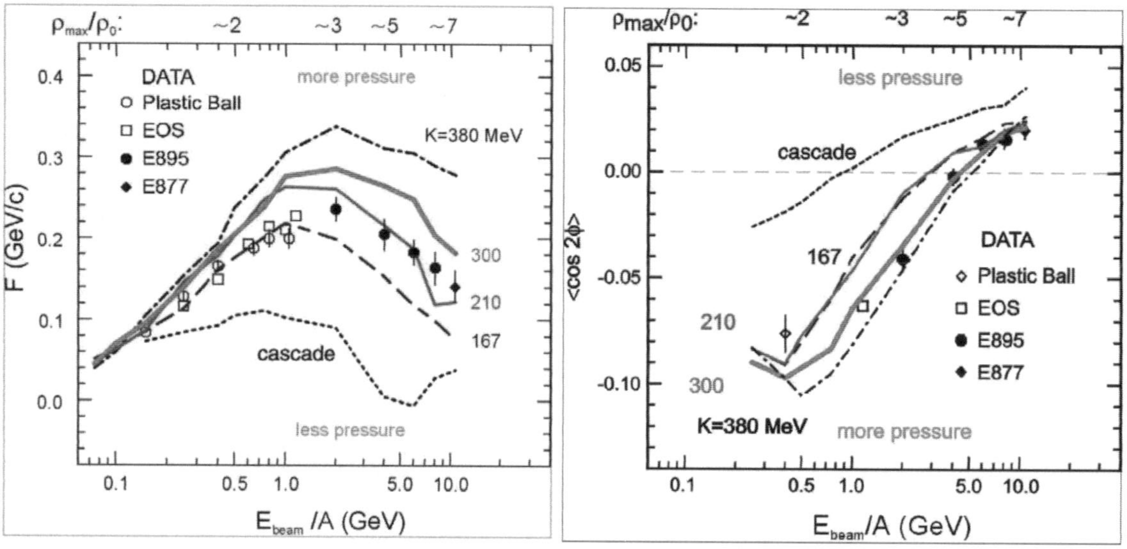

Figure 1. Excitation function of sideward (left) and elliptic flow (right) of protons for Au+Au collisions. The symbols represent experimental data, while the line are simulations assuming different equations of state.

2.2. *Multi-strange Hyperons*

The yields and phase-space distributions of multi-strange hyperons (Ξ, Ω) are promising tools to study the properties and the degrees of freedom of QCD matter at extreme nuclear densities. In particular, these particles may serve as unambiguous messengers of a high-density phase: for hyperon production binary reactions proceeds like $pp \to \Xi^- K^+ K^+ p$ or $pp \to \Omega^- K^+ K^+ K^0 p$ and have accordingly high thresholds of 3.7 and 7.0 GeV, respectively. In dense baryonic matter, however, Ξ^- and Ω^- can also be created via strangeness exchange reactions like $\Lambda\Lambda \to \Xi^- p$ and $\Lambda\Xi^- \to \Omega^- n$ or $\Lambda K^- \to \Xi^- \pi^0$, with the Λ and the K^- previously produced in independent reactions such as $pp \to K^+ \Lambda p$ and $pp \to K^+ K^- pp$, which require only 1.6.and 2.5 GeV, respectively. Alternatively, three-body collisions involving Λ's or kaons open new production channels for Σ and Ω with respect to pp reactions. The production of multi-strange hyperons is thus expected to be enhanced at high densities, and their yield to be sensitive to the baryon density reached in the fireball. Moreover, the energy distributions of multi-strange hyperons provide information on the fireball temperature and the radial flow at the time, when they are emitted. Therefore, systematic measurements of Ξ^- and Ω^- production as function of beam energy and size of the colliding nuclei offer the possibility to study the nuclear matter equation of state, or baryon density fluctuations as they are expected to occur when the system undergoes a first-order phase transition or approaches the QCD critical endpoint.

Existing data on the production of multi-strange hyperons in nuclear collisions at SIS-100 beam energies are scarce [7]. In particular, no data on production below $40A\ GeV$ ($\sqrt{s_{NN}} = 8.8\ AGeV$) exist, except for 312 Ξ^- measured at 6 AGeV at the AGS [8]. A systematic measurement of multi-strange hyperons as diagnostic probes of dense nuclear matter at SIS-100 energies has thus a substantial discovery potential.

Energy [AGeV]	Σ^-	Ω^-	$\bar{\Lambda}^-$	$\bar{\Sigma}^-$	$\bar{\Omega}^+$
4.0	9.0×10^6	1.8×10^5	3.6×10^3	5.3×10^3	9.0×10^2
6.0	2.6×10^7	5.0×10^5	2.4×10^5	1.4×10^{34}	2.8×10^3
8.0	4.0×10^7	1.4×10^6	3.6×10^6	2.0×10^5	6.0×10^4
10.7	5.4×10^8	2.2×10^6	6.8×10^6	3.8×10^5	1.2×10^4

Table 1. Expected hyperon yields [9] measured per week at a minimum-bias interaction rate of $2\times10^4 s^{-1}$. The minimum bias multiplicity is assumed to be 25% of the one in central collisions.

A prerequisite for these measurements is sufficient statistics in finite time. **Table 1** shows that already at 2×10^4 minimum-bias interactions the accumulated data exceed previous measurements in terms of yield.

2.3. Search for Double Hypernuclei

Up to now, only a handful double hypernuclei have been seen, mostly in emulsion experiments. Hypernuclei, i.e., nuclei containing at least one hyperon in addition to nucleons, offer the fascinating perspective to explore the third, strange dimension of the chart of nuclei. Their investigation provides information on the hyperon-nucleon and even on the hyperon-hyperon interactions, which play an important role in neutron star models.

The conventional production mechanism is rather complicated, e.g., via the following chain reaction employing a kaon beam: $K^- + N \rightarrow K^+ + \Xi^-, \Xi^- + ^{12}C \rightarrow {}^{6}_{\Lambda\Lambda}He + ^4He + t$. Hence the production cross sections are exceedingly small. On the other hand, heavy-ion collisions offer the possibility to produce (double) hypernuclei via coalescence of $\Lambda's$ with nucleons or light fragments in the final state of the reaction. In high-energy nucleus-nucleus collisions, Λ hyperons are produced abundantly. Their maximum yield is observed at beam energies between 30 and $40AGeV$(about 50 in central Pb + Pb collisions). At $10\ AGeV$ still 15 $\Lambda's$ are produced. The coalescence probability of Λ with light nuclei is highest at low energies yielding, according to hadron gas model calculations [10, see also 14], $2\times10^{-2}\ {}^{3}_{\Lambda}He$, $10^{-6}\ {}^{5}_{\Lambda\Lambda}He$ and $3\times10^{-8}\ {}^{6}_{\Lambda\Lambda}He$ per central collision at $8 - 10\ AGeV$ (cf. **Figure 2**). This would give, with the foreseen CBM interactions rates about 120 ${}^{5}_{\Lambda\Lambda}He/week$ and still 3.6 ${}^{6}_{\Lambda\Lambda}He/week$. For the detection of the double hypernuclei, the decay chain has to be reconstruction in the tracking detectors, e.g., ${}^{5}_{\Lambda\Lambda}He \rightarrow {}^{5}_{\Lambda}He + \pi^- \rightarrow {}^{4}He + p + \pi^-$.

Such measurements would represent a breakthrough in hypernucleus physics, as up to now only 6 candidates for double-lambda hypernuclei events have been found, and only one ${}^{6}_{\Lambda\Lambda}He$ event has been unambiguously identified.

Journal of Physics: Conference Series **736** (2016) 012003 doi:10.1088/1742-6596/736/1/012003

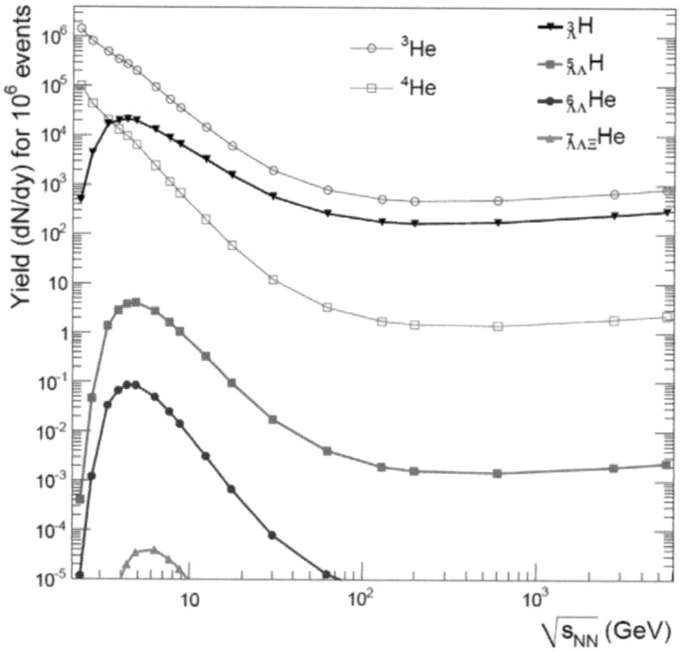

Figure 2. Energy dependence of hypernuclei yields at mid-rapidity for 10^6 central $Au + Au$ events as calculated with the statistical model [10]. The predicted yields of ^3He and ^4He are included for comparison.

2.4. Search for Meta-Stable Multi-Strange Objects

Metastable objects with strangeness, e.g. strangelets and (strange) di-baryons, were proposed long ago as collapsed states of matter, consisting of either baryons or quarks [11,12]. Up to date, none of these objects have been observed. Their existence or absence is an open issue in high-energy physics (cf. [13]).

High-energy nuclear collisions, with kaons and $\Lambda's$ being abundantly produced in a single event, could provide a tool to create such composites objects with multiple units of strangeness. Recent calculations within a hybrid (microscopic transport + hydrodynamics) model predict multiplicities/event of about $3 \cdot 10^{-2}$ for the $(\Xi^0\Lambda)_b$ in central collisions at 35A GeV [14]. Owing to the finite lifetime of such an object its decay, e.g., $(\Xi^0\Lambda)_b \to \Lambda\Lambda$, should be detectable via displaced vertices ($c\tau \approx 1 - 5\, cm$) [15]. Indeed, simulations (cf. **Figure 3**) show that $(\Xi^0\Lambda)_b's$ can be reconstructed with a signal-to-background ratio above 200 for 10^{12} central events [16,17].

Most of the strange di-baryon searches were done via the (K^+, K^-) reaction (cf. [18], and Refs. therein), which is not directly comparable with the production of di-baryons in heavy-ion collisions. The nearest equivalent to heavy-ion reactions is a result reported by the KTeV collaboration [19] at Fermilab, were lightly bound H^0-dibaryons produced in pN-collisions were searched for. However, the acceptance of the experimental apparatus was optimal for a decay length $c\tau = 160\, cm$, and thus the reported limit are not applicable for the search range of $c\tau \approx 1 - 5\, cm$ accessible in the CBM experiment.

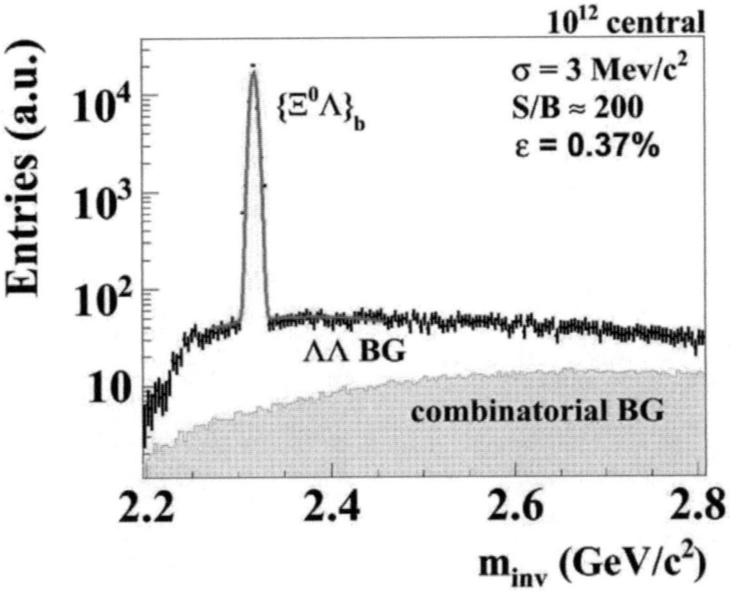

Figure 3. Reconstructed invariant-mass distribution of $\Lambda\Lambda$ candidates

3. Experimental Setup

The experimental setup of the CBM experiment is shown in **Figure 4**. From left to right one has the CBM superconducting magnet with the silicon spectrometer inside, which consists of the Micro Vertex detector (MVD) and the Silicon Tracking System (STS). It is followed by a Ring Imaging Cherenkov (RICH, light blue), four layers of Transition Radiation Detector (TRD, blue) and a Time-of-Flight (TOF, yellow) wall. Downstream of the experiment is a Projectile Spectator Detector (PSD). The muon spectrometer (MuCh) is in its parking position. For the observables discussed above primarily the STS in combination with the TOF wall is relevant.

Figure 4. View of the CBM experimental setup. For details see text.

3.1. STS

The silicon tracking system is intended for the reconstruction of the trajectories of charged particles produced in the interaction and the measurement of their momenta. It is installed between the yokes of a superconducting dipole magnet, providing a field integral of $1\,Tm$ and an aperture of $140\,cm$ in vertical and $280\,cm$ in horizontal direction.

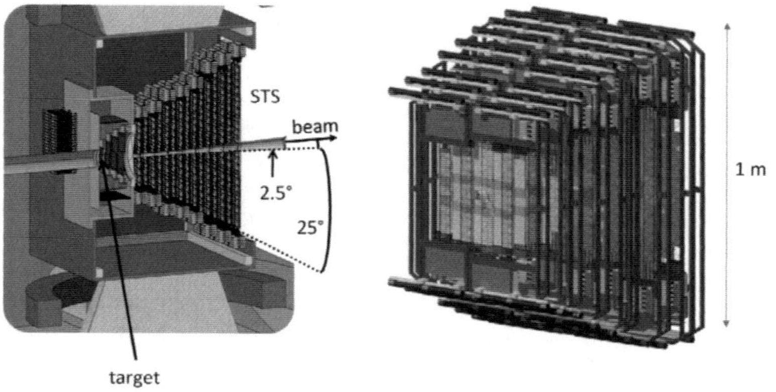

Figure 5. Left: conceptual design of the Silicon Tracking System, showing a cut through the detector in the dipole magnet. Right: engineering model of the STS, with the eight tracking stations to be mounted on the main support frame.

The STS is realized as an array of eight tracking stations constructed from thin double-sided silicon micro-strip detectors. The roughly 1,200 sensors will be mounted on a light-weight carbon structure, the signals from the inner part being routed to the front-end electronics at the periphery of the station by ultra-thin cables. The aim is to keep the material budget below 1% radiation length per station. **Figure 5** shows the conceptual design of the detector system and the current engineering model.

3.2. TOF

The identification of hadrons in CBM is achieved through the measurements of their time-of-flight. The TOF detector is based on the technology of resistive plate chambers (RPC), providing a time resolution of about $60\,ps$. RPC modules will be assembled in a wall with an active area of about $120\,m^2$, the granularity being adapted to cope with the variation of hit rates, which range from $2\,kHz/cm^2$ at the periphery to $25\,kHz/cm^2$ in the inner sections.

4. Status of Preparations

The development of detector and technical systems for CBM is in an advanced stage [20] and will be largely completed by the end of 2016. For most of the systems Technical Design Reports are approved or under evaluation. Series production of detector components will start 2016/17 and is planed to be completed 2019/20, followed by the installation in the CBM cave and commissioning without beam.

First beams from SIS-100 into the CBM cave are not expected before 2022. In view of this delay, the CBM collaboration is investigating possibilities to install and commission parts of detector systems at other experimental facilities. Examples are the installation of a part of the TOF wall at the STAR experiment at RHIC or the co-operation of CBM with BM@N at NICA in the construction of the silicon detectors. Furthermore, CBM intends to install a reduced set-up at a beam line at the SIS-18 accelerator at GSI in order to test the interplay of systems and the full data chain including data processing in real-time.

5. Summary

Heavy-ion collisions in the SIS-100/300 energy range are an ideal tool for the production of hadronic matter at neutron star core densities, and hence offer the unique opportunity to investigate fundamental properties of strongly interacting systems and its constituents: the nuclear matter equation of state, exotic new phases such as quarkyonic matter, in-medium modifications of hadrons as a signature for chiral symmetry restoration, hypernuclei and multi-strange objects, charm production at threshold beam energies, and charm propagation in nuclear matter.

References

[1] P. B. Demorest, T. Pennucci, S. M. Ransom, M. S. E. Roberts, and J. W. T. Hessels, *Nature* **467**, 1081 (2010).
[2] I. Sagert, L. Tolos, D. Chatterjee, J. Schaffner-Bielich and C. Sturm, *Phys. Rev.* C **86** , 045802 (2012)
[3] M. Orsaria, H. Rodrigues, F. Weber, and G. A. Contrera, *Phys. Rev.* D **87**, 023001 (2013)
[4] G. Colucci and A. Sedrakian, *arXiv*:1302.6925 (2013)
[5] B. Frimann et al. (eds.), *"The CBM Physics Book"*, Springer Lecture Notes in Physics 814 (2011)
[6] P. Senger and V. Friese, https://www.gsi.de/documents/DOC-2011-Aug-29.html
[7] C. Blume, *J. Phys.* G **31** (2005) S57
[8] P. Chung et al., *Phys. Rev. Lett.* **91**, 202301 (2003)
[9] A. Andronic, *private communication*
[10] A. Andronic et al., *Phys. Lett.* B **697** (2011) 203
[11] A. R. Bodmer, *Phys. Rev.* D **4** (1971) 1601
[12] R. L. Jaffe, *Phys. Rev. Lett.* **38** (1977) 195
[13] H. Clement, *Prog. Part. Nucl. Phys.* **67**, 486 (2012)
[14] J. Steinheimer et al, *Phys. Lett.* B**714** (2012) 85
[15] J. Schaffner-Bielich, R. Mattiello and H. Sorge, *Phys. Rev. Lett.* **84** (2000) 4305
[16] I.Vassiliev, *private communication*
[17] I. Vassiliev, I. Kisel, and D. Malakhov, *CBM Progress Report* 2009
[18] R.W. Stotzer et al, *Phys. Rev. Lett.* **79**, 3646 (1997)
[19] A.Alavi-Harati et al., *Phys. Rev. Lett.* **84**, 2593 (2000)
[20] V. Friese, Ch. Sturm, A. Toia, A, (eds.), 2015 CBM Progress Report, https://repository.gsi.de/record/186952

Journal of Physics: Conference Series **736** (2016) 012004 doi:10.1088/1742-6596/736/1/012004

Jets In Heavy Ion Collisions with CMS

Sevil Salur for the CMS Collaboration

Department of Physics and Astronomy Rutgers, the State University of New Jersey 136 Frelinghuysen Road Piscataway, NJ 08854-8019

E-mail: salur@physics.rutgers.edu

Abstract.
Jet physics in heavy ion collisions is a rich field which has been rapidly evolving since the first observations of medium interactions at RHIC through back-to-back hadron correlations and at LHC via reconstructed jets. In order to completely characterize the final state via jet-medium interactions and distinguish between competing energy loss mechanisms, complementary and robust jet observables are investigated. Latest developments of jet finding techniques and their applications to heavy ion environments are discussed with an emphasis given on experimental results from CMS experiment.

1. Introduction

Well identified decay products of partonic interactions at large momentum transfers, also called hard probes, are used to study the structure and dynamics of the QGP that is produced in heavy ion collisions [1, 2, 3]. Indirect measurements of energy loss in the medium ("jet quenching") have been made via observables of nuclear modification factors of single particles and the correlations of leading particles [4, 5, 6, 7].

Jets, the most common hard probe of QCD, are collimated sprays of hadronic decay products of hard-scattered partons. They are experimental signatures of quarks and gluons and are expected to reflect kinematics and topology of the initial hard scattered partons. However due to its scale dependence, the definition of the parton is ambiguous, resulting in different jet definitions corresponding to various jet reconstruction algorithms. Therefore, for an acceptable jet algorithm, the main requirement is to merge measured particles to form a jet in a procedure that matches the theoretical calculations without a strong dependence on modeling [8]. For example, the algorithm must be insensitive to soft and collinear radiation i.e., infrared and collinear safety with a seedless implementation for data analysis.

Jets in hadronic colliders are not new phenomena. Successful observation and reconstruction of jets in high transverse energy events at the CERN proton antiproton collider dates back to 1980's [9]. As can be seen in the left panel of Fig. 1 as a publication summary, more than 60% of ATLAS and CMS analyses use jet definitions since the commissioning of LHC due to the fact that if an analysis require the existence of a quark, a gluon or their absence in an event, jet reconstruction tools have to be used. The search for new physics in high luminosity pp collisions at the LHC requires the precise measurement of jet energies over large backgrounds caused by pile up. This has motivated the development of many powerful set of discriminative jet tools including pile up subtraction strategies to correct the jet energy [10, 11]. These algorithms are also applicable in heavy ion environments. Since the beginning of the LHC heavy ion program,

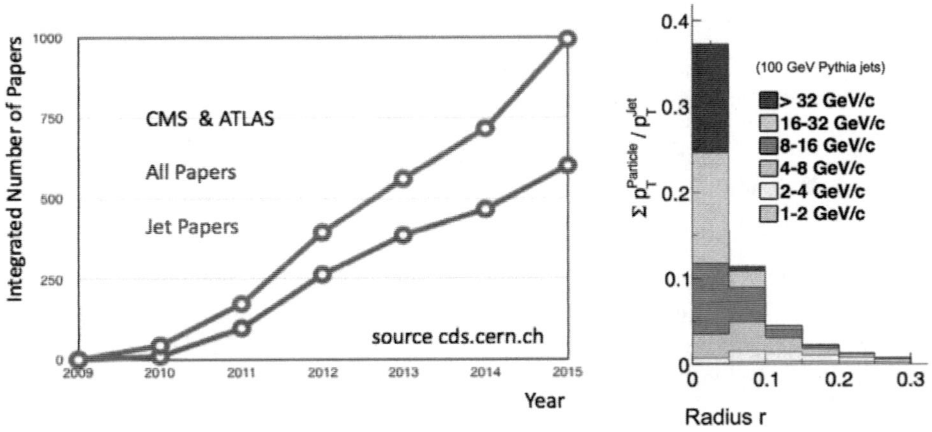

Figure 1. Left: Number of published papers and papers that require a jet reconstruction in their analysis from ATLAS and CMS Collaborations since the start of the LHC. Right: A cartoon distribution of the constituents of an average Pythia Jet that is reconstructed with a transverse momentum of 100 GeV.

jets were used as a way to probe the non-perturbative QCD physics of jet quenching by utilizing pairs of jet-jet and photon-jet measurements. These first measurements showed that the angular correlations of jet pairs were conserved with large transverse momenta imbalance [12, 13, 14, 15]. In this article, we summarize the more recent jet measurements in heavy ion collisions that are collected during the Run 1 period of LHC data taking with the CMS experiment.

2. Experimental Methods

CMS is a multipurpose detector system that was built with many different layers each serving a purpose in particle identification and measurement [16]. By design it is extremely well suited to measure hard scattering processes. Jet reconstruction is performed with anti-k_T sequential reconstruction algorithm that is encoded in the FastJet framework [17]. A small resolution parameter R is used in heavy ion collisions to reduce the deterioration of the jet energy resolution due to fluctuations of the PbPb background from soft interactions. Particle Flow (PF) objects that are reconstructed by combining information from various sub-detectors, most importantly by combining tracks with clusters in electromagnetic and hadronic calorimeters [18] and only calorimetric measurements from HCAL and ECAL are used as inputs to the jet reconstruction algorithms. The reconstructed jet energies are corrected by using a factorized multi-step approach used for all jet analyses in CMS [19]. An algorithm that is a variant of an iterative "noise/pedestal subtraction" technique is used to estimate the heavy ion background event-by-event [20]. And jet energy corrections are derived from PYTHIA [21] simulations without PbPb underlying events.

3. Measurements

3.1. Jet Morphology

Most of the momentum of a jet on average is expected be found close to the jet axis and are carried by high transverse momentum particles. The large angle component of an average jet contains only a small fraction of the overall jet energy, which are carried by softer particles. This can be seen on the right panel of Fig 1 for Pythia jets. Angular and momentum structures of

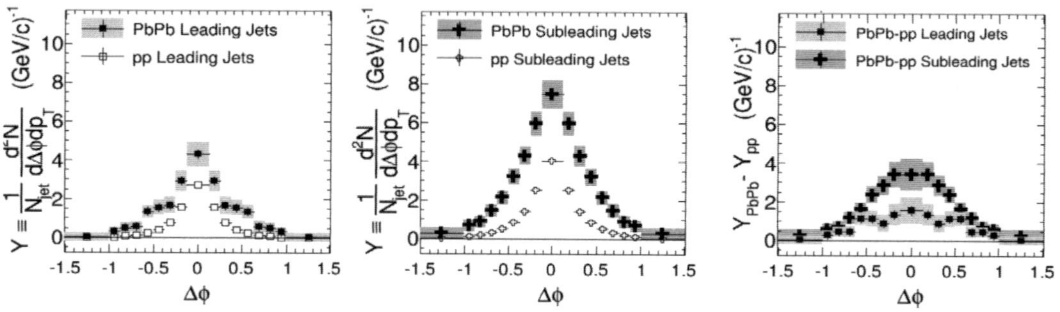

Figure 2. The $\Delta\phi$ distributions of charged-particle background-subtracted yields correlated to leading (left), subleading (middle) in 0-10% PbPb and pp. The right panel show the difference between PbPb and pp for both leading and subleading jets [25].

jets are expected to change in heavy ion collisions due to jet quenching. The internal structure of reconstructed jets are observed to be modified towards an excess of particles far from the jet axis [22]. The momentum structure of reconstructed jets are also observed to be modified towards an excess of particles at low transverse momentum [23]. Further change in the jet morphology was observed in the jet-jet events through the projection of missing transverse momentum of reconstructed charged tracks onto the leading jet axis [13]. These measurements reveal that the missing energy is also recovered in the form of soft particles at large angles with respect to the jet axes. Recently the missing transverse momentum analysis is extended to large annular regions for various jet definitions [24]. This new study suggests possible modifications of leading jets towards getting narrower, and/or the subleading jets towards getting broader due to jet quenching effects. More recently, jets are correlated with charged tracks to study the broadening of jet energy in further detail. The projection of jet-track correlations measured for charged tracks with transverse momentum between 1 and 2 GeV on the $\Delta\phi$ is shown in Fig. 2 [25]. While the distributions look similar in peripheral PbPb and pp events [25] the differences of the correlations strengths are most pronounced in the 0-10% central events for both leading and subleading jets as can be seen in Fig. 2. The integrals of the excess yields (PbPb minus pp) as a function of the transverse momentum of tracks for two PbPb event centralities are shown in Fig. 3. For both centralities the excess yield was found to be largest at the lowest charged track transverse momentum (1-2 GeV) in the most central (0-10%) PbPb data. For peripheral (50-100%) PbPb collisions, correlated charged particle yields are only slightly larger than those for the pp reference. The excess diminishes for higher momentum tracks and becomes similar to the pp reference for charged tracks transverse momentum larger than 4 GeV.

3.2. Parton Dependence

The jet quenching is expected to depend on the flavor of the initiating parton. For example, under the assumption that radiative energy loss is the dominant mechanism, gluon jets should be quenched more strongly than light quark jets, due to their larger color factor for gluon emission. On the other hand, jets initiated by heavy quarks, particularly bottom quarks, are expected to radiate less than light ones due to the so-called dead cone effect [26]. To tag jets and extract information regarding their initiating partons' flavor has been performed by CMS experiment. The following subsections summarizes the current status of these investigations.

3.2.1. Heavy Flavor Tagged Jets Jets formed from heavy flavor quark fragmentation can be tagged by the presence of displaced secondary vertices, either by direct reconstruction of these

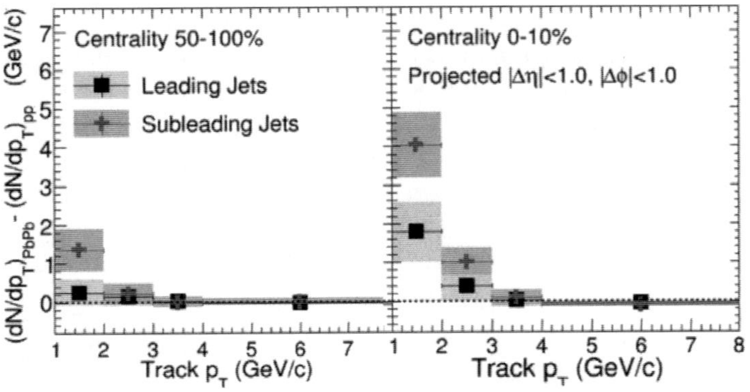

Figure 3. The charge track transverse momentum dependence of the difference between the correlations strengths of PbPb and pp events for both leading and subleading jets [25]

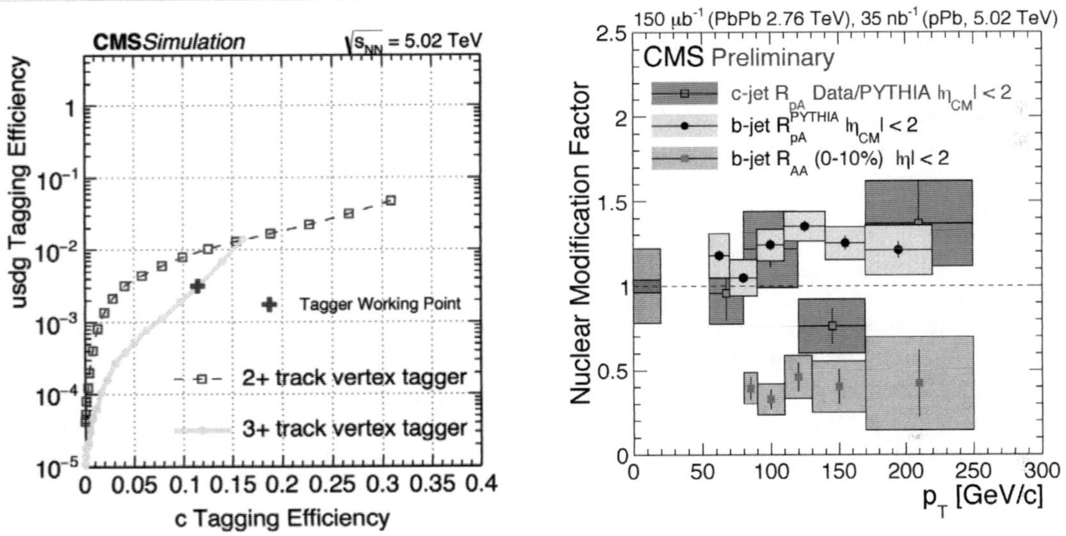

Figure 4. Left: Efficiency of light vs c tagged jet quarks for 2 track vs 3 track vertex tagger [30]. Right: Nuclear modification factors from CMS for heavy flavour associated jets [30, 31, 32].

vertices or by the impact parameter of tracks originating from these vertices [28, 29]. Information from these tracks and vertices are typically combined into a quantity which optimizes their discrimination between heavy and light flavor jets. A variation of this method by including additional discriminative variables such as requiring at least three tracks at the secondary vertex can be used to tag charm quark jets [30]. The left panel of Fig. 4 shows how the purity of charm quark tagging can be improved with the requirement of three tracks at the secondary vertex. The right panel of Fig. 4 shows the current compilation of all heavy flavor associated jet measurements at LHC with CMS experiment. A strong suppression in the nuclear modification of bottom quark associated jets is observed in central PbPb collisions while the nuclear modification calculation is consistent with unity in pPb collisions at a larger centre of mass energy. Similarly the yield of charm quark jets in pPb collisions is observed to be consistent with that predicted by PYTHIA within the systematic uncertainties of the measurement as can be seen with open squares in the right panel of Fig. 4. The inclusive bottom quark jet production at these high transverse momentum appears to have a negligible mass effect, and that the attenuation of bottom quark tagged jets appears to be comparable to the one observed for light quark jets [33]. The situation appears to be different for isolated-photon and B meson tagged bottom quark jets as they are expected to be produced at the early stages of the collision [34]. Additional tagging of bottom quark jets with B mesons or photons are predicted to have a more direct connection to the physics of bottom quark jet energy loss. With high statistics Run 2 data, it is possible to explore multi tags such as c/b jets with additional D/B mesons and photons.

3.2.2. Quark vs Gluon Tagged Jets A large collection of experimental measurements show that hadronic jets initiated by gluons exhibit differences with respect to jets from light-flavor quarks [35, 36, 37, 38, 39]. These observations are primarily grouped into three categories. The first one is the charged particle multiplicity being higher in gluon jets than in light-quark jets. The second one is the considerably softer fragmentation functions of gluon jets than that of quark jets. And finally the gluon jets appeared to be less collimated than quark jets. These differences have been already exploited to differentiate gluon and quark jets in pp collisions [40]. The simplest and most studied variable that is used experimentally is the multiplicity, i.e. the total number of constituent candidates of reconstructed jet. Since gluon hadronization is expected to produce jets which are 'wider' than jets induced by quark hadronization, jet shapes are studied with jet width variables. Since quark jets have harder fragmentation functions compared to gluon jets they are therefore more likely to produce jets with hard constituents that carry a significant fraction of the jet energy. This can be studied with the $\mathrm{p_T^D}$ variable, defined as $\mathrm{p_T^D} = \sqrt{\Sigma_i \mathrm{p_{T,i}^2}}/\Sigma_i \mathrm{p_{T,i}}$. Based on studies of single-variable discrimination power, a likelihood-product discriminator can be defined, built on the product of the studied variables. The left panel of Fig. 5 shows the shape comparisons of the $\mathrm{p_T^D}$ variable used in the discriminator. The data (black markers) are compared to the MADGRAPH/PYTHIA simulation, on which the different components are shown: quarks (blue), gluon (red) and unmatched/pileup (grey). The expected discriminator performance is shown in the right panel of Fig. 5, in terms of light-quark efficiency and gluon rejection. While this study shows a good background rejection and signal efficiency, its stability against pile-up events is under investigation. It might be possible to utilize these tools that are developed for pp to PbPb events especially when combined with other taggers and/or variables.

3.2.3. Z Boson Tagged Jets During the 2015 running of the LHC, the PbPb data at 5.02 TeV per nucleon-pair, corresponding to an integrated luminosity of 404 μb^{-1} and 25.8 pb^{-1} of pp data at the same energy were collected by the CMS experiment. With this data, the so-called "Golden Probe" for jet tomography of the QGP, i.e., coincidences of Z^0 and jet pairs became experimentally accessible [41, 42, 43]. While this channel have served as an essential calibrator of jet energy in TeV pp collisions, in heavy ion collisions they can be used to calibrate in-medium parton energy loss as they carry no color charge and are expected to escape the medium unattenuated. An event display with a back-to-back Z Boson and a jet candidate from Run 2 data taking is shown in the left panel of Fig. 6. The right panel of Fig. 6 shows the invariant mass spectra of Z^0 bosons through their muonic decay channel in pp and PbPb collisions at $\sqrt{s_{NN}} = 5.02$ TeV [1].

[1] The preliminary measurement of the Z Boson tagged jets in PbPb collisions at $\sqrt{s_{NN}} = 5.02$ TeV recently became available [44]. These measurements show that while angular correlations between Z bosons and jets are mostly preserved in central PbPb collisions, the transverse momentum ratio of jets to their Z boson pairs appears to be shifted to lower values with respect to the observations in pp collisions as expected from jet quenching.

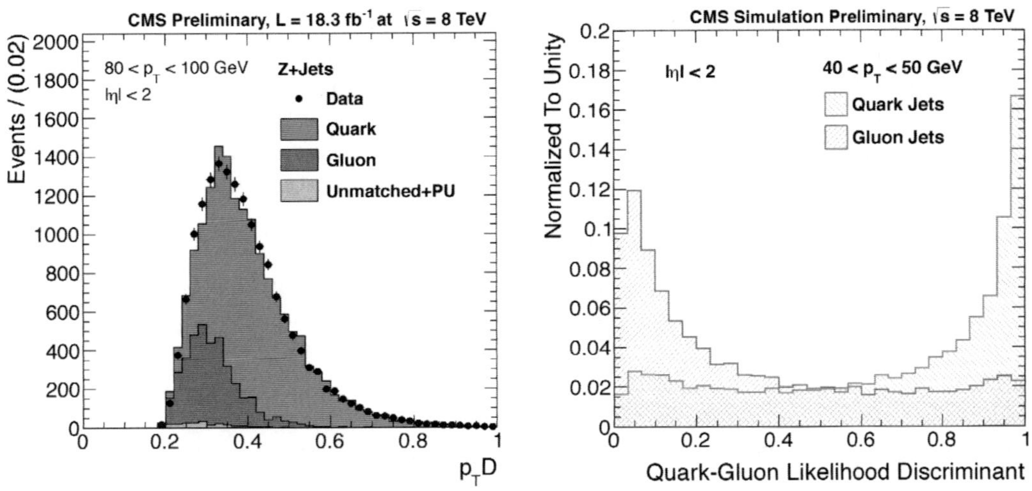

Figure 5. Left: Data-MC comparisons of the pull variable [40]. Right: Shape comparison, in simulated QCD dijet events, of the likelihood discriminator for jets [40].

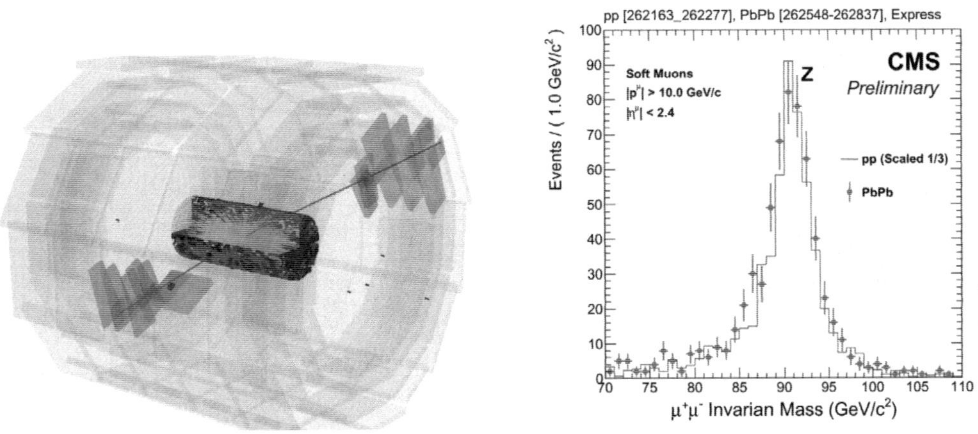

Figure 6. Left: An event with a Z+jet candidate at $\sqrt{s_{NN}} = 5$ TeV PbPb collisions. Right: Invariant mass distribution of Z bosons in pp and PbPb collisions at $\sqrt{s_{NN}} = 5$ TeV.

4. Summary

The jet quenching is studied by complimentary and robust measurements with the data that are collected by the CMS experiment during Run 1 of LHC. A strong increase in the fraction of back-to-back jet pairs with unbalanced transverse momentum with no angular de-correlations are observed. The missing energy is recovered in the form of soft particles at large angles with respect to the jet axis. Jet shapes and fragmentation functions appears to be modified in the PbPb collisions. Through studies of bottom and charm quark tagged jets, it appears that jet quenching does not have a strong dependence on parton mass and flavor of the jet for the studied momentum. These results are all consistent and important for characterizing the jet quenching and shower evolution in the presence of a hot and dense nuclear medium. The first data taking period of Run 2 is also recently completed with success. This provides high statistics PbPb and pp collisions at $\sqrt{s_{NN}} = 5.02$ TeV. Many powerful set of discriminative jet tools that are developed for high-precision studies in pp events could now be explored with this data and future data taking periods of upcoming LHC runs to explore the jet quenching and dependence on its underlying parton flavour. These measurements are essential to distinguish between competing energy loss mechanisms to determine key features of QCD and to extract quantitative properties of QGP.

5. Acknowledgments

This material is based upon work supported by the National Science Foundation under Grant No. 1352081.

References

[1] M. Gyulassy, I. Vitev, X. N. Wang, B. W. Zhang, "Jet quenching and radiative energy loss in dense nuclear matter", Quark Gluon Plasma P123-191 (2003), [nucl-th/0302077].
[2] M. Gyulassy and M. Plumer, "Jet quenching as a probe of dense matter", Nucl. Phys. A 527 (1991) 641.
[3] X. N. Wang and M. Gyulassy, "Gluon shadowing and jet quenching in A + A collisions at $\sqrt{s_{NN}} = 200$-GeV", Phys. Rev. Lett. 68 (1992) 1480.
[4] STAR Collaboration, "Evidence from d + Au measurements for final state suppression of high p(T) hadrons in Au+Au collisions at RHIC", Phys. Rev. Lett. 91 (2003) 072304.
[5] STAR Collaboration, "Direct observation of dijets in central Au+Au collisions at $\sqrt{s_{NN}} = 200$-GeV", Phys. Rev. Lett. 97 (2006) 162301.
[6] PHENIX Collaboration, "Common suppression pattern of eta and pi0 mesons at high transverse momentum in Au+Au collisions at $\sqrt{s_{NN}} = 200$-GeV", Phys. Rev. Lett. 96 (2006) 202301.
[7] CMS Collaboration, "Study of high-p T charged particle suppression in PbPb compared to pp collisions at $\sqrt{s_{NN}} = 2.76$ TeV", Eur. Phys. J. C, 72 (2012) 1945.
[8] G. C. Blazey et al., "Run II jet physics", Batavia, QCD and weak boson physics in Run II (1999) 47, [arXiv:hep-ex/0005012].
[9] UA1 Collaboration, "Observation of Jets in High Transverse Energy Events at the CERN Proton - anti-Proton Collider", Phys. Lett. B123 (1983) 115-122.
[10] M. Cacciari, G. P. Salam and G. Soyez, "The catchment area of jets", JHEP 0804 (2008) 005.
[11] M. Cacciari and G. P. Salam, "Pileup subtraction using jet areas", Phys. Lett. B 659 (2008) 119.
[12] ATLAS Collaboration, "Observation of a Centrality-Dependent Dijet Asymmetry in Lead-Lead Collisions at $\sqrt{s_{NN}} = 2.76$ TeV with the ATLAS Detector at the LHC", Phys. Rev. Lett. 105 (2010) 252303.
[13] CMS Collaboration, "Observation and studies of jet quenching in PbPb collisions at $\sqrt{s_{NN}} = 2.76$ TeV", Phys. Rev. C 84, (2011) 024906.
[14] CMS Collaboration, "Jet momentum dependence of jet quenching in PbPb collisions at $\sqrt{s_{NN}} = 2.76$ TeV", Phys. Lett. B, 712, 3 (2012) 176-197.
[15] CMS Collaboration, "Studies of jet quenching using isolated-photon + jet correlations in PbPb and pp collisions at $\sqrt{s_{NN}} = 2.76$ TeV", Phys. Lett. B 718 (2013) 773.
[16] CMS Collaboration, "CMS Physics Technical Design Report: Addendum on High Density QCD with Heavy Ions", J. Phys. G: Nucl. Part. Phys. 34 (2007) 2307.
[17] M. Cacciari, G. P. Salam and G. Soyez, "The anti-kt jet clustering algorithm", JHEP 0804 (2008) 063.
[18] CMS Collaboration, "Commissioning of the Particle-Flow Reconstruction in Minimum-Bias and Jet Events from pp Collisions at 7 TeV", CMS Physics Analysis Summary CMS-PAS-PFT-10-002, (2010).
[19] CMS Collaboration, "Determination of Jet Energy Calibration and Transverse Momentum Resolution in CMS", JINST 06 (2011) P11002.
[20] O. Kodolova, I. Vardanian, A. Nikitenko et al., "The performance of the jet identification and reconstruction in heavy ions collisions with CMS detector", Eur. Phys. J. C 50 (2007) 117.
[21] T. Sjostrand, S. Mrenna, and P. Skands, "PYTHIA6.4 Physics and Manual", JHEP 05 (2006) 026.
[22] CMS Collaboration, "Modification of jet shapes in PbPb collisions at $\sqrt{s_{NN}} = 2.76$ TeV", Phys. Lett. B, 730 (2104) 243-263.
[23] CMS Collaboration, "Measurement of jet fragmentation in PbPb and pp collisions at $\sqrt{s_{NN}} = 2.76$ TeV", Phys. Rev. C 90, 024908.
[24] CMS Collaboration, "Measurement of transverse momentum relative to dijet systems in PbPb and pp collisions at $\sqrt{s_{NN}} = 2.76$ TeV", Jour. of High Energy Phys., 01 (2016) 006.

[25] CMS Collaboration, "Correlations between jets and charged particles in PbPb and pp collisions at $\sqrt{s_{NN}} = 2.76$ TeV", J. of High Energy Phys., 02 (2016) 156.
[26] Dokshitzer, Yuri L. and Kharzeev, D.E, "Heavy quark colorimetry of QCD matter", Phys. Lett. B519, (2001), 199.
[27] CMS Collaboration, "Suppression of non-prompt J/ψ, prompt J/ψ, and $Y(1S)$ in PbPb collisions at $\sqrt{s_{NN}} = 2.76$ TeV", JHEP 1205 (2012) 063.
[28] CMS Collaboration, "Inclusive b-jet production in pp collisions at $\sqrt{s} = 7$ TeV", JHEP 1204 (2012) 084.
[29] CMS Collaboration, "Identification of b-quark jets with the CMS experiment", JINST 8 (2013) P04013.
[30] CMS Collaboration,"Charm-tagged jet production in pPb collisions at 5.02 TeV and pp collisions at 2.76 TeV", CMS-PAS-HIN-012 (2015).
[31] CMS Collaboration,"Evidence of b-Jet Quenching in PbPb Collisions at $\sqrt{s_{NN}} = 2.76$ TeV", Phys. Rev. Lett., 113 (2014) 132301.
[32] CMS Collaboration,"Transverse momentum spectra of inclusive b jets in pPb collisions at $\sqrt{s_{NN}} = 5.02$ TeV", Phys. Lett. B, 754 (2016) 59-80.
[33] CMS Collaboration, "Nuclear modification factor of high transverse momentum jets in PbPb collisions at $\sqrt{s_{NN}} = 2.76$ TeV", CMS-PAS-HIN-12-004 (2012).
[34] H. Xing, J. Huang, Z. Kang, I. Vitev, "Quenching of inclusive and tagged b-jets at the LHC", Nucl. Phys. B Proc. Supp. 00 (2015) 1-4.
[35] OPAL Collaboration, "A Study of differences between quark and gluon jets using vertex tagging of quark jets", Z. Phys. C58 (1993) 387-404.
[36] OPAL Collaboration, "A Model independent measurement of quark and gluon jet properties and differences", Z. Phys. C68 (1995) 179-202.
[37] DELPHI Collaboration, "Energy dependence of the differences between the quark and gluon jet fragmentation", Z. Phys. C70 (1996) 179-196.
[38] ALEPH Collaboration, "Quark and gluon jet properties in symmetric three jet events", Phys. Lett. B384 (1996) 353-364.
[39] ALEPH Collaboration, "Studies of quantum chromodynamics with the ALEPH detector", Phys. Rept. 294 (1998) 1-165.
[40] CMS Collaboration, "Performance of quark/gluon discrimination using pp collision data at $\sqrt{s} = 8$ TeV", CMS-PAS-JME-13-002 (2013).
[41] X.-N. Wang, Z. Huang, and I. Sarcevic, "Jet Quenching in the Direction Opposite to a Tagged Photon in High-Energy Heavy-Ion Collisions", Phys. Rev. Lett. 77 (1996) 231.
[42] X.-N. Wang and Z. Huang, "Medium-induced parton energy loss in γ+jet events of high-energy heavy-ion collisions", Phys. Rev. C 55 (1997) 3047.
[43] I. Vitev and B. W. Zhang, "The Physics of Z^0/γ^*-tagged jets at the LHC", Phys. Rev. C 83 (2011) 034902.
[44] CMS Collaboration, "Study of Z+jet correlations in PbPb and pp collisions at $\sqrt{s_{NN}} = 5.02$ TeV", CMS-PAS-HIN-15-013 (2016).

Journal of Physics: Conference Series **736** (2016) 012005 doi:10.1088/1742-6596/736/1/012005

Identified Light and Strange Hadron Spectra at $\sqrt{s_{NN}}$ = 14.5 GeV with STAR at RHIC BES I

James Daniel Brandenburg
Rice University, Houston, TX, USA
E-mail: James.D.Brandenburg@rice.edu

Abstract. With the recently measured Au+Au collisions at $\sqrt{s_{NN}}$=14.5 GeV, RHIC completed its first phase of the Beam Energy Scan (BES) program. The main motivation of the BES program is the search for a conjectured critical point and possible first order phase transition. Amongst the various collision energies of 7.7, 11.5, 19.6, 27, and 39 GeV, that have been previously presented by STAR, collisions at 14.5 GeV will provide data set in the relatively large chemical potential gap between the 11.5 and 19.6 GeV center-of-mass energies. In this contribution, we report new STAR measurements of Au+Au at $\sqrt{s_{NN}}$=14.5 GeV that include identified light particle R_{CP} and spectra, as well as measurements of the strange hadrons (K^0_s, Λ, Ξ, and Ω). The spectra from both light and strange particles cover a significant range of the intermediate transverse momentum ($2 < p_T < 5$ GeV/c) in all beam energies. We will discuss the physics implications of these observables and whether hadronic or partonic interactions dominate the collision dynamics at a given center-of-mass energy.

1. Introduction

The Relativistic Heavy Ion Collider's (RHIC) Beam Energy Scan (BES) is a program designed to scan the QCD phase diagram. One of the 3 primary goals of the BES program is to search for the onset of signatures attributed to the formation of a Quark Gluon Plasma (QGP)[1]. As part of the Beam Energy Scan Phase I (BES I) program Au+Au collisions at $\sqrt{s_{NN}}$ = 7.7, 11.5, 19.6, 27.0, and 39.0 GeV were collected in 2010 and 2011. The BES I program was then completed with the recently collected Au+Au collisions at $\sqrt{s_{NN}}$ = 14.5 GeV in 2014. Collisions at these energies cover a significant range in the temperature vs. baryon chemical potential (μ_B) phase space and provide a unique dataset for studying the QCD phase diagram. The final set of Au+Au collisions collected at $\sqrt{s_{NN}}$ = 14.5 GeV is of particular interest because it provides data in the relatively large μ_B gap between $\sqrt{s_{NN}}$ = 11.5 and 19.6 GeV [2]. Particle yields for the light hadrons ($\pi^{\pm}$, p, and $\bar{p}$) and the strange hadrons (K^0_s, Λ, Ξ, and Ω) are obtained by employing the particle identification techniques discussed in Sec. 2. The identified particle yields can be used to construct several experimental observables. In this contribution we report a systematic study of the nuclear modification factor and baryon-to-meson ratio for the BES I energies and discuss their physics implications.

The nuclear modification factor, or $R_{\rm AA}$, is the ratio of particle yields in central heavy ion collisions to the appropriately scaled particle yields in proton-proton (pp) collisions. When pp data at the corresponding collision energy and conditions are not available, as was the case in the BES I, it is common to measure the nuclear modification factor *via* the proxy $R_{\rm CP}$. The $R_{\rm CP}$ is the ratio of particle yields in central heavy ion collisions to the appropriately scaled particle

Journal of Physics: Conference Series **736** (2016) 012005 doi:10.1088/1742-6596/736/1/012005

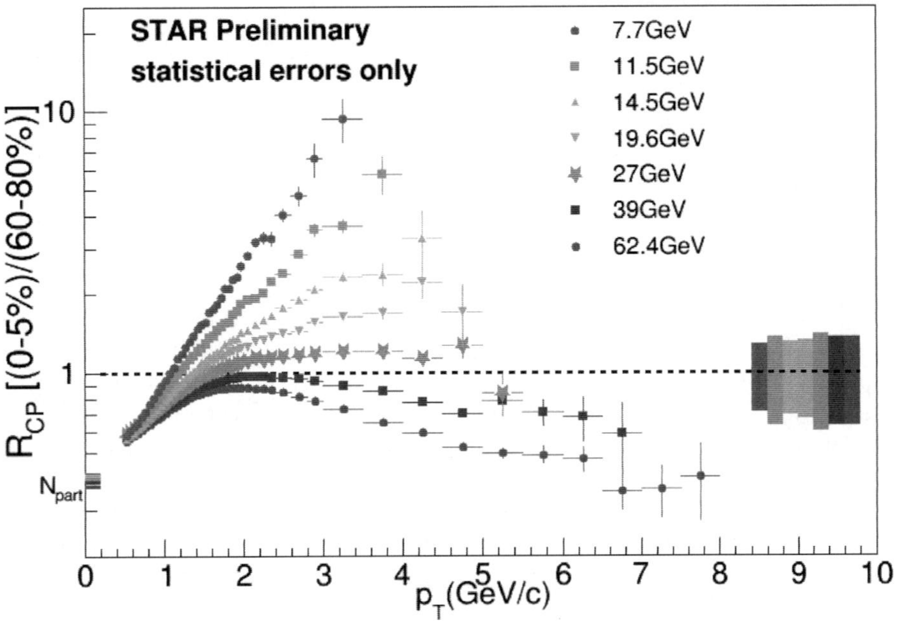

Figure 1. The charged hadron $R_{\rm CP}$ for events with a centrality of 0-5% over those with a centrality of 60-80% for the BES I energies and $\sqrt{s_{NN}} = 62.4$ GeV. The colored error bars on the right side of each plot are the correlated uncertainty due to the N_{coll} scaling for each $\sqrt{s_{NN}}$ energy calculated from MC Glauber. Horizontal bars represent the $p_{\rm T}$ range used for measuring each point in the spectra. Vertical bars are used to show statistical uncertainty only.

yields in peripheral heavy ion collisions. $R_{\rm CP}$ can be calculated from a single minimum bias dataset and is less sensitive to cold nuclear matter effects than $R_{\rm AA}$. However, $R_{\rm CP}$ generally comes with large correlated systematic uncertainties due to the determination, using Monte Carlo (MC) Glauber, of the number of binary collisions (N_{coll}) in peripheral events [3, 4].

Measurements of the nuclear modification factor ($R_{\rm CP}$) for inclusive charged particles in Au+Au collisions at $\sqrt{s_{NN}} = 200$ GeV have shown significantly suppression ($R_{\rm CP} << 1$) at high $p_{\rm T}$ ($p_{\rm T} > 5$ GeV/c)[5]. The observation of strong suppression at high $p_{\rm T}$ in Au+Au collisions at $\sqrt{s_{NN}} = 200$ GeV is attributed to partonic energy loss due to interactions in a hot and dense medium such as the QGP. For this reason strong suppression is considered to be a key signature of QGP formation. In Fig. 1 the $R_{\rm CP}$ of charged hadrons is shown for Au+Au collisions at each BES I energy and for $\sqrt{s_{NN}} = 62.4$ GeV. The $R_{\rm CP}$ of charged hadrons shows a smooth transition from enhancement ($R_{\rm CP}>1$) at the lower $\sqrt{s_{NN}}$ energies to suppression ($R_{\rm CP}<1$) at $\sqrt{s_{NN}} = 62.4$ GeV. This observation in the charged hadrons motivates a systematic study of the identified particle $R_{\rm CP}$. Specifically measurement of the $R_{\rm CP}$ of identified particles will help provide more stringent limits on models and help distinguish competing physics effects.

The two observables discussed in this contribution are constructed from the various identified particle spectra. For this reason the procedures used to conduct particle identification will be briefly discussed. First, the detectors used for particle identification will be introduced. Second, the particle identification techniques used for these results will be presented. Finally, the $R_{\rm CP}$ and baryon-to-meson ratios will be presented and their physics implications will be discussed.

Journal of Physics: Conference Series **736** (2016) 012005 doi:10.1088/1742-6596/736/1/012005

2. Particle Identification with the STAR Detector

The Solenoidal Tracker at RHIC (STAR) [6] is used for the analyses in this paper. The Time Projection Chamber (TPC) [7] is the central detector in STAR and provides charged particle tracking, momentum reconstruction, and particle identification information as dE/dx. The TPC has nominal acceptance in the region $|\eta| < 1.0$. In addition to the TPC, the Time of Flight (TOF) [8] detector is also used for identification of the $\pi^{\pm}$, p, and $\bar{p}$. The TOF detector provides particle identification information through its measurement of $1/\beta$. The nominal acceptance of the TOF detector is in the region $|\eta| < 0.9$. Both the TPC and the TOF detectors provide full azimuthal coverage of 2π.

The measurement of the $\pi^{\pm}$, p, and $\bar{p}$ identified yields is conducted using a method that combines the dE/dx information from the TPC and the $1/\beta$ information from the TOF. In this method the full 2D distribution of the tracks' dE/dx vs. $1/\beta$ is reduced into eight complimentary 1D distributions. Two of these distributions are the 1D projections of all tracks onto the dE/dx and $1/\beta$ axes. The remaining six distributions are subsets of the first two with a relative enhancement for the particle of interest, either π, K, or p. The first three distributions are formed by cutting around the dE/dx peak for each of the π, K, and p and then projecting onto the $1/\beta$ axis. The final three distributions are formed by cutting around the $1/\beta$ peak for each of the π, K, and p and then projecting onto the dE/dx axis. These eight 1D distributions, two "unenhanced" and six "enhanced", capture the vast majority of the information contained in the full 2D distribution while reducing the dataset's overall complexity. These eight distributions are then fit simultaneously to extract the yields for each particle species. In the simultaneous fit, the enhanced distributions are primarily used to constrain shape parameters, while the two unenhanced distributions are used to determine the particle yields.

The strange hadrons, K_s^0, Λ, Ξ, and Ω, are identified using tracks reconstructed in the TPC from each particle's charged decay products to identify the secondary decay vertex. Topological cuts are then applied to these charged daughter tracks to reduce the combinatorial background and to optimize the acceptance of the signal. The particle yields are then extracted by fitting the invariant mass distribution from the daughter particles with combinatorial background subtracted. This topological method of particle identification provides a robust method of extracting K_s^0, Λ, Ξ, and Ω yields over a significant portion of the $p_{\rm T}$ range for the BES I energies.

The spectra for the light hadrons, $\pi^{\pm}$, p, and $\bar{p}$, and the strange hadrons, K_s^0, Λ, Ξ, and Ω are corrected for the TPC's detection efficiency and acceptance using simulated events embedded into real data events. The simulated events are produced using STARSIM; a GEANT3 based simulation of the STAR detector geometries [9]. The $\pi^{\pm}$, p, and $\bar{p}$ are further corrected for the detection efficiency and acceptance of the TOF detector using a data-driven method. Finally the $\pi^{\pm}$, p, $\bar{p}$, and Λ are corrected for weak-decay feeddown contributions and the $\pi^{\pm}$, p, and $\bar{p}$ are corrected for knock-out contributions.

3. Results

With the recent addition of the identified light and strange hadron spectra from Au+Au collisions at $\sqrt{s_{NN}} = 14.5$ GeV, we now have a complete set of identified particle spectra from the BES I energies. These spectra provide a wealth of data from which we can extract observables sensitive to the collision dynamics. In this contribution, we are specifically interested in the $R_{\rm CP}$ and baryon-to-meson ratios; two observables identified as key QGP signatures. With the complete set of identified particle spectra it is possible to systematically study the $\sqrt{s_{NN}}$ dependence of $R_{\rm CP}$ and the baryon-to-meson ratios for the BES I energies.

In Fig. 2 the $R_{\rm CP}$ calculated from 0-5% central events over 60-80% central events for $\pi^{\pm}$, p, and $\bar{p}$ is shown. In this figure, each plot shows the $R_{\rm CP}$ for a single particle species for all six BES I energies to highlight the $\sqrt{s_{NN}}$ dependence. The $R_{\rm CP}$ for $\pi^{\pm}$, p, and $\bar{p}$ are all below

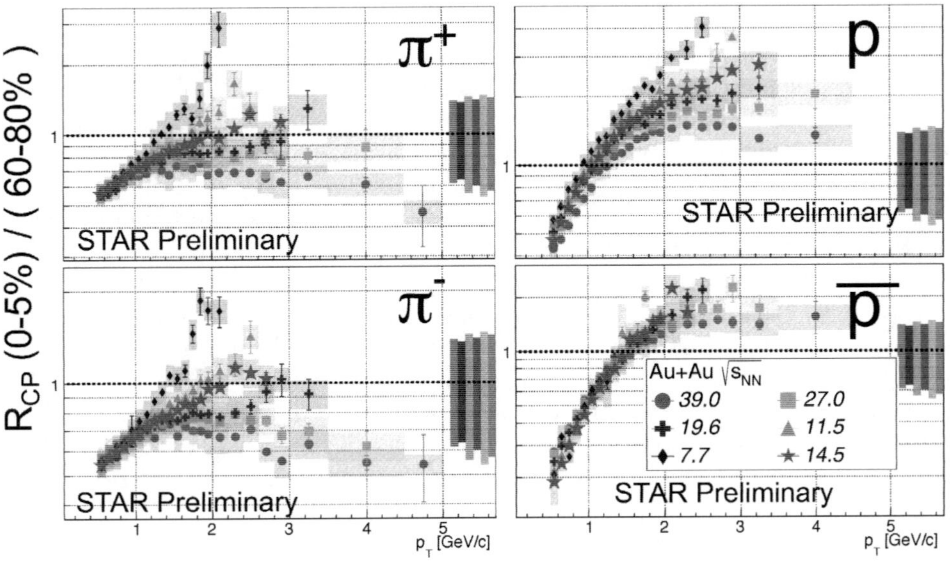

Figure 2. The R_{CP} for Au+Au collisions at the BES I energies constructed from events with a centrality of 0-5% over those with a centrality of 60-80%. The colored error bars on the right side of each plot are the correlated uncertainty due to the N_{coll} scaling for each $\sqrt{s_{NN}}$ energy calculated from MC Glauber. Horizontal bars represent the p_{T} range used for measuring each point in the spectra. Vertical bars are used to show statistical uncertainty while shaded boxes are used to show systematic uncertainties.

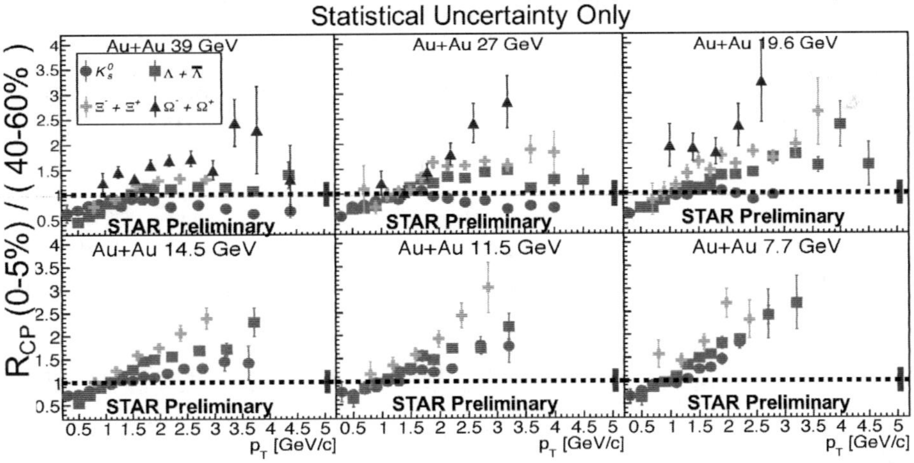

Figure 3. The R_{CP} for Au+Au collisions at the BES I energies for events with a centrality of 0-5% over those with a centrality of 40-60%. The gray error bar on the right side of each plot is the correlated uncertainty due to the N_{coll} scaling for each $\sqrt{s_{NN}}$ energy calculated from MC Glauber. Vertical bars are used to show statistical uncertainty only.

unity at low p_{T} (< 1.0 GeV/c). In this region number of binary collision scaling (N_{coll}) is not expected to hold. Instead number-of-participant scaling (N_{part}) is expected to better describe

Journal of Physics: Conference Series **736** (2016) 012005 doi:10.1088/1742-6596/736/1/012005

the production at low p_{T} where soft processes dominate.

From Fig. 2 we can observe that the R_{CP} for $\bar{p}$ shows very little $\sqrt{s_{NN}}$ dependence, while the R_{CP} of p shows some non-trivial splitting at intermediate p_{T} as a function of $\sqrt{s_{NN}}$. The R_{CP} of both p and $\bar{p}$ show enhancement at the highest measured p_{T} for each energy. Unlike p and $\bar{p}$, the R_{CP} of π^+ and π^- show significant dependence on $\sqrt{s_{NN}}$ for $p_{\mathrm{T}} > 1.0$ GeV/c.

While high p_{T} suppression is attributed to partonic energy loss, other effects such as coalescence, radial flow and cold nuclear matter effects, similar to the Cronin effect, can lead to enhancement in the R_{CP} [10]. Previous measurements have shown that there is significant radial flow even at the BES I energies [11]. The presence of radial flow will contribute to the enhancement observed in the p and $\bar{p}$ R_{CP} at intermediate p_{T}. The pions are less affected by radial flow and coalescence and, therefore, the pion R_{CP} should be a more sensitive probe for observing suppression due to partonic energy loss.

The uncertainty on the N_{coll} scaling calculated with MC Glauber is shown on the right side of each plot. The scaling uncertainty makes it difficult to determine at exactly which energy the π^+ and π^- R_{CP} transitions from suppression to unity to enhancement. However, since the scaling uncertainty is correlated from one energy to the next, it is clear that the relative amount of suppression increases as a function of $\sqrt{s_{NN}}$.

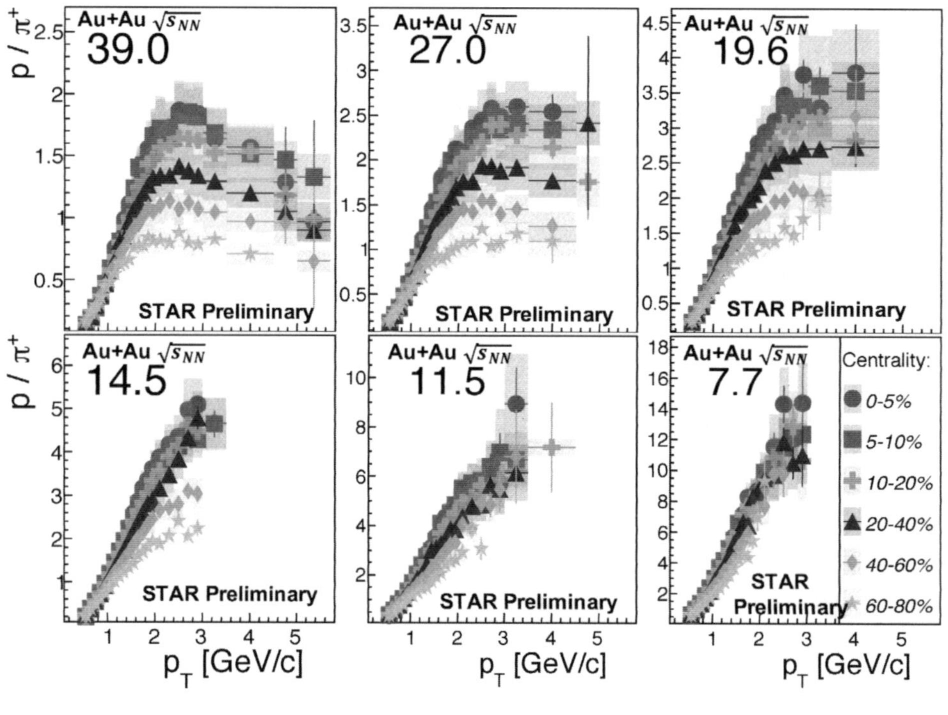

Figure 4. The p / π^+ ratio for Au+Au collisions at $\sqrt{s_{NN}} = 7.7$, 11.5, 14.5, 19.6, 27.0, and 39.0 GeV. Horizontal bars represent the p_{T} range used for measuring each point in the spectra. Vertical bars are used to show statistical uncertainty while shaded boxes are used to show systematic uncertainties.

In Fig. 3 the R_{CP} calculated from 0-5% central events over 40-60% central events for K^0_s, Λ, Ξ, and Ω is shown. In this figure, each plot shows the R_{CP} for a single $\sqrt{s_{NN}}$ energy for all particle to highlight the particle species dependence at each energy. At $\sqrt{s_{NN}} = 14.5$ GeV and above the R_{CP} shows significant dependence on the particle species. However, for the two lowest $\sqrt{s_{NN}}$ energies, the R_{CP} for K_s^0, Λ, and Ξ are roughly consistent. This is evidence that hadronic degrees of freedom are dominant at the lowest $\sqrt{s_{NN}}$ energies, while partonic degrees of freedom may play a larger roll as the $\sqrt{s_{NN}}$ energy is increased.

Previous STAR measurements in Au+Au collisions at $\sqrt{s_{NN}} = 200$ GeV observed that the baryon-to-meson ratio (e.g. Λ/K_s^0) in central events at intermediate p_{T} is significantly larger than the same ratio measured in pp collisions[12, 13, 14]. The models capable of reproducing this effect generally include coalescence and recombination when simulating the momentum distribution of the particles[15]. Since modeling the baryon-to-meson ratios require the use of coalescence and recombination, the baryon-to-meson ratio is an observable that is considered sensitive to the formation of a QGP.

Measurements of the p/π^+ ratios are shown in Fig. 4 for several centralities for each BES I energy. The Λ/K_s^0 ratios are shown in Fig. 5 (previously shown in [16]) for several centralities for each BES I energy. In both the p/π^+ and Λ/K_s^0 ratios we observe a significant centrality dependence at intermediate to high p_{T} for $\sqrt{s_{NN}}$ energies of 14.5 GeV and above. However, no significant centrality dependence is observed for $\sqrt{s_{NN}} = 11.5$ GeV and below. This may be evidence that recombination and coalescence are more dominant production mechanisms at higher $\sqrt{s_{NN}}$ energies, while fragmentation is a more dominant production mechanism at lower $\sqrt{s_{NN}}$.

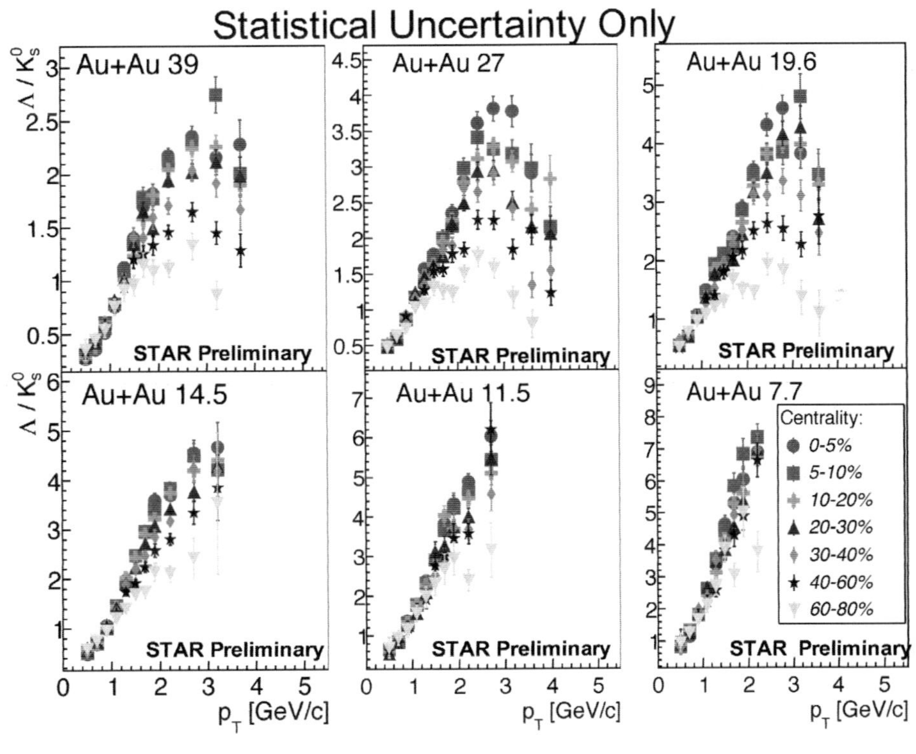

Figure 5. The Λ/K_s^0 ratio for Au+Au collisions at the BES I energy. Vertical bars are used to show statistical uncertainty only.

Journal of Physics: Conference Series **736** (2016) 012005 doi:10.1088/1742-6596/736/1/012005

4. Summary
In this contribution a systematic study of the particle-identified R_{CP} and baryon-to-meson ratio is presented for Au+Au collisions at the BES I energies of $\sqrt{s_{NN}}$ = 7.7, 11.5, 14.5, 19.6, 27.0, and 39.0 GeV. The measurements presented here would benefit from an increased p_T reach, especially for the lowest collision energies. The Beam Energy Scan Phase II (BES II) program at RHIC will provide significantly more data at these energies and will allow these results to be improved in the future.

5. Acknowledgements
This work is supported by the US Department of Energy under grant DE-FG02-10ER41666.

[1] Aggarwal, M *et al* 2010 *arXiv* **1007.2613**
[2] Cleymans J, Oeschler H, Redlich K and Wheaton S 2006 *Phys. Rev. C* **73**
[3] Miller M L, Reygers K, Sanders S J and Steinberg P 2007 *Annu. Rev. Nucl. Part. Sci.* **57** 205–43
[4] Abelev, B I *et al* 2009 *Phys. Rev. C* **79**
[5] Adams, J *et al* 2003 *Phys. Rev. Lett.* **91**
[6] Ackermann, K H *et al* 2003 *Nucl. Instr. and Meth. A* **499** 624–32
[7] Anderson, M *et al* 2003 *Nucl. Instr. and Meth. A* **499** 659–78
[8] Llope, W J *et al* 2004 *Nucl. Instr. and Meth. A* **522** 252–73
[9] Brun R, Bruyant F, Maire M, McPherson A C and Zanarini P 1987 **CERN-DD-EE-84-1**
[10] Cronin J W, Frisch H J, Shochet M J, Boymond J P, Pirou P A and Sumner R L 1975 *Phys. Rev. D* **11** 3105–23
[11] Odyniec G 2013 *J. Phys.: Conf. Ser.* **455**
[12] Abelev, B I *et al* 2007 *Physics Letters B* **655** 104–13
[13] Caines, H *et al* 2006 *Phys. Rev. Lett.* **97**
[14] Lamont, M A *et al* 2006 *J. Phys. G: Nucl. Part. Phys.* **32**
[15] Cuautle E and Ayala A 2014 *J. Phys.: Conf. Ser.* **509**
[16] Ashraf M U 2016 *Journal of Physics: Conference Series* **668**

Journal of Physics: Conference Series **736** (2016) 012006 doi:10.1088/1742-6596/736/1/012006

Photon and neutral meson measurements with the ALICE experiment at the LHC

A Vauthier (for the ALICE Collaboration)
LPSC, Université Grenoble-Alpes, CNRS/IN2P3 - 53, Avenue des Martyrs 38026 Grenoble Cedex

E-mail: astrid.vauthier@cern.ch

Abstract. The measurements of photons and neutral mesons, such as π^0 and η, allow to explore the QCD matter created in heavy-ions collisions. In pp collisions, observables such as differential particle production cross sections can be used to test perturbative QCD calculations and constrain PDFs. In Pb-Pb collisions, neutral meson spectra address the medium induced suppression, whereas low transverse momentum direct photons allow to obtain information on the QCD medium temperature or anisotropic flow.

We present an overview of photon and neutral meson measurements, employing three different methods which use both ALICE calorimeters, as well as the central tracking system via the photon conversion method. These experimental results are compared with theoretical predictions.

1. Introduction

Quantum chromodynamics (QCD), which describes the strong interaction, predicts a phase transition above a given critical energy density from ordinary nuclear matter - composed of hadrons - to a hot and dense plasma of deconfined partons: the Quark-Gluon Plasma (QGP). The measurement of photons and neutral mesons allows to test perturbative QCD (pQCD) with pp collisions, but also to study the properties of the QGP, which is created in ultra-relativistic heavy-ion collisions.

Measured neutral mesons and photons give different information depending on the p_{T} range of the measurements. At low p_{T}, neutral mesons can give insight about collective effects via the modification of their p_{T} spectra and thermal photons, which are radiated by the fireball, allow to access bulk properties. On the other hand, at high p_{T}, neutral mesons coming from the fragmentation of partons created in hard processes during the early stages of the collision allow us to study medium induced parton energy loss. On the contrary, photons produced in hard scattering processes, do not interact strongly with the QGP and provide unmodified information on the early stage of the collisions.

2. Photon and neutral meson detection in the ALICE experiment

The ALICE experiment is dedicated to the study of the QGP at the LHC [1].

Photons are detected using three complementary methods: electromagnetic shower detection with both ALICE electromagnetic calorimeters (EMCal and PHOS) and the photon conversion method (PCM) within the ALICE tracking system, using the Inner Tracking System (ITS) and the Time Projection Chamber (TPC).

Journal of Physics: Conference Series **736** (2016) 012006 doi:10.1088/1742-6596/736/1/012006

The EMCal calorimeter is composed of sampling layers of Pb (1.4 mm thickness) and scintillator (1.7 mm thickness). The detector energy resolution is: $\frac{\sigma_E}{E} = \frac{0.05}{E} \oplus \frac{0.11}{\sqrt{E}} \oplus 0.017$. The EMCal is located at 4.4 m from the ALICE interaction point and covers 1.4 pseudo-rapidity units and 100° in azimuth angle during Run I [2].

The PHOS calorimeter is composed of $PbWO_4$ crystals. The granularity of the detector is higher than for the EMCal and its energy resolution is: $\frac{\sigma_E}{E} = \frac{0.018}{E} \oplus \frac{0.033}{\sqrt{E}} \oplus 0.011$. The PHOS is located at 4.6 m from the ALICE interaction point and its acceptance covers 0.26 units in pseudo-rapidity and 60° in azimuth angle during Run I [3].

The PCM method allows to measure photons by reconstructing the e^+e^- pairs coming from the conversion of photons in material. The low conversion probability, of about 8 % [4], is compensated by the large acceptance of the tracking system: 1.8 units in pseudo-rapidity and full coverage in azimuth (φ).

The π^0 and η neutral mesons are identified by reconstructing the invariant mass ($M_{\gamma\gamma}$) of all photon pairs in an event. The PCM method allows to reach very low p_T ($p_T > 0.3$ GeV/c) for the π^0 reconstruction with an excellent resolution, while EMCal and PHOS allow to reach higher p_T (up to 20 GeV/c).

All three methods provide independent measurements with different sources of systematic uncertainties. Combining their measurements therefore allows to minimize those.

3. Neutral meson measurement in pp collisions

Measurements in pp collisions are necessary as a reference for Pb-Pb collisions and will test the pQCD predictions.

Figure 1 shows the π^0 p_{T}-differential invariant cross section obtained from PCM and PHOS combined measurements. The invariant cross section is presented for several center of mass energies $\sqrt{s} = 0.9$, 2.76, 7 and 8 TeV (the last one is a preliminary PHOS only measurement) [5]. The invariant cross sections are compared with PYTHIA and next-to-leading order pQCD (NLO pQCD) calculations with MSTW parton distribution functions (PDF) and DSS14 fragmentation functions (FF) [6]. A good overall agreement is found between data points and calculations, although some tension appears at high p_{T} and center of mass energy.

Figure 2 presents the η p_{T}-differential invariant cross sections also obtained with PCM and PHOS combined measurements. The invariant cross sections are shown for three center of mass energies $\sqrt{s} = 0.9$, 2.76 and 7 TeV. The measurements are compared to NLO pQCD calculations with CTEQ6M5 PDF and the AES fragmentation function [7]. The observed stronger discrepancy between data and calculations compared to π^0 is due to the PDF and fragmentation function chosen as shown by Fig. 2.

The p_{T} dependence of the η/π^0 ratio, presented in Fig. 3, for center of mass energies $\sqrt{s} = 2.76$ and 7 TeV [5], allows us to compare strange and light quark production. This ratio flattens, above 4 GeV/c, for both center of mass energies around the value compatible with the NLO pQCD calculations using CTEQ6M5 as PDF and DSS (π^0) and AES (η) for the fragmentation functions. Although this set of PDF and fragmentation function doesn't reproduce the individual spectra, their bias seems to mainly cancel out in the ratio.

4. Neutral meson measurement in Pb-Pb collisions

The nuclear effects present in Pb-Pb collisions can be quantified with the nuclear modification factor:

$$R_{AA} = \frac{\frac{d^2N_{AA}}{dp_{\mathrm{T}}d\eta}}{< T_{AA} > \frac{d^2\sigma_{pp}}{dp_{\mathrm{T}}d\eta}} \tag{1}$$

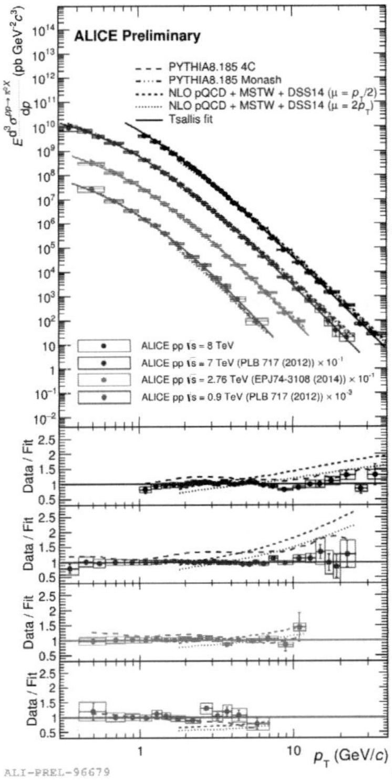

Figure 1. π^0 invariant cross section in pp collisions at $\sqrt{s}$ = 0.9, 2.76, 7 and 8 TeV, obtained from PCM and PHOS combined measurements. The results are compared to NLO pQCD (MSTW + DSS) and PYTHIA calculations.

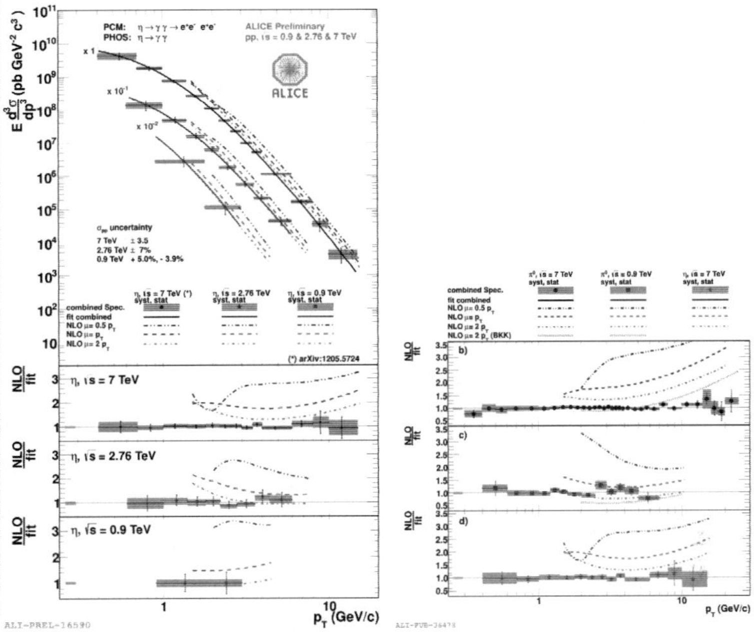

Figure 2. Left: η invariant cross section in pp collisions at $\sqrt{s}$ = 0.9, 2.76 and 7 TeV, obtained from PCM and PHOS combined measurements. The results are compared to NLO pQCD (CTEQ6M5 + AES) and PYTHIA calculations. Right: Comparison of π^0 invariant cross section with NLO pQCD (CTEQ6M5 + AES). The same discrepancy is observed for η and π^0.

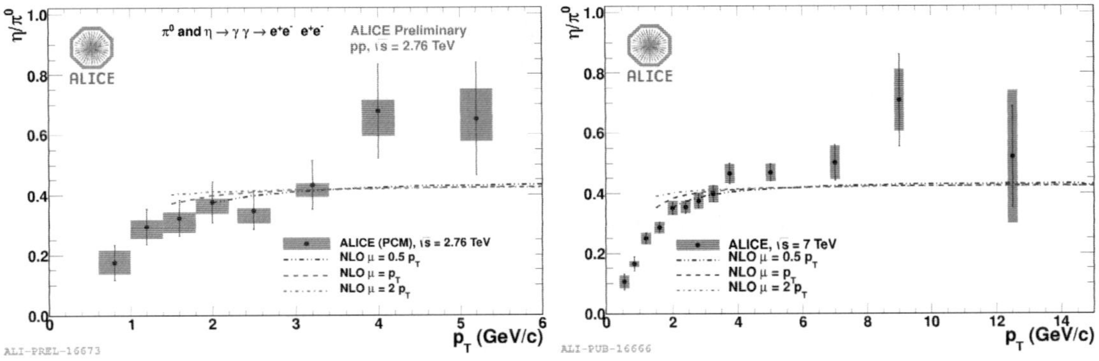

Figure 3. p_{T} dependence of the η/π^0 ratio for pp collisions at $\sqrt{s}$ = 2.76 (left) and 7 TeV (right) together with results from NLO pQCD calculations.

Journal of Physics: Conference Series **736** (2016) 012006 doi:10.1088/1742-6596/736/1/012006

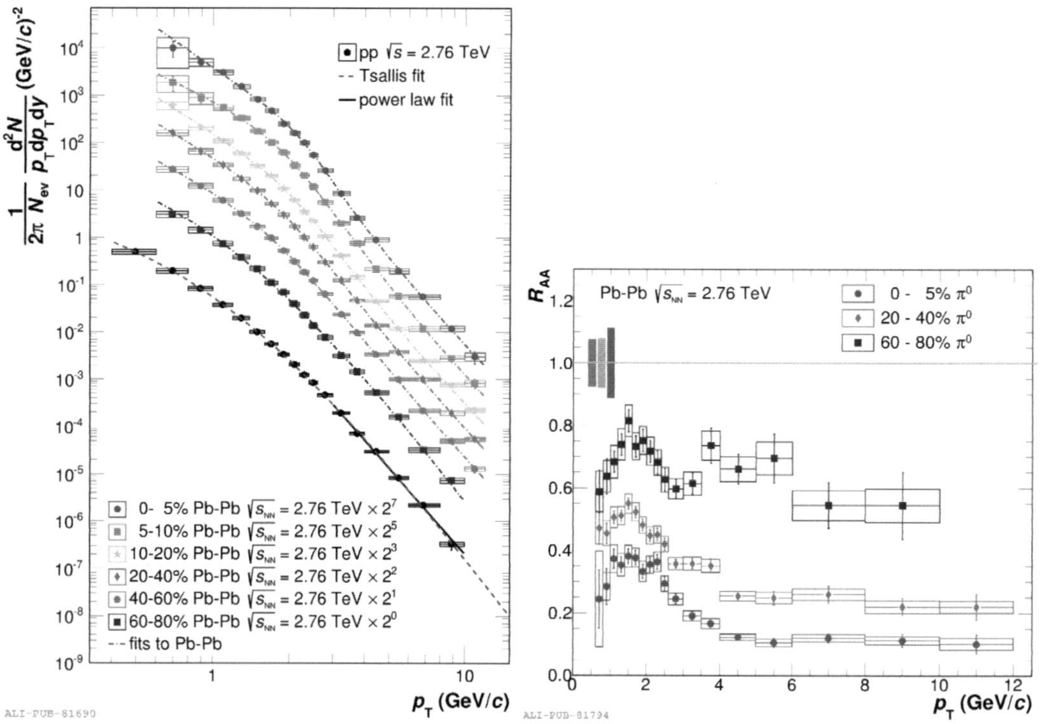

Figure 4. Left: π^0 invariant yield in Pb-Pb collisions at $\sqrt{s_{\mathrm{NN}}} = 2.76$ TeV, from PCM and PHOS combined measurement for six centrality classes. Right: π^0 R_{AA} p_{T} dependence for three centrality classes.

where $< T_{AA} >=< N_{coll} > /\sigma_{pp}^{inel}$, $< N_{coll} >$ is the number of binary collisions in Pb-Pb collisions and σ_{pp}^{inel} is the inelastic cross section for pp collisions.

Figure 4, left panel, presents the PCM and PHOS combined measurement of the invariant yield of π^0 from the 2010 data [8]. The invariant yield is shown for six centrality classes. The right panel presents the nuclear modification factor obtained for three centrality classes: central (0-5%), semi-central (20-40%) and peripheral (60-80%). A suppression at high p_T is observed and is attributed to an increasing energy loss in the medium with centrality. Those results are in agreement, at high p_{T}, with the R_{AA} of other particle species [9].

The energy dependence of the π^0 nuclear modification factor as a function of p_{T} is presented in Fig. 5 [8]. The ALICE result for central collisions is compared to the SPS result [10] at $\sqrt{s} = 17.3$ GeV and to the PHENIX results at $\sqrt{s} = 39, 62.4$ and 200 GeV [11][12]. With the growth of the center of mass energy, two competitive effects appear: a higher energy density which enhances the suppression of the high-p_{T} particle yields, and a harder parton spectrum which drives the nuclear modification factor closer to unity. The net effect observed is a decrease of the nuclear modification factor with an increasing center of mass energy.

Figure 6 presents the first η (right panel) measurement in Pb-Pb collisions at the LHC together with π^0 (left panel) invariant yields from EMCal and PCM combined measurements with 2011 data for central (0-10 %) and semi-central (20-50 %) collisions. The 2011 dataset allows to extend the p_T reach up to 20 GeV/c, with respect to the 2010 dataset. NLO pQCD calculations with MSTW for the PDF and DSS14 (π^0) and DSS07 (η), scaled by $< N_{coll} >$ are used as a the pp reference up to 20 GeV/c and appear as the grey lines on both figures to highlight the known suppression in Pb-Pb collisions compared to pp collisions.

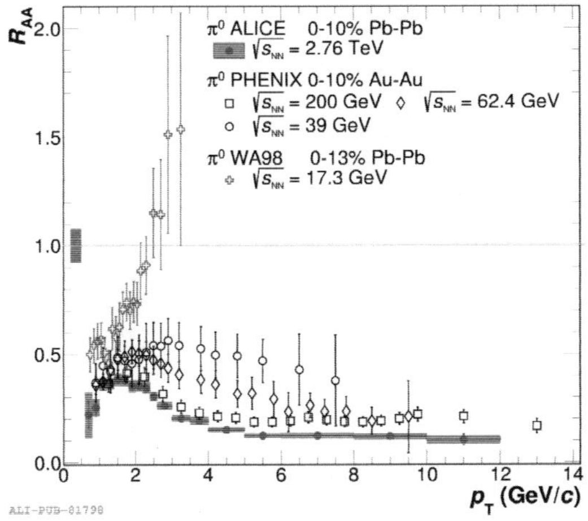

Figure 5. $\sqrt{s_{NN}}$ evolution of the π^0 R_{AA} p_T dependence.

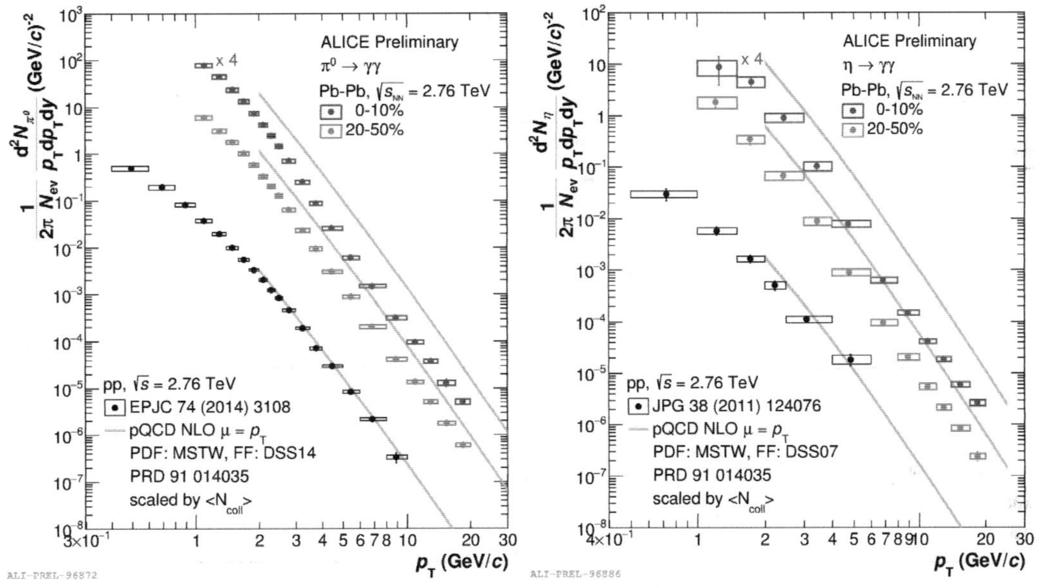

Figure 6. Left: π^0 invariant yield from EMCal and PCM combined measurements in Pb-Pb collisions at $\sqrt{s_{NN}}$ = 2.76 TeV with the 2011 data. Right: η invariant yield from EMCal and PCM combined measurements in Pb-Pb collisions with 2011 data. The grey lines correspond to NLO pQCD calculations.

The p_T dependence of the η/π^0 ratio obtained from 2011 data is shown in Fig. 7. This ratio is compared to the η/π^0 ratio measured in 7 TeV pp collisions [5] (left), to the $K^\pm/\pi^\pm$ ratio measured in Pb-Pb collisions [13] (middle) and to NLO pQCD with added energy loss [14] (right). The η/π^0 ratio seems compatible, within the large uncertainties, with η/π^0 ratio in pp collisions and $K^\pm/\pi^\pm$ ratio in Pb-Pb collisions. The pQCD calculation reproduces qualitatively the data.

5. π^0-hadron correlations

Partons that are produced in hard processes at the early stage of the collisions will hadronize in the QCD vacuum in pp collisions. In Pb-Pb collisions, the parton will pass through the QGP

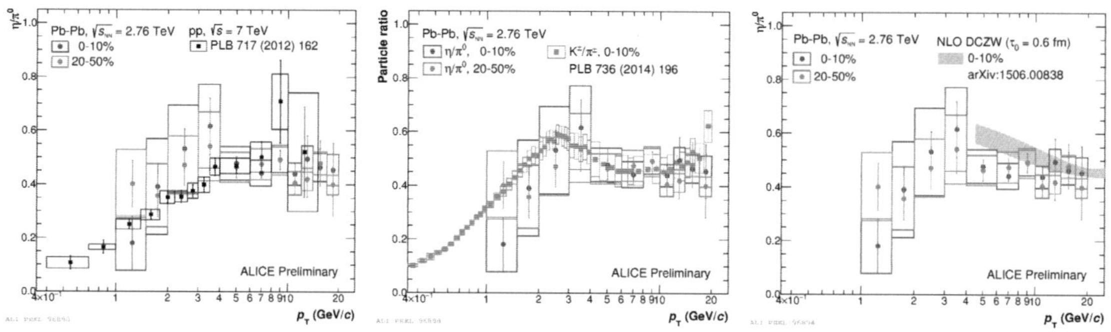

Figure 7. The η/π^0 ratio obtained for Pb-Pb collisions with 2011 data, shown in all three panels, is compared to the η/π^0 ratio in pp collisions (left), to the $K^{\pm}/\pi^{\pm}$ ratio in Pb-Pb collisions (middle) and to NLO pQCD calculations with added energy loss (right).

and will interact with the medium and therefore lose energy and then will hadronize.

The measurement of π^0-hadron correlations is performed by associating high p_T π^0 which comes from a parton emitted in a hard process with all the hadrons of the same event. On the away side (opposite side of the π^0), correlated hadrons mainly come from the recoiling parton that went through the medium in Pb-Pb collisions, leading to an energy loss of this parton. Hence the comparison of π^0-hadron correlations in pp and Pb-Pb collisions allows to highlight the parton energy loss mechanism.

The angular correlations $\Delta\varphi = \varphi^{\pi^0} - \varphi^{hadron}$ for pp and Pb-Pb collisions are presented in Fig. 8. The underlying event is not subtracted in the figures presented. The two peaks observed at $\Delta\varphi = 0$ and $\Delta\varphi = \pi$ come from the two jets emitted back-to-back in the hard process. The medium induced per-trigger yield modification factor, I_{AA} is built from the yield Y of one of the peaks (i.e. its integral after subtraction of the underlying event):

$$I_{AA}(p_T^{\pi^0}, p_T^{\text{assoc}}) = \frac{Y^{AA}(p_T^{\pi^0}, p_T^{\text{assoc}})}{Y^{pp}(p_T^{\pi^0}, p_T^{\text{assoc}})} \tag{2}$$

The variation of the yield, Y, with $p_{\text{T}}^{\text{assoc}}$ (transverse momentum of the associated hadrons) is presented in Fig. 9 for the near and away side peaks, for a given $p_{\text{T}}^{\text{trig}}$ bin (transverse momentum of the π^0). An enhancement of the yield is observed in Pb-Pb collisions compared to pp collisions on the near side (where the π^0 is). On the away side, where the hadrons from the fragmentation of the recoiling parton are found, a suppression is observed, which is attributed to parton energy loss in the QCD medium.

6. Direct photon measurement in Pb-Pb collisions

The measurement of direct photons in Pb-Pb collisions is crucial to better estimate bulk properties, as the temperature or energy density. Direct low p_{T} photons can be extracted by subtracting the decay photons from the inclusive photon spectrum. This is achieved by the calculation of the double ratio $R_\gamma = \frac{\gamma_{\text{inc}}}{\pi^0} / \frac{\gamma_{\text{decay}}}{\pi^0_{\text{param}}}$ where γ_{inc} stands for the inclusive photon spectrum, γ_{decay} for the simulated decay photon cocktail spectrum and π^0_{param} for the parametrization of the measured π^0 spectrum. Figure 10 (left panel), shows that R_γ is above unity, in particular in central collisions over the full measured p_{T} range, which reflects the contribution from direct photons [4]. At high p_{T} ($p_{\text{T}} > 4$ GeV/c) and all centralities an excess compatible with pQCD and JETPHOX predictions [15] is observed. At low p_{T} and central

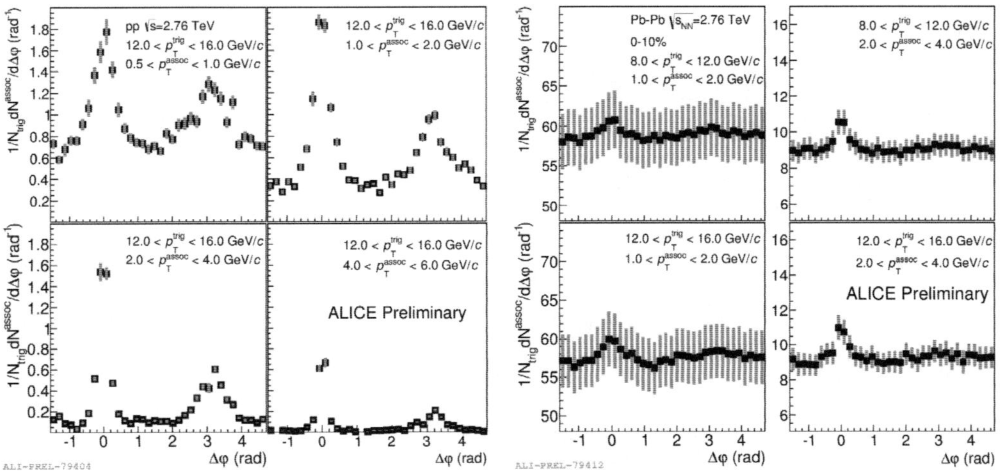

Figure 8. Relative azimuthal angle distribution of correlated π^0-hadron pairs for pp collisions (left) and Pb-Pb collisions (right). The angular distributions are presented for several $p_{\mathrm{T}}^{\mathrm{assoc}}$ bins. The away side peak is hardly visible in Pb-Pb collisions due to a large contribution from the underlying event.

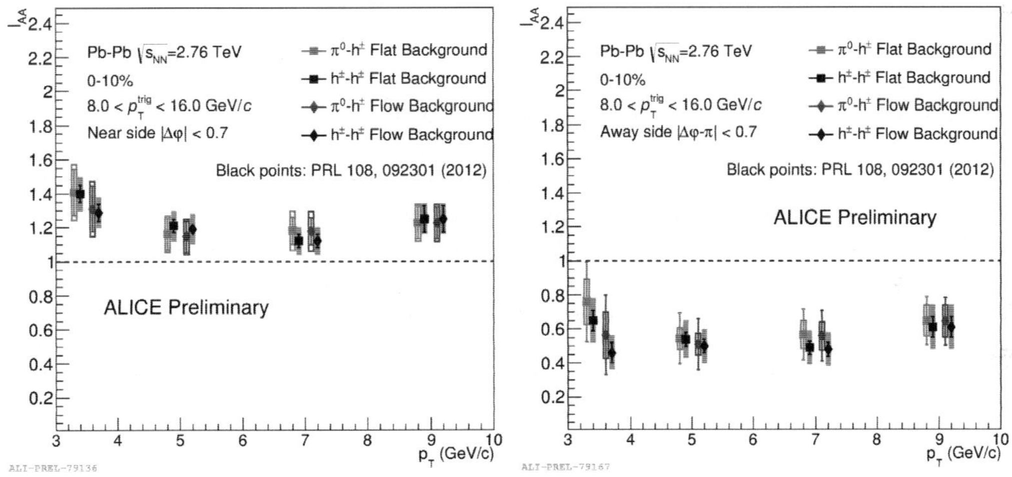

Figure 9. I_{AA} on the near side (left) and on the away side (right) peaks for a given $p_{\mathrm{T}}^{\mathrm{trig}}$ bin (for explanation see text).

collisions a 2.6σ excess relative to pQCD is observed (Fig. 10 right panel), which is attributed to the presence of direct thermal photons radiated by the QGP.

Figure 10, right panel, presents the p_{T} spectrum extracted for direct photons where an excess of about 2.6σ of direct photons attributed to thermal photons radiated by the QGP can be seen at low p_{T} for central collisions.

7. Conclusions

Neutral mesons and photons are measured by ALICE in a wide p_T range with complementary methods.

The neutral meson cross sections have been presented for pp collisions. They are fairly well

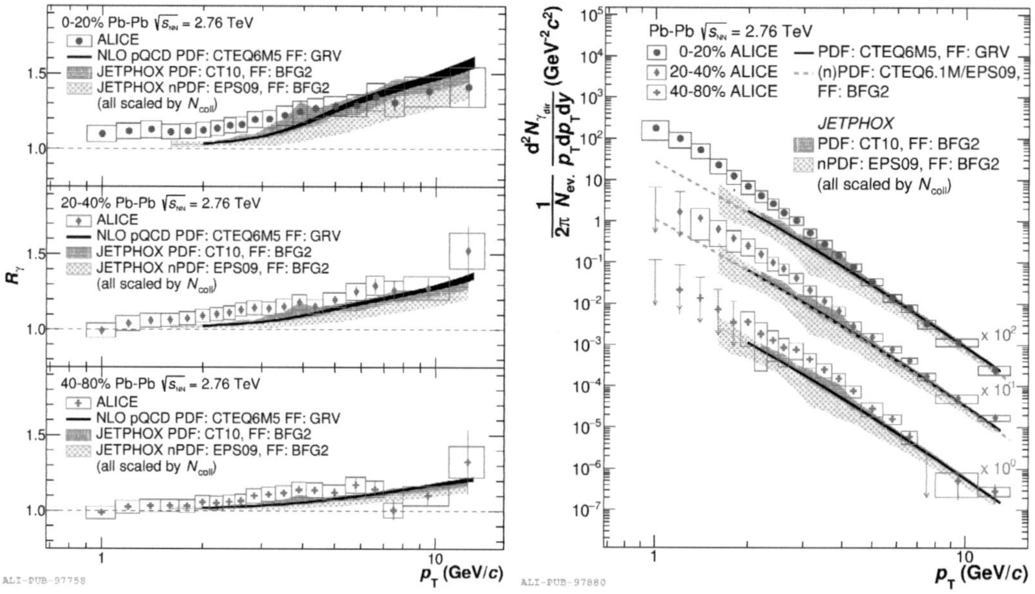

Figure 10. Left: R_γ as a function of $p_{\rm T}$ (see text). Right: direct photon $p_{\rm T}$ spectrum for three centrality classes in Pb-Pb collisions at $\sqrt{s_{NN}} = 2.76$ TeV.

described by NLO pQCD calculations and PYTHIA in the intermediate p_T region (< 10 GeV/c). The η/π^0 ratio is also well described by NLO pQCD calculations.

In Pb-Pb collisions, the π^0 invariant yield measurement and the corresponding R_{AA} show an increasing π^0 suppression with increasing centrality and center of mass energy. The η measurement in Pb-Pb collisions, which is the first η measurement at the LHC, allows to access the η/π^0 ratio in Pb-Pb collisions. Within the large uncertainties this ratio is compatible with η/π^0 ratio in pp collisions and also the $K^\pm/\pi^\pm$ ratio in Pb-Pb collisions.

π^0-hadron correlations have been presented for pp and Pb-Pb collisions. The medium induced per-trigger yield modification factor, I_{AA}, shows a suppression in Pb-Pb collisions on the away side, which is attributed to parton medium-induced energy loss.

Finally, an excess of low p_T photons has been observed in Pb-Pb collisions, and it can be attributed to thermal photons radiated by the QGP.

References

[1] Antinori F and Wiedemann U 2004 *Journal of Physics G: Nuclear and Particle Physics* **30** 1515
[2] Cortese P *et al.* (ALICE) 2008 *CERN-LHCC-2008-014, CERN-ALICE-TDR-014*
[3] Dellacasa G *et al.* (ALICE) 1999 *CERN-LHCC-99-04*
[4] Adam J *et al.* (ALICE) 2016 *Physics Letters B* **754** 235 – 248 ISSN 0370-2693
[5] Abelev B *et al.* (ALICE) 2012 *Phys. Lett.* **B717** 162–172 (*Preprint* 1205.5724)
[6] de Florian D, Sassot R, Epele M, Hernández-Pinto R J and Stratmann M 2015 *Phys. Rev. D* **91**(1) 014035
[7] Aidala C A, Ellinghaus F, Sassot R, Seele J P and Stratmann M 2011 *Phys. Rev. D* **83**(3) 034002
[8] Abelev B B *et al.* (ALICE) 2014 *Eur. Phys. J.* **C74** 3108 (*Preprint* 1405.3794)
[9] Abelev B B *et al.* (ALICE) 2014 *Phys. Lett.* **B736** 196–207 (*Preprint* 1401.1250)
[10] Aggarwal M M *et al.* (WA98 Collaboration) 2008 *Phys. Rev. Lett.* **100**(24) 242301
[11] Adare A *et al.* (PHENIX Collaboration) 2012 *Phys. Rev. Lett.* **109**(15) 152301
[12] Adare A *et al.* (PHENIX Collaboration) 2008 *Phys. Rev. Lett.* **101**(23) 232301
[13] Abelev B *et al.* 2014 *Physics Letters B* **736** 196 – 207 ISSN 0370-2693
[14] Dai W, Chen X F, Zhang B W and Wang E 2015 *Physics Letters B* **750** 390 – 395 ISSN 0370-2693
[15] Klasen M, Klein-Bosing C, Konig F and Wessels J P 2013 *JHEP* **10** 119 (*Preprint* 1307.7034)

Journal of Physics: Conference Series **736** (2016) 012007 doi:10.1088/1742-6596/736/1/012007

STAR D^0 meson v_2 measurement

Liang He, For the STAR Collaboration
Department of Physics and Astronomy, Purdue University, West Lafayette, IN 47907, USA
E-mail: he202@purdue.edu

Abstract. Experiments at the Relativistic Heavy Ion Collider (RHIC) indicate that a novel QCD matter consistent with a strongly-coupled Quark-Gluon Plasma (sQGP) has been created. Heavy quarks are predominantly created in initial hard scatterings, and thus experience the entire evolution of the system including the sQGP phase. The sensitivity of heavy quarks to the transport properties of sQGP is encoded in experimental observables such as elliptic flow (v_2). In this presentation, we present the first measurement of v_2 of D^0 mesons reconstructed with the STAR Heavy Flavor Tracker (HFT) in Au+Au collisions at $\sqrt{s_{\rm NN}}$=200 GeV, for a wide transverse momentum range. The measured D^0 v_2 in 0-80% most central collisions is finite and suggests that the charm quark exhibits collective behavior. Several theoretical calculations with dimensionless spacial diffusion coefficient ($2\pi T D_s$) for charm quarks in the range of ~2-12 can simultaneously reproduce our D^0 v_2 result as well as the previously published D^0 nuclear modification factor.

1. Introduction

Heavy quarks (charm and bottom) are predominantly produced during hard scatterings at the early stage of heavy-ion collisions [1]. They thus experience the entire evolution of the medium created in these high energy nuclear collisions. Compared to the nuclear modification factor R_{AA}, the anisotropic flow (v_2) of heavy quarks at low transverse momenta(p_T) is less affected by several cold nuclear matter effects such as shadowing and initial-state energy loss [2]. It can thus be used to probe the strongly-coupled bulk medium. Since the masses of heavy quarks are significantly larger than the medium energy scale, Brownian motion may be one of the approaches to describe the propagation of heavy quarks in the medium. It allows to establish quantitative connections between the heavy-quark spacial diffusion coefficient($2\pi T D_s$) and the anisotropic flow [1].

Recent measurements at RHIC and LHC show that yields of high p_T charmed hadron are considerably suppressed in central collisions suggesting strong interactions between charm quarks and the bulk medium [3, 4, 5]. The D^0 v_2 measured by ALICE [6, 7] is comparable with that of light quarks. At RHIC energy the charm quark flow so far can only be inferred from measurements of semi-leptonic decays, which however suffer from large uncertainties from the experimental background, charm hadron decay kinematics and ambiguity of charm or bottom decays. A precise measurement of charm hadron v_2 over a wide momentum range will provide valuable insights into transport properties of sQGP.

2. Experimental setup

The data used for this analysis were collected by the STAR experiment at RHIC in Au+Au collisions at $\sqrt{s_{\rm NN}}$ = 200 GeV in 2014. About 780 million events with minimum bias (MB) trigger are used in this analysis. D^0 are reconstructed through the hadronic decay channels $D^0(\bar{D^0}) \to K^{\mp}\pi^{\pm}$, with $c\tau \approx 122.9$ μm and branching ratio of $3.88 \pm 0.05\%$.

Tracks are reconstructed with the Time Projection Chamber (TPC) and Heavy Flavor Tracker (HFT). The TPC covers the full azimuth at mid-rapidity ($-1 < \eta < 1$). The HFT, a newly installed silicon detector, has about the same coverage as TPC. Figure 1 shows the track pointing resolution of kaon, pion and proton along the transverse direction and beam direction. It is better than 50 μm for tracks with $p_T > 750$ MeV/c. The tracks in this analysis are required to have at least 20 space points in the TPC and at least one hit on each of three layers of the HFT.

The particle identification for D^0 daughters is done with the TPC and the Time of Flight (TOF) detector. The TPC provides energy loss measurement dE/dx. Particles are required to be within 2 (for $K^{\pm}$) or 3 (for $\pi^{\pm}$) standard deviations from the expected value. If the TOF information is available, the $1/\beta$ is required to be within 0.03 from the expected value.

The decay vertex is reconstructed as the mid point of DCA between $K^{\pm}$ and $\pi^{\mp}$. Topological cuts are used to suppress the combinatorial background. The topological cuts are:

- decay length (distance from decay vertex to primary vertex)
- DCA between $K^{\pm}$ and $\pi^{\mp}$
- DCA of $K^{\pm}$ ($\pi^{\mp}$) to primary vertex
- DCA of D^0 candidates to primary vertex

The TMVA(Toolkit for MultiVariate Analysis) is used to optimize cuts to achieve best signal significance for each $p_T(K\pi)$ bin. The optimized cuts are listed in Table 1.

$D^0 p_T$ (GeV/c) / cuts	0-1	1-2	2-3	3-5	5-10
decay length (μm) >	145	181	212	247	259
DCA(K, π)(μm) <	84	66	57	50	60
DCA(K, PV)(μm) >	103	91	95	79	58
DCA(π, PV)(μm) >	110	111	86	81	62
DCA($V0, PV$)(μm) >	61	49	38	38	40

Table 1. Topological cuts used for D^0 reconstruction.

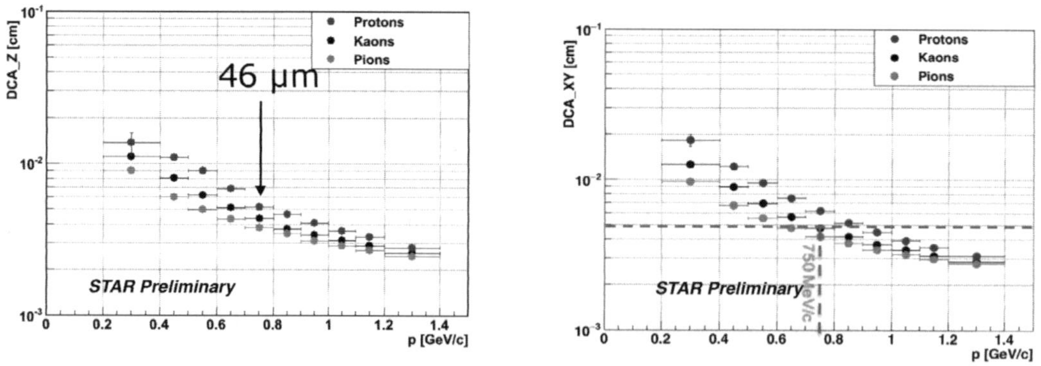

Figure 1. Track pointing resolution along beam direction (left) and transverse direction (right) as a function of the particle momentum.

3. Azimuthal anisotropy

Two different methods are used to measure D^0 v_2: event plane method and two-particle correlation method.

In the event plane method, the second order event plane Ψ_2 is reconstructed using TPC tracks and corrected for the non-uniform detector efficiency [8]. A minimum pseudorapidity (η) gap of 0.15 around the D^0 candidate is applied to the tracks used for event plane reconstruction to suppress non-flow effect. For different p_T bins the yields of D^0 candidates are measured in azimuth relative to the event plane Ψ. In order to avoid bias from D^0 reconstruction efficiency and event plane resolution, D^0 condidates are weighted by a factor of $1/(\epsilon * R)$, where ϵ is the D^0 reconstruction efficiency and R is the event plane resolution [8]. The background is estimated from mixed events [9] scaled to the like-sign yield within the invariant mass range of 1.6-2.1 GeV/c^2. For $1 < p_T < 5$ GeV/c, the D^0 yield is obtained by fitting invariant mass distribution of $K\pi$ unlike-sign pairs with a Gaussian function (representing D^0 signal) plus a first-order polynomial function (for background). In $5 < p_T < 10$ GeV/c bin, the D^0 yield is obtained by subtracting scaled counts in side-band regions (1.71-1.80 and 1.93-2.02 GeV/c^2) from counts in the signal region (1.82-1.91 GeV/c^2). Figure 2 shows the D^0 yield as a function of $\phi - \Psi$ for 3 $< p_T < 4$ GeV/c, where ϕ is the azimuthal angle of D^0. $v_2^{observed}$ is obtained by fitting the data with $A(1 + 2v_2 \cos(2(\phi - \Psi)))$. The observed v_2 is scaled by $\langle 1/R \rangle$ to obtain the final v_2 [10].

In the correlation method [11, 12], D^0 v_2 is calculated through the D^0-hadron azimuthal cumulant $V_2^{D^0 \cdot h} \equiv < \cos(2\phi_{D^0} - 2\phi_h) >$. D^0 v_2 can be obtained via $v_2^{D^0} = V_2^{D^0 \cdot h}/v_2^h$. v_2^h is measured using the same method, as $v_2^h = \sqrt{V_2^{h \cdot h}}$. To suppress the non-flow contribution, $|\Delta\eta| > 0.15$ between D^0 and correlated hadrons is applied in D^0-hadron azimuthal cumulant calculation. The hadron-hadron cumulant uses particles on opposite sides of the TPC in η. D^0 candidate v_2 and background v_2 are measured with the same method. D^0 candidates are unlike-sign $K\pi$ pairs within D^0 mass window ($\pm 3\sigma$); and background is represented by like-sign $K\pi$ pairs under D^0 peak and side bands ($4 - 9\sigma$ around D^0 peak using both like-sign and unlike-sign $K\pi$ pairs). Figure 2 shows the azimuthal cumulant of both D^0 candidate and background. With v_2 of D^0 candidate and background, as well as the yield of D^0 signal from the fitting, the signal v_2 is calculated as $v_2^{signal} = (N_{cand} \cdot v_{2cand} - N_{bg} \cdot v_{2bg})/N_{signal}$, where N_{cand} is the count under D^0 peak and $N_{bg} = N_{cand} - N_{signal}$.

Figure 3 presents the D^0 v_2 measured with the event plane and correlation methods. The results from two different methods show good agreement within systematic uncertainties. The grey band is an estimation of non-flow contribution. It is deduced from D^*-hadron correlation measured with p+p data taken by STAR in the year 2012, and corrected for different feed-down effects for D^0 and D^* using simulation. The systematic uncertainty on D^0 v_2 is dominated by background treatment. In the event plane method, the systematic uncertainty is estimated by varying several aspects of fit and side band methods mentioned above. In the correlation method, the uncertainty is estimated as the largest difference between using like-sign pairs and side bands. The yield difference from using different fitting functions (Gaussian plus exponential) is also considered as a contribution to systematic uncertainty in the correlation method.

4. Discussion

Figure 4 compares the D^0 v_2 with K_s [13], Ω and ϕ [14]. In order to eliminate the effects of different particle masses and different number of constituent quarks (NCQ), Figure 4 plots v_2/NCQ as a function of $(m_T - m_0)/NCQ$, where m_0 is the rest mass of particle and $m_T = \sqrt{p_T^2 + m_0^2}$. v_2/NCQ of K_s, Ω and ϕ fall into same trend. There is an indication that D^0 v_2 is systematically lower than those of the light flavor particles, which might suggest that charm quark is not fully thermalized in the medium.

Figure 4, right panel, shows the comparison of D^0 v_2 to different model calculations.

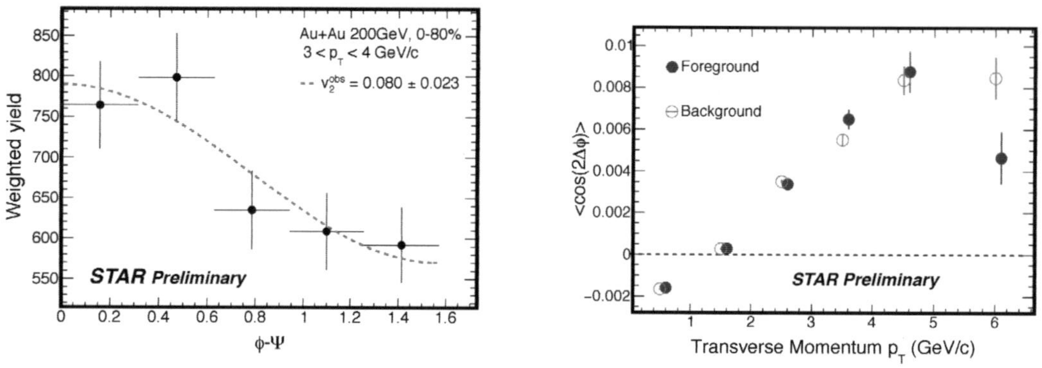

Figure 2. Left:D^0 yield as a function of $\phi - \Psi$ in $3 < p_T < 4$ GeV/c bin; Right:Azimuthal cumulant of D^0 candidates and background as a function of p_T.

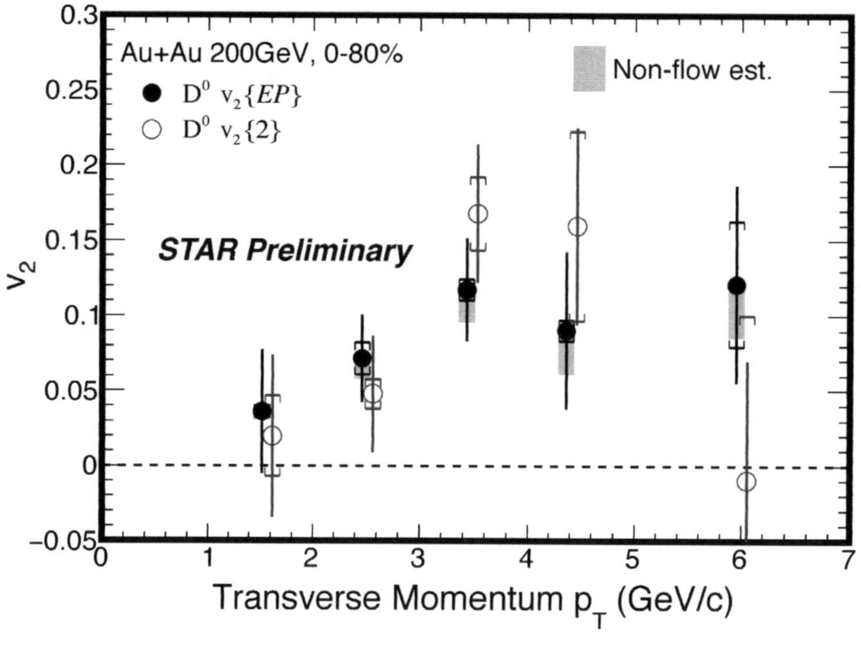

Figure 3. Comparison of v_2 using event plane method and two-particle correlation method

The models use different treatments of charm-medium interaction and have different effective diffusion coefficient $2\pi TD_s$. The SUBATECH model [15, 16] uses pQCD with Hard Thermal Loop (HTL) describing soft processes. The TAMU model [17, 18] is a non-perturbative model, using T-matrix with internal energy potential. Two curves of TAMU models are shown: the blue curve includes charm diffusion while the pink curve does not. Table 2 summarizes the $\chi^2/n.d.f$ measuring the difference between the measured D^0 v_2 and those from different models. The model developed by Duke group [19, 20] has a fixed diffusion coefficient $2\pi TD_s = 7$. The Parton-Hadron-String Dynamic (PHSD) model [21] is a transport model. From the comparison of D^0 v_2 with models, the measurement agrees with most of the models with diffusion coefficient

Table 2. The fitting result of different models to D^0 v_2 measurement.

Model	diffusion coefficient	$\chi^2/n.d.f$
SUBATECH	2-4	2.8/5
TAMU c quark diff.	2-7	2.1/5
TAMU no c quark diff.	N/A	7.4/5
DUKE	7	9.3/5
PHSD	5-12	0.46/4

ranging from 2-12.

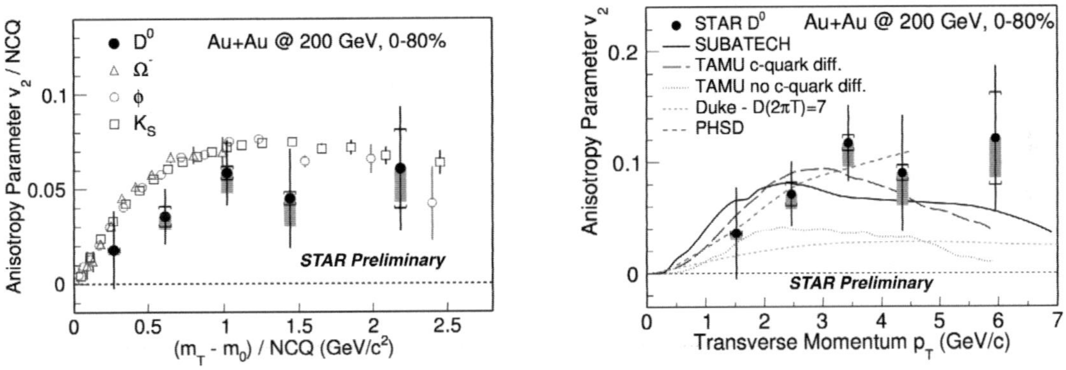

Figure 4. Left: Comparison of v_2 of D^0 with light flavor particles; Right: Comparison of D^0 v_2 with different models.

5. Summary

D^0 v_2 is measured by STAR with the newly installed HFT, for 0-80% most central Au+Au collisions at $\sqrt{s_{\rm NN}} = 200$ GeV. The measurement is carried out in a p_T range from 1 to 10 GeV/c. The measured D^0 v_2 is finite. The smaller D^0 v_2 compared to the light hadrons indicates that charm quarks may not be fully thermalized. Different models with charm diffusion coefficient $2\pi T D_s$ ranging from 2 to 12 are consistent with the measured D^0 v_2. The model comparisons suggest that charm quarks have collective behavior.

As part of continuing heavy flavor program in STAR, 2 billion Au+Au minimum bias events have been recorded in 2016 with thinner aluminum cables used for the innermost silicon layer of the HFT. An expected factor of 2-3 improvement in the D^0 significance at $p_T = 1$ GeV/c will allow for the study of the centrality dependence of D^0 v_2 in heavy ion collisions.

References

[1] Rapp R and van Hees H 2008 (*Preprint* 0803.0901)
[2] Andronic A *et al.* 2015 (*Preprint* 1506.03981)
[3] Adamczyk L *et al.* (STAR) 2014 *Phys. Rev. Lett.* **113** 142301
[4] Abelev B *et al.* (ALICE) 2012 *JHEP* **09** 112
[5] Adam J *et al.* (ALICE) 2015 (*Preprint* 1509.06888)
[6] Abelev B *et al.* (ALICE) 2013 *Phys. Rev. Lett.* **111** 102301

Journal of Physics: Conference Series **736** (2016) 012007 doi:10.1088/1742-6596/736/1/012007

[7] Abelev B B *et al.* (ALICE) 2014 *Phys. Rev.* **C90** 034904
[8] Poskanzer A M and Voloshin S A 1998 *Phys. Rev. C* **58** 1671–1678
[9] Drijard D, Fischer H G and Nakada T 1984 *Nucl. Instrum. Meth.* **A225** 367
[10] Masui H and Schmah A 2012 (*Preprint* 1212.3650)
[11] Borghini N, Dinh P M and Ollitrault J Y 2001 *Phys. Rev.* **C63** 054906
[12] Bilandzic A, Snellings R and Voloshin S 2011 *Phys. Rev.* **C83** 044913
[13] Abelev B I *et al.* (STAR) 2008 *Phys. Rev.* **C77** 054901
[14] Adamczyk L *et al.* (STAR) (*Preprint* 1507.05247)
[15] Gossiaux P B, Aichelin J, Gousset T and Guiho V 2010 *J. Phys.* **G37** 094019
[16] Nahrgang M, Aichelin J, Bass S, Gossiaux P B and Werner K 2015 *Phys. Rev.* **C91** 014904
[17] He M, Fries R J and Rapp R 2012 *Phys. Rev.* **C86** 014903
[18] He M, Fries R J and Rapp R 2013 *Phys. Rev. Lett.* **110** 112301
[19] Cao S, Qin G Y and Bass S A 2013 *Phys. Rev.* **C88** 044907
[20] Cao S, Qin G Y and Bass S A 2015 *Phys. Rev.* **C92** 024907
[21] Berrehrah H, Gossiaux P B, Aichelin J, Cassing W, Torres-Rincon J M and Bratkovskaya E 2014 *Phys. Rev.* **C90** 051901

Journal of Physics: Conference Series **736** (2016) 012008 doi:10.1088/1742-6596/736/1/012008

Tomography of the QGP by heavy quarks

T Song[1,2], **H Berrehrah**[1], **E L Bratkovskaya**[1,2,3], **D Cabrera**[1], **W Cassing**[4], **J M Torres-Rincon**[1], **L Tolos**[1,5]

[1] Frankfurt Institute for Advanced Studies, University of Frankfurt, Frankfurt, Germany
[2] Institute for Theoretical Physics, University of Frankfurt, Frankfurt, Germany
[3] GSI Helmholtzzentrum für Schwerionenforschung GmbH, Darmstadt, Germany
[4] Institute for Theoretical Physics, University of Giessen, Giessen, Germany
[5] Institut de Ciencies de l'Espai (IEEC/CSIC), Universitat Autonoma de Barcelona, Spain

E-mail: E.Bratkovskaya@gsi.de

Abstract. The dynamics of partons and hadrons in ultra-relativistic nucleus-nucleus collisions is analyzed within the Parton-Hadron-String Dynamics (PHSD) transport approach, which is based on a dynamical quasiparticle model for the partonic phase (DQPM) including a dynamical hadronization scheme while reproducing lattice QCD results in thermodynamic equilibrium for the equation-of-state as well as transport coefficients like shear and bulk viscosities, the electric conductivity or the charm diffusion coefficient of the hot QCD medium. In this contribution we report on the recent results on the charm dynamics and elliptic flow in Au+Au collisions at RHIC and Pb+Pb reactions at the LHC as well as on the single electron spectra from $D-$ and $B-$meson semileptonic decays in Au+Au collisions at $\sqrt{s_{NN}}$ =200 and 62.4 GeV. We find that the PHSD approach well describes the R_{AA} and elliptic flow v_2 of open charm mesons in Au+Au collisions at $\sqrt{s_{NN}}$ = 200 GeV (from STAR) and 2.76 TeV (from ALICE) as well as the elliptic flow of single electrons at $\sqrt{s_{NN}}$ = 200 and 62.4 GeV (from PHENIX), however, the large R_{AA} at $\sqrt{s_{NN}}$ = 62.4 GeV is not reproduced at all which might indicate a new 'PHENIX puzzle'.

1. Introduction

The dynamics of the early universe in terms of the 'Big Bang' may be studied experimentally by ultra-relativistic nucleus-nucleus collisions at Relativistic-Heavy-Ion-Collider (RHIC) or Large-Hadron-Collider (LHC) energies in terms of 'tiny bangs' in the laboratory. With sufficiently strong parton interactions, the medium in the collision zone can be expected to achieve local equilibrium after some initial delay and exhibit approximately hydrodynamic flow [1, 2, 3]. In these collisions a new state of strongly interacting matter is created, being characterized by a very low shear viscosity η to entropy density s ratio, η/s, close to a nearly perfect fluid [4, 5]. Lattice QCD (lQCD) calculations [6, 7] indicate that a crossover region between hadron and quark-gluon matter should have been reached in these experiments.

Since the hot and dense matter produced in relativistic heavy-ion collisions appears only for a couple of fm/c, it is a big challenge for experiment to investigate its properties. The heavy flavor mesons are considered to be promising probes in this respect since the production of heavy flavor requires a large energy-momentum transfer. Thus it takes place early in the heavy-ion collisions, and - due to the large energy-momentum transfer - should be described by perturbative quantum chromodynamics (pQCD). The produced heavy flavor then interacts with the hot dense matter

Journal of Physics: Conference Series **736** (2016) 012008 doi:10.1088/1742-6596/736/1/012008

(of partonic or hadronic nature) by exchanging energy and momentum. As a result, the ratio of the measured number of heavy flavors in heavy-ion collisions to the expected number in the absence of nuclear or partonic matter, which is the definition of $R_{\rm AA}$, is suppressed at high transverse momentum, and the elliptic flow of heavy flavor is generated by the interactions in noncentral heavy-ion collisions. Although it had been expected that the $R_{\rm AA}$ of heavy flavor is less suppressed and its elliptic flow is smaller as compared to the corresponding quantities for light hadrons, the experimental data show that the suppression of heavy-flavor hadrons at high transverse momentum and its elliptic flow v_2 are comparable to those of light hadrons [8, 9]. This is a puzzle for heavy-flavor production and dynamics in relativistic heavy-ion collisions as pointed out by many groups [10, 11, 12]. For recent reviews we refer the reader to Refs. [13, 14].

Since the heavy-flavor interactions are closely related to the dynamics of the partonic or hadronic degrees-of-freedom due to their mutual interactions, a proper description of the relativistic heavy-ion collisions and their bulk dynamics is necessary. In this study we employ the parton-hadron-string dynamics (PHSD) approach, which differs from the conventional Boltzmann-type models [15, 16] in the aspect that the degrees-of-freedom for the QGP phase are off-shell massive strongly-interacting quasi-particles that generate their own mean-field potential [17]. The masses of the dynamical quarks and gluons in the QGP are distributed according to spectral functions whose pole positions and widths, respectively, are defined by the real and imaginary parts of their self-energies [18]. The partonic propagators and self-energies, furthermore, are defined in the dynamical quasiparticle model (DQPM) in which the strong coupling and the self-energies are fitted to lattice QCD results [19]. We recall that the PHSD approach has been successfully applied to p+A and Au+Au (Pb+Pb) collisions from SPS to LHC energies with respect to single-particle spectra, collective flows and electromagnetic observables [20, 21, 22].

2. Charm and bottom dynamics in PHSD

2.1. Initial heavy flavor production and fragmentation

In the PHSD the charm and bottom quark pairs are produced through initial hard nucleon-nucleon scattering in relativistic heavy-ion collisions. We employ the PYTHIA event generator [23] to produce the heavy-quark pairs and modify slightly their transverse momentum and rapidity such that they are similar to those from the FONLL calculations [24]. We recall that in the case of heavy-ion collisions at the top RHIC energy of $\sqrt{s}$ = 200 GeV the transverse momentum and the rapidity of the charm quark are reduced by 10 % and 16 % [25], respectively, while those of the bottom quark are unmodified. The corrections employed at other ivariant energies can be found in Refs. [26, 27]. Accordingly, our tuned PYTHIA generator gives very similar charm and bottom distributions as those from FONLL calculations [24, 28], which fixes the input from pQCD in our approach.

The produced charm and bottom quarks in hard nucleon-nucleon interactions are hadronized in p+p collisions by emitting soft gluons, which is denoted by 'fragmentation'. We use the fragmentation function of Peterson which reads as [29]

$$D_Q^H(z) \sim \frac{1}{z[1-1/z-\epsilon_Q/(1-z)]^2}, \tag{1}$$

where z is the momentum fraction of the hadron H fragmented from the heavy quark Q while ϵ_Q is a fitting parameter which is taken to be $\epsilon_Q = 0.01$ for charm [25] and 0.004 for bottom [30]. Since the charm quark is much heavier than the soft emitted gluons, it takes a large momentum fraction in the fragmentation into a $D-$meson. This is even more pronounced in the case of a bottom quark.

The excited $D^*(B^*)$ mesons first decay into $D(B)+\pi$ or $D(B)+\gamma$, and finally the $D-$ and $B-$mesons produce single electrons through the semileptonic decay [31]. This completes the

Journal of Physics: Conference Series **736** (2016) 012008 doi:10.1088/1742-6596/736/1/012008

production and decay of heavy-flavor mesons in p+p collisions.

2.2. Shadowing in heavy-ion reactions

The production cross section for heavy-quark pairs in a nucleon-nucleon collision is calculated by convoluting the partonic cross section with the parton distribution functions (PDFs) of the nucleon [32]. However, the parton distribution functions are modified in a hadronic medium, i.e. suppressed at small momentum fraction x, slightly enhanced for $x \approx 0.1$ and then suppressed again for larger x (EMC effect). This is denoted as 'shadowing (antishadowing)' effect and is incorporated in the PHSD by employing the EPS09 package from Ref. [32] as in most of the related approaches [14]. The details of the implementation are given in Ref. [26].

2.3. Heavy-quark interactions in the QGP

In PHSD the baryon-baryon and baryon-meson collisions at high-energy produce strings. If the local energy density is above the critical energy density (~ 0.5 GeV/fm^3), the strings melt into quarks and antiquarks with masses determined by the temperature-dependent spectral functions from the DQPM [19]. Massive gluons are formed through flavor-neutral quark and antiquark fusion in line with the DQPM. In contrast to normal elastic scattering, off-shell partons may change their mass after the elastic scattering according to the local temperature T in the cell (or local space-time volume) where the scattering happens. This automatically updates the parton masses as the hot and dense matter expands, i.e. the local temperature decreases with time. The same holds true for the reaction chain from gluon decay to quark+antiquark ($g \to q + \bar{q}$) and the inverse reaction ($q + \bar{q} \to g$) following detailed balance.

The heavy quarks and antiquarks produced in early hard collisions - as described above - interact with the dressed lighter off-shell partons in the partonic phase. The cross sections for the heavy-quark scattering with massive off-shell partons have been calculated by considering explicitly the mass spectra of the final state particles in Refs. [33, 34, 35]. The elastic scattering of heavy quarks in the QGP is treated consistently in the PHSD by including the non-perturbative effects of the strongly interacting quark-gluon plasma (sQGP) constituents, i.e. the temperature-dependent coupling $g(T/T_c)$ as well as the effective propagators with broad spectral functions (imaginary parts) from the DQPM [19]. We recall that the divergence encountered in the t-channel scattering is cured self-consistently, since the infrared regulator is given by the finite DQPM gluon mass and width. We note that heavy flavor interactions in the QGP – as described by the DQPM charm scattering cross sections – differ substantially form the pQCD scenario [36], however, the spacial diffusion constant for charm quarks $D_s(T)$ is consistent with the lQCD data [26, 35]. For further details we refer the reader to Refs. [33, 34, 35].

2.4. Heavy-quark coalescence

The heavy-quark hadronization in heavy-ion collisions is realized via 'dynamical coalescence' and fragmentation. Here 'dynamical coalescence' means that the probability to find a coalescence partner is defined by Monte Carlo in the vicinity of the critical energy density $0.4 \leq \epsilon \leq 0.75$ GeV/fm^3 as explained below. We note that such a dynamical realization of heavy-quark coalescence is in line with the dynamical hadronization of light quarks in the PHSD and differs from the 'spontaneous' coalescence used in the early work [25] when heavy-quarks have been forced to hadronize at a critical energy density $\epsilon_c \approx 0.5$ GeV/fm^3 via coalescence or fragmentation by Monte Carlo. Indeed, the dynamical realization gives some window in energy density to find the proper light partner and leads to an enhancement of the heavy-quark fraction that hadronizes via coalescence.

In PHSD all antiquarks neighboring in phase space are candidates for the coalescence partner of a heavy quark. From the distances in coordinate and momentum spaces between the heavy quark and light antiquark (or vice versa), the coalescence probability is given by [27]

$$f(\rho, \mathbf{k}_\rho) = \frac{8g_H}{6^2} \exp\left[-\frac{\rho^2}{\delta^2} - \mathbf{k}_\rho^2\delta^2\right], \tag{2}$$

where g_H is the degeneracy of the heavy meson, and

$$\rho = \frac{1}{\sqrt{2}}(\mathbf{r}_1 - \mathbf{r}_2), \quad \mathbf{k}_\rho = \sqrt{2}\,\frac{m_2\mathbf{k}_1 - m_1\mathbf{k}_2}{m_1 + m_2}, \tag{3}$$

with m_i, $\mathbf{r}_i$ and $\mathbf{k}_i$ denoting the mass, position and momentum of the quark or antiquark i in the center-of-mass frame, respectively. The width parameter δ is related to the root-mean-square radius of the produced heavy meson through

$$\langle r^2 \rangle = \frac{3}{2}\frac{m_1^2 + m_2^2}{(m_1 + m_2)^2}\delta^2. \tag{4}$$

Since this prescription gives a larger coalescence probability at low transverse momentum, the radius is taken to be 0.9 fm for a charm quark as well as for a bottom quark [25]. We also include the coalescence of charm quarks into highly excited states, $D_0^*(2400)^0$, $D_1(2420)^0$, and $D_2^*(2460)^{0,\pm}$ and the coalescence of bottom quarks into $B_1(5721)^{+,0}$, $B_2^*(5747)^{+,0}$, and $B(5970)^{+,0}$ which are, respectively, assumed to immediately decay to D (or D^*) and π and to B (or B^*) and π after hadronization as described in Ref. [25].

Summing up the coalescence probabilities from all candidates, whether the heavy quark or heavy antiquark hadronizes by coalescence or not, and which quark or antiquark among the candidates will be the coalescence partner, is decided by Monte Carlo. If a random number is above the sum of the coalescence probabilities, it is tried again in the next time step till the local energy density is lower than 0.4 GeV/fm^3, i.e. in the dominantly hadronic phase. The heavy quark or heavy antiquark, which does not succeed to hadronize by coalescence, then hadronizes through fragmentation as in p+p collisions.

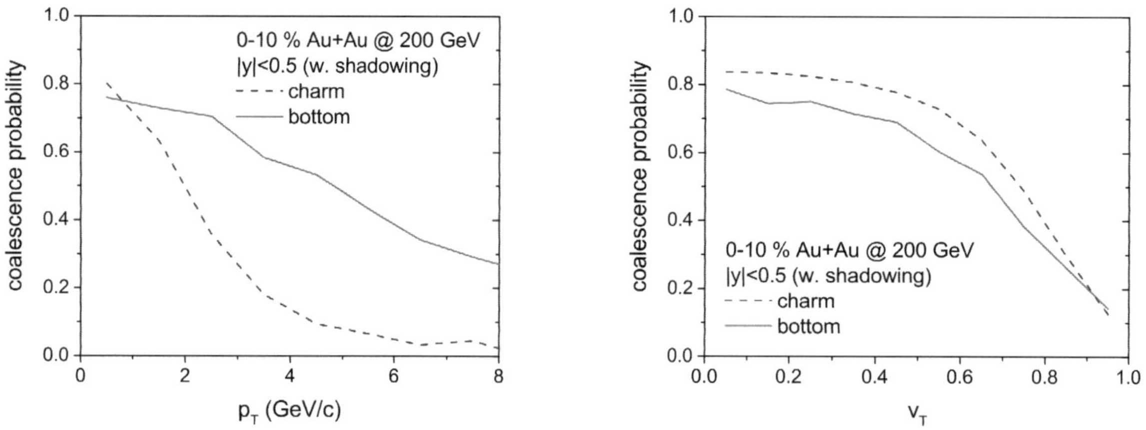

Figure 1. Coalescence probability of charm and bottom quarks at midrapidity ($|y| < 0.5$) as functions of transverse momentum (lhs) and of transverse velocity (rhs) in 0-10 % central Au+Au collisions at $\sqrt{s_{\rm NN}}$ =200 GeV taking into account the shadowing effect.

Fig. 1 shows the coalescence probabilities of charm and bottom quarks at midrapidity ($|y| < 0.5$) as functions of transverse momentum (lhs) and of transverse velocity (rhs) in 0-10 % central Au+Au collisions at $\sqrt{s_{\rm NN}}$ =200 GeV. Since a heavy quark with a large transverse momentum has a smaller chance to find a coalescence partner close by in phase space, the coalescence probability decreases with increasing transverse momentum. It appears from the left panel of Fig. 1 that the coalescence probability of a bottom quark is larger than that of a charm quark. It emerges, however, because the bottom quark is much heavier than the charm quark. The right panel of Fig. 1 clearly shows that the coalescence probability is similar for a bottom or charm quark, when it is expressed as a function of the transverse velocity v_T (for $|y| < 0.5$).

2.5. Interactions of charm and bottom mesons with the hadronic medium

After the hadronization of heavy quarks and their subsequent decay into D, D^*, B and B^* mesons, the final stage of the evolution concerns the interaction of these states with the hadrons conforming the expanding bulk medium. A realistic description of the hadron-hadron scattering —potentially affected by resonant interactions— includes collisions with the states $\pi, K, \bar{K}, \eta, N, \bar{N}, \Delta, \bar{\Delta}$. Such a description of their interactions has been developed in Refs. [37, 38, 39] using effective field theory. Moreover, after the application of an effective theory, one has to implement to the tree-level scattering amplitudes a unitarization method to better control the behavior of the cross sections at moderates energies with potentially resonant scattering amplitudes.

The details of the interaction for the four heavy states follows quite in parallel by virtue of the "heavy-quark spin-flavor symmetry". It accounts for the fact that if the heavy masses are much larger than any other typical scale in the system, like Λ_{QCD}, temperature and the light hadron masses, then the physics of the heavy subsystem is decoupled from the light sector, and the former is not dependent on the mass nor on the spin of the heavy particle. This symmetry is exact in the ideal limit $m_Q \to \infty$, with m_Q being the mass of the heavy quark confined in the heavy hadron. In the opposite limit $m_Q \to 0$, one can exploit the chiral symmetry of the QCD Lagrangian to develop an effective realization for the light particles. This applies to the pseudoscalar meson octet $(\pi, K, \bar{K}, \eta)$. Although both symmetries are broken in nature (as in our approach, when implementing physical masses), the construction of the effective field theories incorporates the breaking of these symmetries in a controlled way. In particular, it provides a systematic expansion in powers of $1/m_H$ (inverse heavy-meson mass) and powers of p, m_l (typical momentum and mass of the light meson). For details the reader is referred to Refs. [37, 38, 39].

3. Results for heavy-ion reactions

So far we have described the interactions of the heavy flavor produced in relativistic heavy-ion collisions with partonic and hadronic degrees-of-freedom as well as their hadronization by coalescence or fragmentation. Since the matter produced in heavy-ion collisions is extremely dense, the interactions with the bulk matter suppresses heavy flavors at high-p_T. On the other hand, the partonic or nuclear matter is accelerated outward (exploding), and a strong flow is generated via the interactions of the bulk particles and the repulsive scalar interaction for the lighter partons. Since the heavy flavor strongly interacts with the expanding matter, it is also accelerated outwards. Such effects of the medium on the heavy-flavor dynamics are expressed in terms of the nuclear modification factor defined as

$$R_{\rm AA}(p_{\rm T}) \equiv \frac{dN_{\rm AA}/dp_{\rm T}}{N^{\rm AA}_{\rm binary} \times dN_{\rm pp}/dp_{\rm T}}, \tag{5}$$

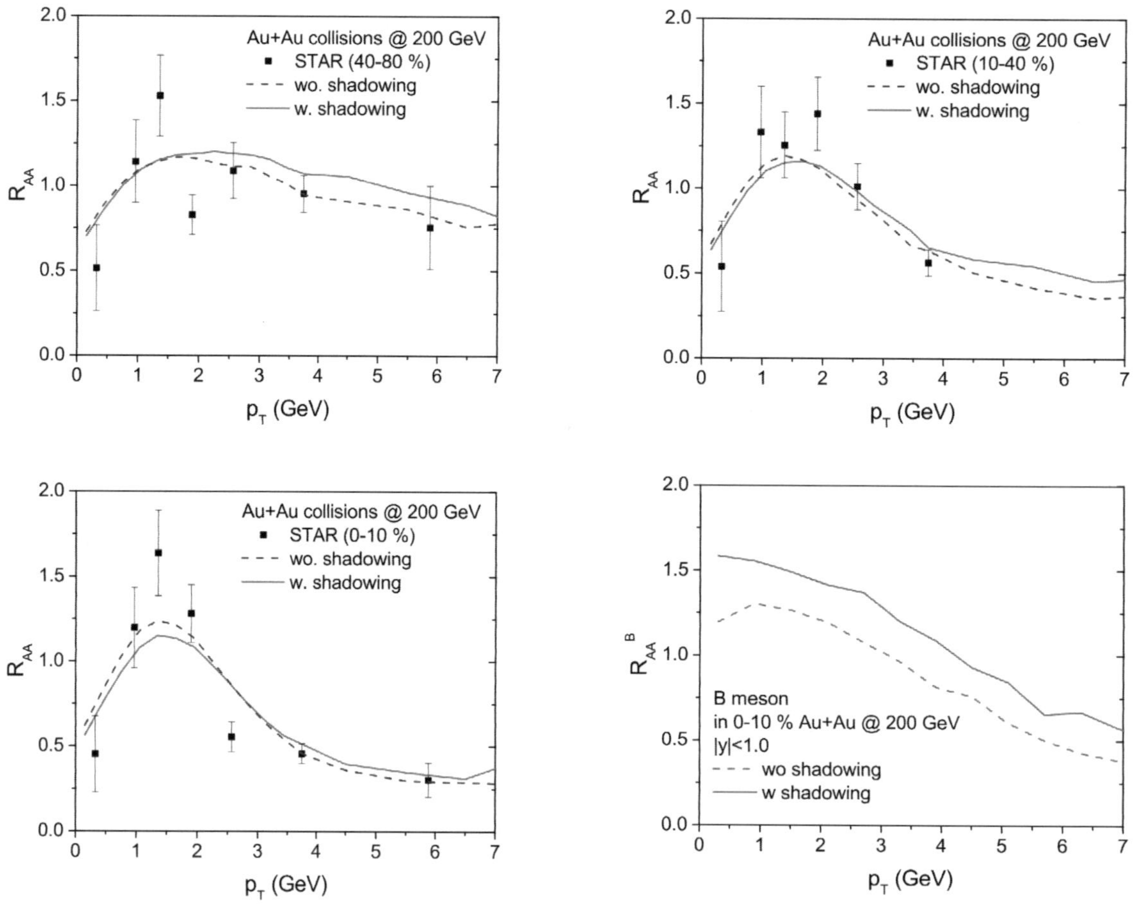

Figure 2. $R_{\rm AA}$ of $D-$mesons with (solid) and without (dashed) shadowing effect in Au+Au collisions at $\sqrt{s_{\rm NN}}$ =200 GeV for different centralities in comparison to the experimental data from the STAR collaboration [40]. (lower right panel) $R_{\rm AA}$ of $B-$mesons with (solid) and without (dashed) shadowing effect in central Au+Au collisions at $\sqrt{s_{\rm NN}}$ =200 GeV.

where $N_{\rm AA}$ and $N_{\rm pp}$ are, respectively, the number of particles produced in heavy-ion collisions and that in p+p collisions, and $N^{\rm AA}_{\rm binary}$ is the number of binary nucleon-nucleon collisions in the heavy-ion collision for the centrality class considered. Note that if the heavy flavor does not interact with the medium in heavy-ion collisions, the numerator of Eq. (5) will be similar to the denominator. For the same reason, an $R_{\rm AA}$ smaller (larger) than one in a specific $\rm p_T$ region implies that the nuclear matter suppresses (enhances) the production of heavy flavors in that transverse momentum region.

Furthermore, in noncentral heavy-ion collisions the produced matter expands anisotropically due to the different pressure gradients in plane and out-of plane. If the heavy flavor interacts strongly with the nuclear matter, then it also follows this anisotropic motion to some extend. The dominant anisotropic flow is expressed in terms of the elliptic flow v_2 which reads

$$v_2({\rm p_T}) \equiv \frac{\int d\phi \cos 2\phi (dN_{\rm AA}/d{\rm p_T} d\phi)}{2\pi dN_{\rm AA}/d{\rm p_T}}, \tag{6}$$

where ϕ is the azimuthal angle of a particle in momentum space with respect to the reaction plane. In the following Subsections, we will present our results on the production of heavy flavors and single electrons in Au+Au (or Pb+Pb) collisions at different invariant energies.

3.1. Au+Au at $\sqrt{s_{\rm NN}}$ *=200 GeV*

We start with Au+Au collisions at the top RHIC energy and focus on the open charm mesons.

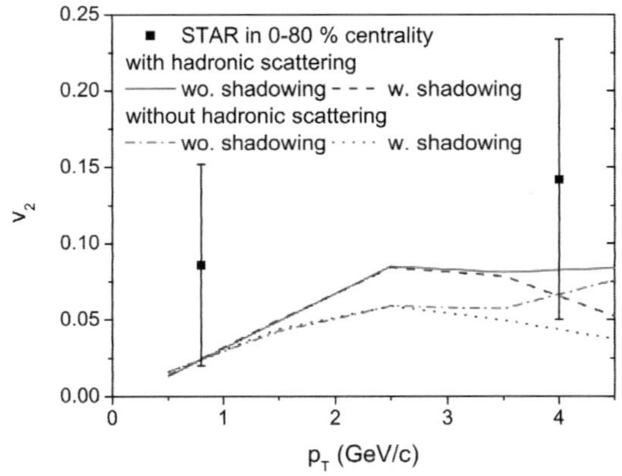

Figure 3. The elliptic flow v_2 of D^0 mesons with and without hadronic scattering and with and without shadowing in Au+Au collisions at $\sqrt{s_{\rm NN}}$ =200 GeV and 0-80% centrality. The experimental data are from the STAR Collaboration [40, 41].

In Fig. 2 we show the $R_{\rm AA}$ of $D-$mesons in different centrality classes with (solid) and without (dashed) shadowing effect in Au+Au collisions at $\sqrt{s_{\rm NN}}$ =200 GeV in comparison to the experimental data from the STAR collaboration [40]. Furthermore, we show in the lower right panel the $R_{\rm AA}$ of $B-$mesons in 0-10 % central Au+Au collisions at $\sqrt{s_{\rm NN}}$ =200 GeV from the PHSD calculations. The shadowing effect is excluded in the dashed lines and is included in the solid lines. We note that the $R_{\rm AA}$ of $D-$mesons without shadowing effect is slightly different from our previous results in Ref. [25], because the elastic backward scattering has been improved and the coalescence of charm quark takes place continuously in time as described above. As shown in Fig. 2, the shadowing effect decreases the charm production by ~8 % in central reactions and increases the bottom production by ~20 %. Apparently, the $R_{\rm AA}$ of $B-$mesons is much larger than that of $D-$mesons at the same transverse momentum. However, this is attributed to the larger bottom mass than charm mass as demonstrated in Fig. 1.

Fig. 3 displays the elliptic flow v_2 of D^0 mesons with and without hadronic scattering and with and without shadowing in 0-80% centrality Au+Au collisions at the top RHIC energy. We can see that the hadronic scattering plays an important role in the elliptic flow v_2 and enhances the v_2 of final D mesons by up to 30% at higher $p_{\rm T}$ essentially due to resonant scatterings [25]. The shadowing effect, furthermore, is ineffective up to $p_T \sim 3$ GeV/c.

3.2. Pb+Pb at $\sqrt{s_{\rm NN}}$ *=2.76 TeV*

Fig. 4 (lhs) shows the ratio $R_{\rm AA}$ of D^0, D^+, and D^{*+} mesons within the rapidity range $|y| < 0.5$ as a function of $p_{\rm T}$ in 0-10 % central Pb+Pb collisions at $\sqrt{s_{\rm NN}}$ =2.76 TeV. Here the charm quark mass is taken to be 1.5 GeV, independent of temperature. The solid and dotted lines are, respectively, the $R_{\rm AA}$ of D mesons with and without (anti-)shadowing. We can see that the $R_{\rm AA}$ of D mesons decreases especially at small transverse momentum due to shadowing. When

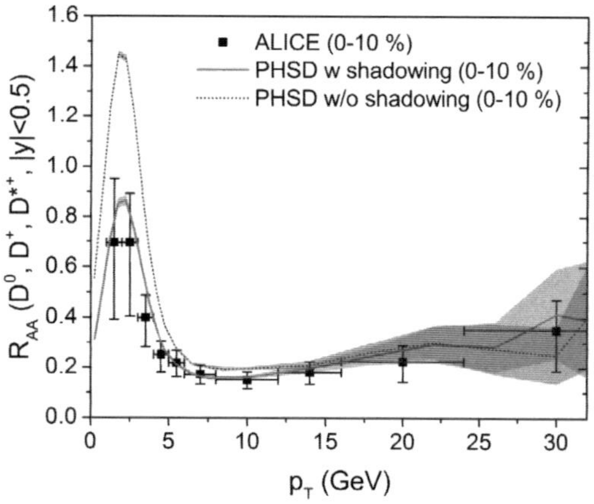

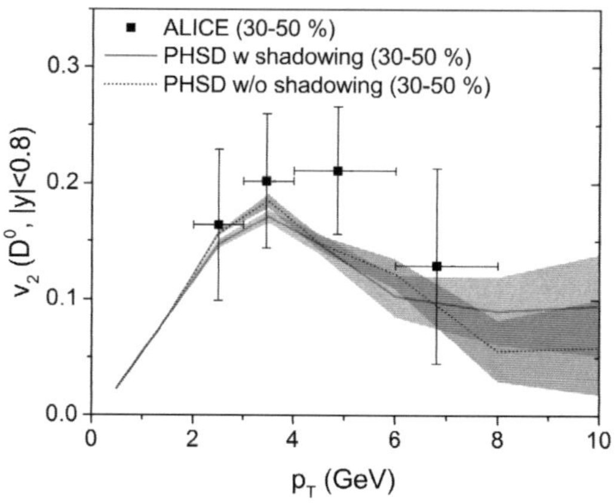

Figure 4. (lhs) The ratio $R_{\rm AA}$ of D^0, D^+, and D^{*+} mesons within $|y| < 0.5$ as a function of p_T in 0-10 % central Pb+Pb collisions at $\sqrt{s_{\rm NN}}$ =2.76 TeV compared with the experimental data from the ALICE collaboration [42]. The solid and dotted lines are, respectively, $R_{\rm AA}$ with and without (anti-)shadowing. The charm quark mass is taken to be 1.5 GeV. (rhs) The elliptic flow v_2 of D^0 mesons within $|y| < 0.8$ in 30-50 % central Pb+Pb collisions at $\sqrt{s_{\rm NN}}$ =2.76 TeV compared with the experimental data from the ALICE collaboration [44]. The solid (red) and dotted (blue) lines are, respectively, v_2 with and without (anti-)shadowing.

including shadowing our results are in a good agreement with the experimental data from the ALICE collaboration [42]. Since the radiative energy loss is not yet included in our study, it seems that the latter is not significant for transverse momenta up to 15 GeV/c (cf. Ref. [43]).

Fig. 4 (rhs) shows the elliptic flows v_2 of D^0 mesons within the rapidity range $|y| < 0.8$ in 30-50 % central Pb+Pb collisions at $\sqrt{s_{\rm NN}}$ =2.76 TeV. The solid and dotted lines are, respectively, the results with and without (anti-)shadowing. Again the charm quark mass is taken to be 1.5 GeV. We can see that the PHSD results roughly reproduce the experimental data from the ALICE collaboration [44]. The (anti-)shadowing effect slightly decreases the elliptic flows, because it reduces the production of low-$p_{\rm T}$ charm which more easily follows the bulk flow. We recall that so far it has been a challenge for theoretical models to reproduce the experimental data and to explain simultaneously the large energy loss of charm quarks ($R_{\rm AA}$) and the strong collectivity (v_2) [13]. According to the studies performed above the microscopic PHSD approach appears to be quite in line with the experimental observations for charm quarks at top RHIC and LHC energies.

3.3. Single electrons from Au+Au reactions at RHIC

The PHENIX Collaboration has measured the $R_{\rm AA}$ and elliptic flow of single electrons which emerge from charm and bottom decays. In this case it is important to include the bottom production and dynamics in the microscopic calculations since the single electrons from bottom decays become dominant at high $p_{\rm T}$ [13].

Fig. 5 shows the $R_{\rm AA}$ of single electrons from $D-$meson and $B-$meson semileptonic decays, which correspond to the dashed and dot-dashed lines, respectively, while the solid lines are the sum of them in 0-10 % central Au+Au collisions at $\sqrt{s_{\rm NN}}$ =200 GeV. The PHSD calculations shown here include the shadowing effect, which enhances the bottom production and suppresses the charm production at low transverse momentum in line with the discussion above. We find that the single electrons from B decay have a larger contribution than that from D decay above

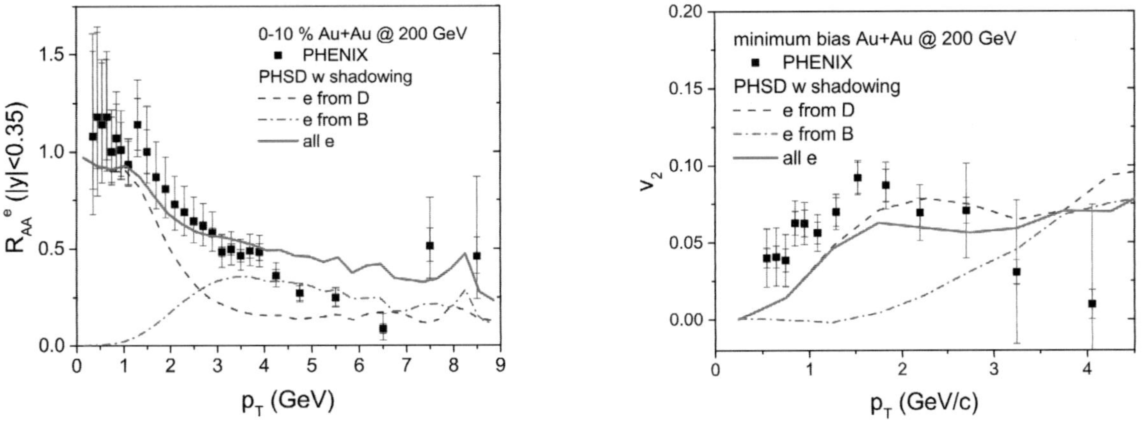

Figure 5. (lhs) R_{AA} of single electrons from the semi-leptonic decay of $D-$mesons (dashed) and of $B-$mesons (dot-dashed) and the sum of them (solid) with the shadowing effect included in 0-10 % central Au+Au collisions at $\sqrt{s_{NN}}$ =200 GeV in comparison to the experimental data from the PHENIX collaboration [48]. (rhs) The elliptic flow v_2 of single electrons from the semi-leptonic decay of $D-$mesons (dashed) and of $B-$mesons (dot-dashed) and of both of them (solid) with the shadowing effect included in minimum-bias Au+Au collisions at $\sqrt{s_{NN}}$ =200 GeV in comparison to the experimental data from the PHENIX collaboration [45].

$p_T \approx 2.7 - 2.8$ GeV. In p+p collisions, the contribution from B decay starts to be larger than that from D decay at about $p_T \approx 4$ GeV as shown in Ref. [25]. The reason for the dominance of B decay at lower transverse momentum in Au+Au collisions is that the R_{AA} of $B-$mesons is larger than that of $D-$mesons at high transverse momentum. In view of the results in Fig. 5 we may conclude that apart from the charm dynamics in Au+Au collisions at the top RHIC energy also the bottom dynamics is reasonably reproduced by the PHSD calculations.

The beam energy scan (BES) program at RHIC has been carried out by colliding Au nuclei at various energies down to $\sqrt{s_{NN}}$ =7.7 GeV. The aim of the program is to find information on the phase boundary and the critical point in the QCD phase diagram as pointed out in Refs. [46, 47]. It is expected that if the trajectories of the produced nuclear matter in the QCD phase diagram pass close to the critical point, some drastic changes of observables could be measured in experiments. Since the PHENIX collaboration recently measured the single electrons from heavy flavor decay at $\sqrt{s_{NN}}$ =62.4 GeV [48], which is much lower than the maximum energy at RHIC, we will also address this system in the present subsection.

Figs. 6 (lhs) and 6 (rhs) show, respectively, the R_{AA} and elliptic flow v_2 of single electrons from the semileptonic decay of heavy flavors in 0-20 % central Au+Au collisions at $\sqrt{s_{NN}}$ =62.4 GeV with the shadowing effect included. As at $\sqrt{s_{NN}}$ =200 GeV, the contribution from $D-$meson decay is important at low transverse momentum and superseeded by the contribution from B decay above 3 GeV in transverse momentum. The contribution from B decay becomes dominant at higher transverse momentum than at $\sqrt{s_{NN}}$ =200 GeV, because the ratio of the scattering cross section for bottom production to that for charm production is much lower at $\sqrt{s_{NN}}$ =62.4 GeV. The ratio is 0.75 % at $\sqrt{s_{NN}}$ =200 GeV and 0.145 % at $\sqrt{s_{NN}}$ =62.4 GeV according to the FONLL calculations [24].

The PHSD results in Fig. 6 (lhs) underestimate R_{AA} besides touching the lower error bars of the experimental data at low and high p_T. Although the shadowing effect enhances R_{AA} at low p_T, there is still a large discrepancy between the experimental data and our results in the range

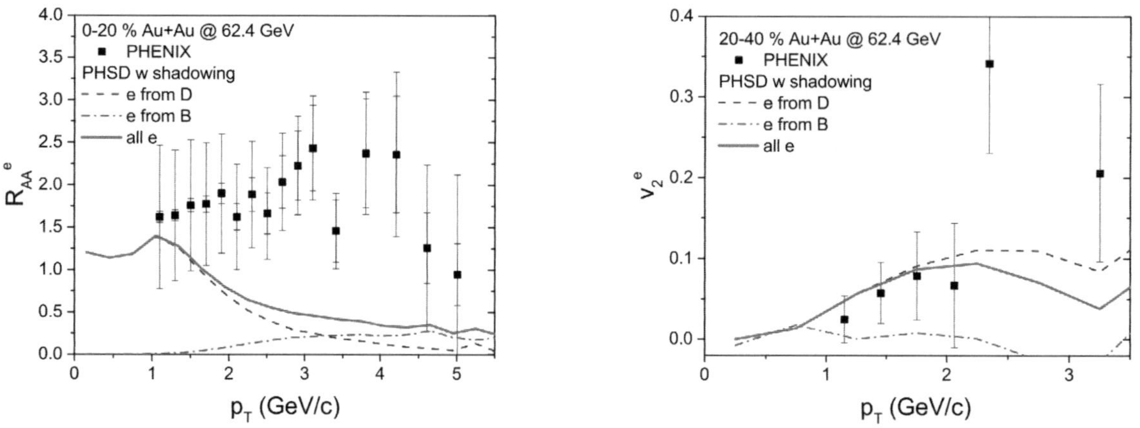

Figure 6. (lhs) R_{AA} of single electrons from the semi-leptonic decay of $D-$mesons (dashed) and of $B-$mesons (dot-dashed) and the sum of them (solid) with the shadowing effect included in 0-20 % central Au+Au collisions at $\sqrt{s_{NN}}$ =62.4 GeV in comparison to the experimental data from the PHENIX collaboration [48]. (rhs) The elliptic flow v_2 of single electrons from the semi-leptonic decay of $D-$mesons (dashed) and of $B-$mesons (dot-dashed) and of both of them (solid) with the shadowing effect in 20-40 % central Au+Au collisions at $\sqrt{s_{NN}}$ =62.4 GeV in comparison to the experimental data from the PHENIX collaboration [48].

of p_T between 2.5 and 4 GeV, which clearly lacks an explanation. We mention that a similar pattern of results has been shown in Ref. [49] and partially been attributed to the Cronin effect. We recall that the Cronin effect for the charm quarks is included in the PHSD via the EPS09 package used for the shadowing effect.

In spite of the difficulty in reproducing R_{AA}, the elliptic flow v_2 of single electrons at $\sqrt{s_{NN}}$ =62.4 GeV is well described by the PHSD approach up to $p_T \approx 2$ GeV/c irrespective whether the shadowing effect is included or not, as shown in Fig. 6 (rhs). The v_2 of single electrons from $B-$meson decay is small at low transverse momentum for the same reason as at $\sqrt{s_{NN}}$ =200 GeV.

4. Summary

In this contribution we have studied the dynamics of charm and bottom mesons as well as electron production through the semileptonic decay of heavy mesons in relativistic heavy-ion collisions at $\sqrt{s_{NN}}$ =62.4, 200, and 2760 GeV within the PHSD transport approach. The ratio of the initial scattering cross section for bottom production to that for charm production at RHIC collision energies is less than 1 %. However, since the p_T spectrum of bottom quarks is harder than that of charm quarks and the single electrons from $B-$meson decay is much more energetic than that from $D-$meson decay, it is essential to take $B-$meson production into account in order to study the single electron production, especially at high p_T.

In analogy to the charm quark pairs, the bottom pairs are produced by using the PYTHIA event generator [23] which is tuned to reproduce the p_T spectrum and rapidity distribution of bottom quark pairs from the FONLL calculations [24]. The (anti)shadowing effect, which is the modification of the nucleon parton distributions in a nucleus, is implemented by means of the EPS09 package [32]. We have found that the (anti)shadowing effect is not so strong at RHIC energies as compared to LHC energies [26] where the shadowing becomes essential to find agreement with the experimental measurements especially at low p_T (cf. Fig. 4 (lhs)).

The charm and bottom partons - produced by the initial hard nucleon-nucleon scattering - interact with the massive quarks and gluons in the QGP by using the scattering cross sections calculated in the Dynamical Quasi-Particle Model (DQPM) which reproduces heavy-quark diffusion coefficients from lattice QCD calculations at temperatures above the deconfinement transition. When approaching the critical energy density for the phase transition from above, the charm and bottom (anti)quarks are hadronized into $D-$ and $B-$mesons through the coalescence with light (anti)quarks. Those heavy quarks, which fail in coalescence until the local energy density is below 0.4 GeV/fm^3, hadronize by fragmentation as in p+p collisions. The hadronized $D-$ and $B-$mesons then interact with light hadrons in the hadronic phase with cross sections that have been calculated in an effective lagrangian approach with heavy-quark spin symmetry. Finally, after freeze-out of the $D-$ and $B-$mesons they produce single electrons through semileptonic decays with the branching ratios given by the Particle Data Group (PDG) [31].

We have found that the coalescence probability for bottom quarks is still large at high p_T compared to charm quarks, and the R_{AA} of $B-$mesons is larger than that of $D-$mesons at the same (high) p_T. However, this can dominantly be attributed to the much larger mass of the bottom quark. If the coalescence probability and the R_{AA} are expressed as a function of the transverse velocity of the heavy quark, both charm and bottom coalescence become similar since both are comoving with the neighbouring light antiquarks (cf. Fig. 1).

Our studies show that the energy loss of D mesons at high p_T can be dominantly attributed to partonic scattering and that the collisional energy loss is dominant at least up to $p_T = 6$ GeV/c at top RHIC energies. The shape of R_{AA} versus p_T reflects the influence of quark coalescence and hadronic rescattering at low p_T while the D-meson elliptic flow v_2 gets enhanced sizeably by hadronic scattering [25]. Furthermore, both the R_{AA} and the elliptic flow of D mesons from the ALICE collaboration are reasonably reproduced (cf. Fig.4). This supports the validity and consistency of the PHSD approach for charm production and propagation in relativistic heavy-ion collisions in close analogy to our previous studies on electromagnetic probes from relativistic heavy-ion collisions [18].

Furthermore, we have found that the shadowing effect suppresses charm production preferentially at small transverse momentum and mid-rapidity, and it helps the PHSD to reproduce the ratio R_{AA} of D mesons from the ALICE collaboration at LHC energies. The shadowing also slightly decreases the elliptic flow of D mesons because it suppresses the production of charm quarks with small transverse momentum which more easily follow the bulk flow.

Finally, we found that the PHSD approach can roughly reproduce the experimental data on single electron production in Au+Au collisions at $\sqrt{s_{NN}}$ =200 – including essential contributions from bottom decay – and the elliptic flow of single electrons at $\sqrt{s_{NN}}$ =62.4 GeV from the PHENIX collaboration. However, the R_{AA} at $\sqrt{s_{NN}}$ =62.4 GeV is clearly underestimated which presently remains as an open puzzle.

Acknowledgments

The authors acknowledge inspiring discussions with J. Aichelin, P. B. Gossiaux, C. M. Ko, O. Linnyk, R. Marty, V. Ozvenchuk, and R. Vogt. This work was supported by DFG under contract BR 4000/3-1, and by the LOEWE center HIC for FAIR. JMTR is supported by the Program TOGETHER from Region Pays de la Loire and the European I3-Hadron Physics program. LT acknowledges support from the Ramon y Cajal Research Programme and contracts FPA2010-16963 and FPA2013-43425-P from Ministerio de Ciencia e Innovaci on, as well as from FP7-PEOPLE-2011-CIG under Contract No. PCIG09-GA-2011-291679. The computational resources have been provided by the LOEWE-CSC.

Journal of Physics: Conference Series **736** (2016) 012008 doi:10.1088/1742-6596/736/1/012008

5. References

[1] Ollitrault J Y 1992 *Phys. Rev.* D **46** 229
[2] Heinz U and Kolb P 2002 *Nucl. Phys.* A **702** 269
[3] Shuryak E V 2009 *Prog. Part. Nucl. Phys.* **62** 48
[4] Shuryak E V 2005 *Nucl. Phys.* A **750** 64
[5] Gyulassy M and McLerran L 2005 *Nucl. Phys.* A **750** 30
[6] Cheng M *et al.* 2008 *Phys. Rev.* D **77** 014511
[7] Aoki Y *et al.* 2009 *JHEP* **0906** 088
[8] Abelev B *et al.* [ALICE Collaboration] 2012 *JHEP* **1209** 112
[9] Abelev B *et al.* [ALICE Collaboration] 2013 *Phys. Rev. Lett.* **111** 102301
[10] Das S K Scardina F Plumari S Greco V 2015 *Phys. Lett.* B **747** 260
[11] Nahrgang M Aichelin J Gossiaux P B Werner K 2016 *Phys. Rev.* C **93** 044909
[12] Aichelin J Gossiaux P B Gousset T 2014 *Phys. Rev.* D **89** 074018
[13] Andronic A *et al.* 2016 *Eur. Phys. J.* C **76** 107
[14] Primo F Rapp R 2016 *arXiv*:1603.00529
[15] Plumari S Puglisi A Scardina F and Greco V 2012 *Phys. Rev.* C **86** 054902
[16] Xu Z and Greiner C 2005 *Phys. Rev.* C **71** 064901
[17] Cassing W and Bratkovskaya E L 2009 *Nucl. Phys.* A **831** 215
Bratkovskaya E L *et al.* 2011 *Nucl. Phys.* A **856** 162
[18] Linnyk O Bratkovskaya E Cassing W 2016 *Prog. Part. Nucl. Phys.* **87** 50
[19] Cassing W 2009 *Eur. Phys. J. ST* **168** 3
Cassing W 2007 *Nucl. Phys.* A **795** 70
[20] Linnyk O Cassing W and Bratkovskaya E L 2014 *Phys.Rev.* C **89** 034908
Linnyk O Konchakovski V P Cassing W and Bratkovskaya E L 2013 *Phys. Rev.* C **88** 034904
Linnyk O Konchakovski V Steinert T Cassing W Bratkovskaya E L 2015 *Phys. Rev.* C **92** 054914
[21] Konchakovski V P *et al.* 2012 *Phys. Rev.* C **85** 044922
[22] Konchakovski V P Cassing W and Toneev V D 2015 *J. Phys. G* **42** 055106
[23] Sjostrand T Mrenna S Skands P Z 2006 *JHEP* **0605** 026
[24] Cacciari M Frixione S Houdeau N Mangano M L Nason P Ridolfi G 2012 *JHEP* **1210** 137
[25] Song T Berrehrah H Cabrera D Torres-Rincon J M Tolos L Cassing W Bratkovskaya E 2015 *Phys. Rev.* C **92** 014910
[26] Song T Berrehrah H Cabrera D Cassing W Bratkovskaya E 2016 *Phys. Rev.* C **93** 034906
[27] Song T Berrehrah H Cabrera D Torres-Rincon J M Tolos L Cassing W Bratkovskaya E 2016 *arXiv*:1605.07887
[28] Cacciari M Nason P Vogt R 2005 Phys. Rev. Lett. **95** 122001
[29] Peterson C Schlatter D Schmitt I Zerwas P M 1983 *Phys. Rev.* D **27** 105
[30] Abdallah J *et al.* [DELPHI Collaboration] 2011 *Eur. Phys. J.* C **71** 1557
[31] Olive K A *et al.* [Particle Data Group Collaboration] 2014 *Chin. Phys.* C **38** 090001
[32] Eskola K J Paukkunen H Salgado C A 2009 *JHEP* **0904** 065
[33] Berrehrah H Bratkovskaya E Cassing W Gossiaux P B Aichelin J Bleicher M 2014 *Phys. Rev.* C **89** 054901
[34] Berrehrah H Gossiaux P B Aichelin J Cassing W Bratkovskaya E 2014 *Phys. Rev.* C **90** 064906
[35] Berrehrah H Bratkovskaya E Steinert T Cassing W 2016 *Int. Journal of Mod. Phys.* E **25** 1642003
[36] Berrehrah H *et al.* 2016 *arXiv*:1604.02343 [hep-ph]
[37] Tolos L Torres-Rincon J M 2013 *Phys. Rev.* D **88** 074019
[38] Torres-Rincon J M Tolos L Romanets O 2014 Phys. Rev. D **89** 074042
[39] Tolos L 2013 *Int. J. Mod. Phys.* E **22** 1330027
[40] Adamczyk L *et al.* [STAR Collaboration] 2014 *Phys. Rev. Lett.* **113** 142301
[41] Tlusty D [STAR Collaboration] 2013 *Nucl. Phys.* A **904-905** 639c
[42] Adam J *et al.* [ALICE Collaboration] 2015 *arXiv*:1509.06888 [nucl-ex]
[43] Cao S Qin G Y Bass S A 2015 *Phys. Rev.* C **92** 054909
[44] Abelev B B *et al.* [ALICE Collaboration] 2014 *Phys. Rev.* C **90** 034904
[45] Adare A *et al.* [PHENIX Collaboration] 2007 *Phys. Rev. Lett.* **98** 172301
[46] Mohanty B [STAR Collaboration] 2011 *J. Phys.* G **38** 124023
[47] Kumar L [STAR Collaboration] 2011 *J. Phys.* G **38** 124145
[48] Adare A *et al.* [PHENIX Collaboration] 2015 *Phys. Rev.* C **91** 044907
[49] He M Fries R J Rapp R 2015 *Phys. Rev.* C **91** 024904

Charm production in high multiplicity pp events

K. Werner$^{(a)}$, **B. Guiot**$^{(a,b)}$, **Iu. Karpenko**$^{(c,d,e)}$, **T. Pierog**$^{(f)}$, **G. Sophys**$^{(a)}$

$^{(a)}$ SUBATECH, University of Nantes – IN2P3/CNRS– EMN, Nantes,France
$^{(b)}$ Universidad Tcnica Federico Santa Mara, Valparaiso , Chile
$^{(c)}$ FIAS, Johann Wolfgang Goethe Universitaet, Frankfurt am Main, Germany
$^{(d)}$ Bogolyubov Institute for Theoretical Physics, Kiev 143, 03680, Ukraine
$^{(e)}$ INFN - Sezione di Firenze, Via G. Sansone 1, I-50019 Sesto Fiorentino (Firenze), Italy
$^{(f)}$Karlsruhe Inst. of Technology, KIT, Campus North, Inst. f. Kernphysik, Germany

Abstract. Studying proton-proton scattering at 7 TeV, the ALICE collaboration found the unexpected result that the D meson multiplicity increases more than linear as a function of the charged particle multiplicity. We try to understand this behavior using the EPOS3 approach. Two issues play an important role in this context: multiple scattering, in particular its impact on multiplicity fluctuations, and the collective hydrodynamic expansion. These data contain therefore valuable information about very basic features of the reaction mechanism in proton-proton collisions.

Recently, several experimental groups investigated the dependence of heavy quark production on the event activity, both for open and hidden charm or bottom, in high energy proton-proton collisions. We will focus here on D meson production, where the term "D meson multiplicity" refers in the following to the average multiplicity of D^+, D^0 and D^{*+}. The ALICE collaboration found a quite unexpected result [1]: When plotting the D meson multiplicity versus the charged particle multiplicity, both divided by the corresponding minimum bias mean values, one obtains a dependence which is very significantly more than linear (where "linear" means $N_D/\langle N_D\rangle = N_{\rm ch}/\langle N_{\rm ch}\rangle$). The effect seems to be bigger for larger transverse momentum (p_t). Both D meson and charged particle multiplicity refer to central rapidities.

It is clear that the experimental observations are very interesting, and provide valuable insight into the very nature of the reaction mechanism in pp scattering, in particular in case of high event activity. So we try in this paper to provide an analysis of the phenomenon in the EPOS3 framework. Two key aspects are: Multiple scattering and collectivity.

EPOS3 [2] is a universal model in the sense that for pp, pA, and AA collisions, the same procedure applies, based on several stages:

Initial conditions. A Gribov-Regge multiple scattering approach is employed ("Parton-Based Gribov-Regge Theory" PBGRT [3]), where the elementary object (by definition called Pomeron) is a DGLAP parton ladder, using in addition a CGC motivated saturation scale [4] for each Pomeron, of the form $Q_s \propto N_{\rm part}\,\hat{s}^{\lambda}$, where $N_{\rm part}$ is the number of nucleons connected the Pomeron in question, and $\hat{s}$ its energy. The parton ladders are treated as classical relativistic (kinky) strings.

Core-corona approach. At some early proper time τ_0, one separates fluid (core) and escaping hadrons, including jet hadrons (corona), based on the momenta and the density of string

Journal of Physics: Conference Series **736** (2016) 012009 doi:10.1088/1742-6596/736/1/012009

segments (First described in [5], a more recent discussion in [2]). The corresponding energy-momentum tensor of the core part is transformed into an equilibrium one, needed to start the hydrodynamical evolution. This is based on the hypothesis that equilibration happens rapidly and affects essentially the space components of the energy-momentum tensor.

Viscous hydrodynamic expansion. Starting from the initial proper time τ_0, the core part of the system evolves according to the equations of relativistic viscous hydrodynamics [2, 6], where we use presently $\eta/s = 0.08$. A cross-over equation-of-state is used, compatible with lattice QCD [7, 8]. The "core-matter" hadronizes on some hyper-surface defined by a constant temperature T_H, where a so-called Cooper-Frye procedure is employed, using equilibrium hadron distributions, see [8]. After hadronization, there occur still hadron-hadron rescatterings, realized via UrQMD [9].

The above procedure is employed for each event (event-by-event procedure).

Heavy quarks (Q) are produced during the initial stage, in the PBGRT formalism, in the same way as light quarks. We have several parton ladders, each one composed of two space-like parton cascades (SLC) and a Born process. The time-like partons emitted in the SLC or the Born process are in general starting points of time like cascades (TLC). In all these processes, whenever quark–antiquark production is possible, heavy quarks may be produced. We take of course into account the modified kinematics in case of non-zero quark masses (we use $m_c = 1.3$, $m_b = 4.2$). D meson production in the EPOS3 framework has been studied extensively, comparing to data and other calculation, in ref. [10].

We try to understand the dependence of the D meson multiplicity on the charged particle multiplicity, first for EPOS basic (without hydro). We study the case, where both multiplicities refer to central rapidities ($|y| \leq 0.5$ for the D mesons, and $|\eta| \leq 1$ for the charged particles). We use the variables $N_{\rm ch}$ for the charged particle multiplicity, and N_{Di} for the D meson multiplicities for different p_t ranges (N_{D1} for $1 < p_t[{\rm GeV/c}] < 2$, and N_{D8} for $8 < p_t[{\rm GeV/c}] < 12$).

In EPOS3, we have in each individual event a certain number of parton ladders (cut Pomerons). Each ladder contributes (roughly, on the average) the same to both charged particle and charm production, so both corresponding multiplicities are proportional to the number $N_{\rm Pom}$ of cut Pomerons: $N_{Di} \propto N_{\rm ch} \propto N_{\rm Pom}$, which leads to a "natural" linear relation between the charged particle multiplicity $N_{\rm ch}$ and the D meson multiplicities N_{Di} (to first approximation).

We define normalized multiplicities, $n = N/\langle N\rangle$, both for charged particles ($n_{\rm ch}$) and D meson multiplicities (n_{Di}). In the following, we consider fixed values $n^*_{\rm ch}$ of normalized charged multiplicities.

We will study the average normalized D meson multiplicity for the largest p_t range, for some given $n^*_{\rm ch}$, which may be expressed in terms of the Pomeron number distribution prob$(N_{\rm Pom}, n_{\rm ch}{}^*)$ at fixed $n^*_{\rm ch}$ and the number $n_{D8}(N_{\rm Pom}, n_{\rm ch}{}^*)$ of D mesons for fixed $N_{\rm Pom}$ and $n_{\rm ch}{}^*$, as

$$n_{D8}(n^*_{\rm ch}) = \sum_{N_{\rm Pom}} {\rm prob}(N_{\rm Pom}, n_{\rm ch}{}^*) \times n_{D8}(N_{\rm Pom}, n_{\rm ch}{}^*) \quad . \tag{1}$$

The two curves representing prob$(N_{\rm Pom}, n_{\rm ch}{}^*)$ and $n_{D8}(N_{\rm Pom}, n_{\rm ch}{}^*)$ are shown in fig. 1 (left). We see in the figure that $n_{D8}(N_{\rm Pom}, n_{\rm ch}{}^*)$ increases strongly towards small $N_{\rm Pom}$ with an increasing slope. Let us compare the expression of eq. (1) with the corresponding sum (as a reference) where we use $n_{D8}(N_{\rm Pom}, n_{\rm ch}{}^*) = n_{\rm ch}{}^*$, which would lead to $n_{D8}(n^*_{\rm ch}) = n^*_{\rm ch}$. For large $N_{\rm Pom}$, the contribution to the sum in eq. (1) will be less than the reference case, but this is more than compensated at small $N_{\rm Pom}$. Therefore, we have $n_{D8}(n^*_{\rm ch}) > n_{\rm ch}{}^*$, which is confirmed by the precise calculation shown in fig. 1 (right) as red point. Also shown is the complete curve $n_{D8}(n_{\rm ch})$ as obtained from EPOS basic. Indeed, we get a more than linear increase.

Journal of Physics: Conference Series **736** (2016) 012009 doi:10.1088/1742-6596/736/1/012009

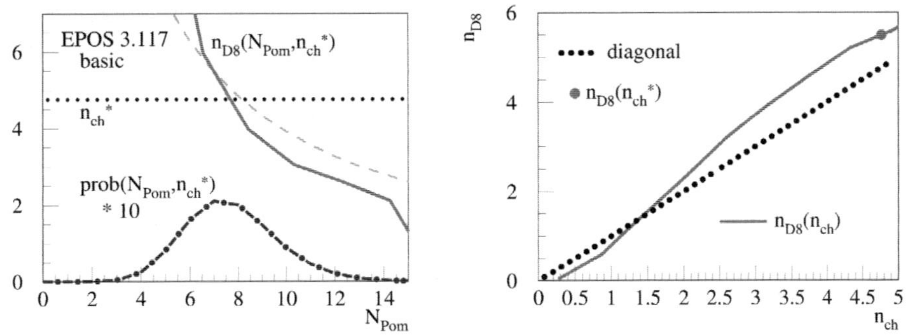

Figure 1. (Color online) Left: Pomeron number distribution at fixed charged multiplicity, prob($N_{\mathrm{Pom}}, n_{\mathrm{ch}}{}^{*}$) (blue line), and number $n_{D8}(N_{\mathrm{Pom}}, n_{\mathrm{ch}}{}^{*})$ of D mesons (large p_t) for fixed N_{Pom} and $n_{\mathrm{ch}}{}^{*}$ as a function of the Pomeron number N_{Pom} (red line). The dotted line represents the constant value $n_{\mathrm{ch}}{}^{*}$.
Right: Average multiplicity $n_{D8}(n_{\mathrm{ch}})$ of D mesons (large p_t) as a function of n_{ch} (red line) and the diagonal ($n_{D1} = n_{\mathrm{ch}}$, dotted line). The red point refers to $n_{D8}(n^{*}_{\mathrm{ch}})$ for the particular value of $n_{\mathrm{ch}}{}^{*}$ used in eq. (1).

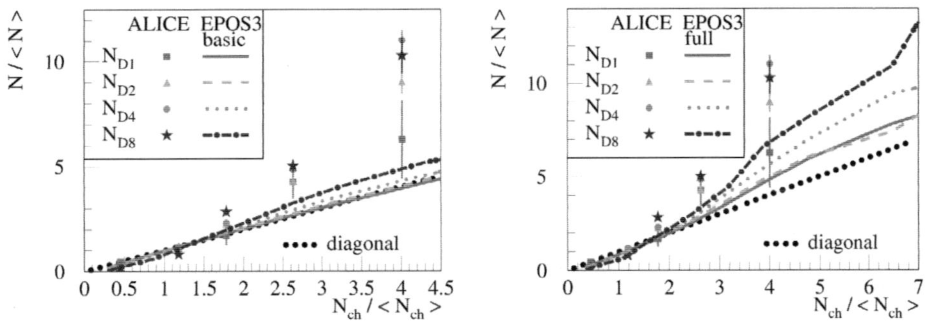

Figure 2. (Color online) D meson multiplicities versus the charged particle multiplicity, both divided by the corresponding minimum bias mean values. The different symbols and the notations N_{D1}, N_{D2}, N_{D4}, N_{D8} refer to different p_t ranges: 1-2, 2-4, 4-8, 8-12 (in GeV), N_{ch} refers to the charged particle multiplicity. We compare our calculations (lines) to ALICE data (points). Left plot : EPOS basic. Right plot : full EPOS.

The results of our calculation agree qualitatively with the trend in the data, namely a more than linear increase, in particular for high transverse momentum D mesons. But the effect is actually too small, as seen in fig. 2 (left), where we plot the D meson multiplicities versus the charged particle multiplicity, both for our calculation and data from ALICE [1].

But anyhow, EPOS basic (w/o hydro) reproduces neither spectra nor correlations, we have to consider the full approach, i.e. EPOS with hydrodynamical evolution (with or without hadronic cascade makes no difference). In fig. 2 (right), we plot again the D meson multiplicities versus the charged particle, EPOS3 compared to data, but here we refer to the calculations based on the full EPOS model (with hydro). We see a significant non-linear increase, much more pronounced as in the case of EPOS basic (without hydro), mainly due to the fact that the multiplicities from full EPOS are considerably below the results from EPOS basic. A much more detailed discussion will be provided in a separate publication.

To summarize: We analyzed the dependence of D meson multiplicities (in different p_t ranges) on the charged particle multiplicity in proton-proton collisions at 7 TeV, using the EPOS3 approach. We find a non-linear increase. Two issues play an important role: Multiplicity fluctuations due to multiple scattering (realized via multiple Pomerons), and the collective hydrodynamic expansion. Multiplicity fluctuations are important since in particular high p_t D meson production at given (large) charged particle multiplicity is very much favored for small Pomeron numbers, which is responsible for the strong increase of the D meson production with multiplicity. In addition, the effect is amplified when turning on the hydrodynamical expansion, due to a reduction of the charged particle multiplicity with respect to the model without hydro.

Acknowlegments This research was carried out within the scope of the GDRE (European Research Group) "Heavy ions at ultrarelativistic energies". B.G. acknowledges the financial support by the TOGETHER project of the Region of "Pays de la Loire". B. G. gratefully acknowledges generous support from Chilean FONDECYT grants 3160493.

[1] ALICE collaboration, arXiv:1505.00664v1
[2] K. Werner, B. Guiot, I. Karpenko, and T. Pierog, Phys.Rev. C89 (2014) 064903, arXiv:1312.1233.
[3] H. J. Drescher, M. Hladik, S. Ostapchenko, T. Pierog and K. Werner, Phys. Rept. 350, 93, 2001
[4] L. McLerran, R. Venugopalan, Phys. Rev. D 49 (1994) 2233; L. McLerran, R. Venugopalan, Phys. Rev. D 49 (1994) 3352; L. McLerran, R. Venugopalan, Phys. Rev. D 50 (1994) 2225.
[5] K. Werner, Phys. Rev. Lett. 98, 152301 (2007)
[6] Iu. Karpenko, P. Huovinen, M. Bleicher, arXiv:1312.4160
[7] S. Borsanyi et al., JHEP 1011 (2010) 077, arXiv:1007.2580
[8] K. Werner, Iu. Karpenko, T. Pierog, M. Bleicher, K. Mikhailov, arXiv:1010.0400, Phys. Rev. C 83, 044915 (2011)
[9] M. Bleicher et al., J. Phys. G25 (1999) 1859; H. Petersen, J. Steinheimer, G. Burau, M. Bleicher and H. Stocker, Phys. Rev. C78 (2008) 044901
[10] B. Guiot, Ph.D. Thesis, University of Nantes, 2014

Journal of Physics: Conference Series **736** (2016) 012010 doi:10.1088/1742-6596/736/1/012010

Latest Heavy Flavor Results from the PHENIX Experiment at RHIC

Alexandre Lebedev for the PHENIX Collaboartion

Department of Physics and Astronomy, Iowa State University, Ames, IA 50011-3160, USA

E-mail: lebedev@iastate.edu

Abstract. Heavy quark production can be used as a stringent test of perturbative QCD in proton-proton collisions, and is a valuable reference for the study of heavy ion collisions. In nucleus-nucleus collisions, the measurement of heavy quark production provides a powerful tool for studying the properties of hot and dense matter created in these collisions. The PHENIX experiment has studied many important observables related to heavy flavor via leptonic measurements. Such observables include the invariant yield and azimuthal anisotropy of electrons from non-photonic sources and prompt single muons, both of which are dominated by decays of D and B mesons. Complimentary to single lepton measurements, PHENIX has measured invariant yield, flow, and polarization of various quarkonia states. Such measurements provide additional insight into heavy flavor production mechanisms. The most recent PHENIX heavy flavor results will be presented, and compared to various theoretical model predictions.

1. $\psi(2S)/J/\psi$ ratios in p+Au and p+Al collisions

The PHENIX experiment has measured previously [1] $\psi(2S)$ nuclear modification factor in d+Au collisions at $\sqrt{s} = 200$ GeV/c. The measurement was done at mid-rapidity. The magnitude and the trend with centraility of $\psi(2S)$ meson suppression turned out to be very different than that of J/ψ [2]. In most central d+Au collisions $\psi(2S)$ suppression was measured to be almost 3 times larger than that of J/ψ. Since the initial state factors in a nucleus, such as shadowing and energy loss, should be the same or very similar for $c\bar{c}$ precursor pairs for both $\psi(2S)$ and J/ψ, something different must be happening in the later stages of the collision.

The new PHENIX measurement of $\psi(2S)$ production in p+p, p+Au and p+Al collisions at $\sqrt{s} = 200$ GeV/c at forward/bacward rapidity may shed light on the relatively large suppression of $\psi(2S)$. Fig. 1 shows the fraction of $\psi(2S)/J/\psi$ as a function of transverse momentum (p_T) in proton-proton collisions. Integrated over all p_T, this fraction is around 2%, which is in agreement with the world data (see Fig. 2). This measurement serves as a baseline for the proton-nucleus measurements.

Fig. 3 shows relative suppression of $\psi(2S)/J/\psi$ as a function of rapidity for p+Au and p+Al collisions, as well as the old d+Au result at mid-rapidity. In the proton-going direction (positive rapidity) suppression of $\psi(2S)$ is, essentially, the same as that of J/ψ, while in the nucleus-going direction (negative rapidity) $\psi(2S)$ are much more suppressed both in p+Au and p+Al collisions. In other words, $\psi(2S)$'s are strongly suppressed in both directions, while J/ψ's are strongly suppressed in proton-going direction, and moderately in nucleus-going direction.

Fig. 4 shows relative suppression of $\psi(2S)/J/\psi$ as a function of transverse momentum. As one can see, the slowest J/ψ, which spend the most time with soft co-movers are completely

Journal of Physics: Conference Series **736** (2016) 012010 doi:10.1088/1742-6596/736/1/012010

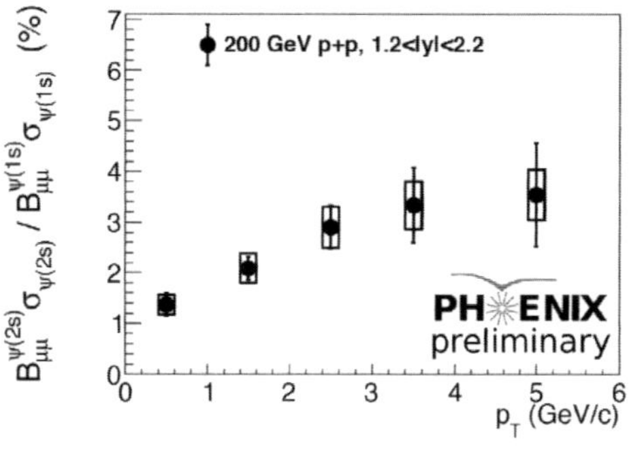

Figure 1. Fraction of $\psi(2S)/J/\psi$ as a function of transverse momentum in proton-proton collisions.

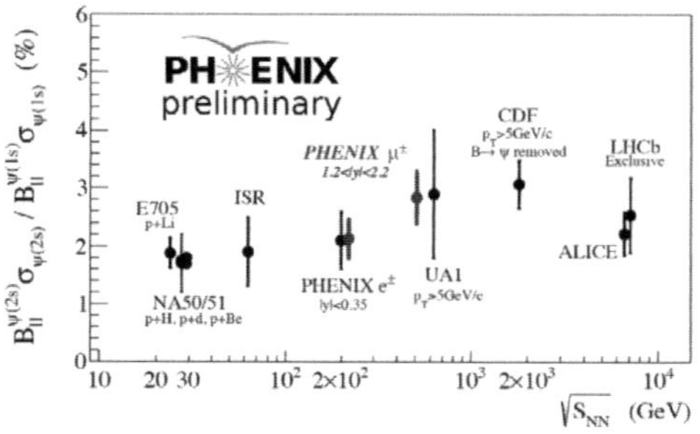

Figure 2. Integrated over p_T $\psi(2S)/J/\psi$ fraction as a function of $\sqrt{s}$.

gone. Thus looks qualitatively different from similar results from the ALICE experiment [9], where little or no dependence on J/ψ transverse momentum is observed. However, both results agree within rather large experimental uncertainties.

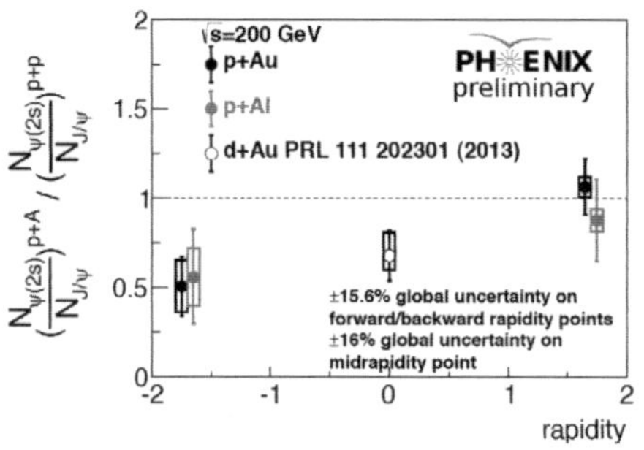

Figure 3. Relative $\psi(2S)/J/\psi$ suppression as a function of rapidity.

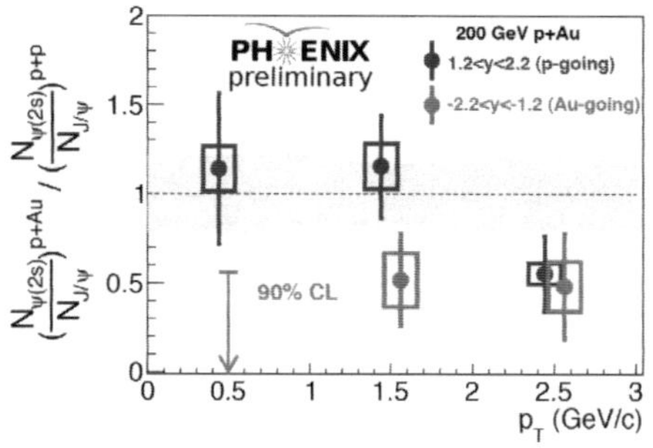

Figure 4. Relative $\psi(2S)/J/\psi$ suppression as a function of transverse momentum.

2. J/ψ suppression in U+U collisions

J/ψ suppression at RHIC energies is much stronger than that at LHC energies for similar energy density created in nucleus-nucleus collisions, which may be an indication of importance of coalescence at LHC. At the same time, at RHIC, J/ψ suppression is almost the same at $\sqrt{s}$ = 39, 62 and 200 GeV, perhaps being strongest at 200 GeV. It can be explained as a balance between color screening and coalescence [3]. So, the question is, when does the coalescence becomes important? Colliding uranium nuclei allows us to go to higher energy density at RHIC and, perhaps, answer this question. Central U+U collisions are expected to have 15-20% higher energy density than central Au+Au collisions, and increased charm production from larger (by ~25%) number of binary collisions (N_{coll}). The former will make color screening stronger, while the latter will result in more coalescence.

The calculation of number of binary collisions for the uranium-uranium collisions is complicated by the non-spheric shape of the uranium nucleus. In this study we used two different shapes. The first one is a "conventional" description of the uranium nucleus

Journal of Physics: Conference Series **736** (2016) 012010 doi:10.1088/1742-6596/736/1/012010

deformation, where the mean radius and diffuseness are taken from electron scattering measurements [7]. The second parametrization [8] takes into account the finite radius of a nucleon and averages over all orientations of axis-of-symmetry. Average radius and diffuseness obtained from this parametrization matches values from electron scattering measurements. The first parametrization has larger surface diffuseness, which results in a less compact nucleus, larger cross-section (by ~12%), and smaller number of binary collisions (by 6-15%).

Figs. 5 and 6 show J/ψ nuclear modification factor (R_{AA}) in uranium-uranium collisions as a function of number of participant nucleons for the first and second parametrizations. On the same figures R_{AA} for Au+Au collisions is shown for comparison. In both cases in most central events J/ψ suppression in U+U collisions appears to be less strong than that in Au+Au collisions.

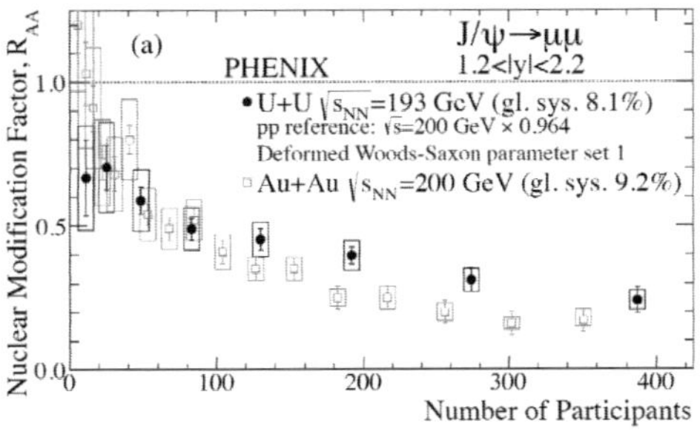

Figure 5. $J\psi$ nuclear modification factor R_{AA} as a function of number of participant nucleons for the first parametrization.

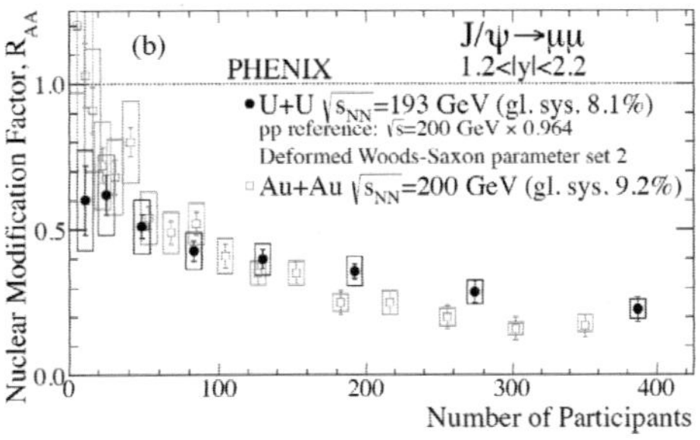

Figure 6. $J\psi$ nuclear modification factor R_{AA} as a function of number of participant nucleons for the second parametrization.

The measured U+U/Au+Au ratio vs. centrality does not depend on number of binary collisions. However, we can predict how this ratio would change if J/ψ production is only dependent on N_{coll}. Fig. 7 shows experimentally measured J/ψ R_{AA} ratio in U+U and Au+Au collisions as a function of centrality. Color curves in this fiigure show how the ratio should change if J/ψ production scales with N_{coll} (dashed curves) or N_{coll}^2 (solid curves). The first parametrization favors N_{coll}^2 dependence, while the first one is consistent with both N_{coll}^2 and N_{coll}, slightly favoring the former in most central collisions. This observation is consistent with the picture in which the increase in charm coalescence becomes more important than the increased color screening when going from Au+Au to U+U collisions.

3. $b\bar{b}$ production in p+p collisions

3.1. Opposite sign di-electrons at $\sqrt{s} = 200$ GeV

Beauty production in p+p collisions at $\sqrt{s} = 200$ GeV was studied using opposite sign di-electrons using simultaneous fit of invariant mass and transverse momentum distributions [4].

The spectrum of di-electrons produced in p+p collisions is well understood in terms of hadronic cocktail (at low mass) and Drell-Yan, charm, and beauty decays (high mass). After the yield from vector and pseudo-scalar mesons, and Drell-Yan is subtracted from the measured di-electron spectrum, we are left with di-elctrons from semi-leptonic decays of charm and beauty. The main idea behind charm/beauty separation is that di-electrons from charm dominate low invariant mass and low p_T region of this two-dimasional space, while di-electrons from beauty dominate either high mass and low p_T, or low mass and high p_T regions. One-dimensional projections of two-dimensional fits to invariant mass axis are shown in Fig. 8.

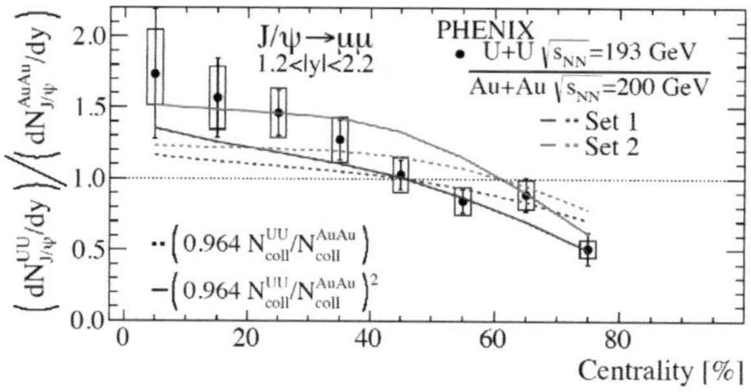

Figure 7. Measured ratio of J/ψ nuclear modification factors in U+U and Au+Au collisions as a function of centrality. Blue curves: first parametrization; red curves: second parametrization. Dashed curves: N_{coll} dependence; solid curves: N^2_{coll} dependence.

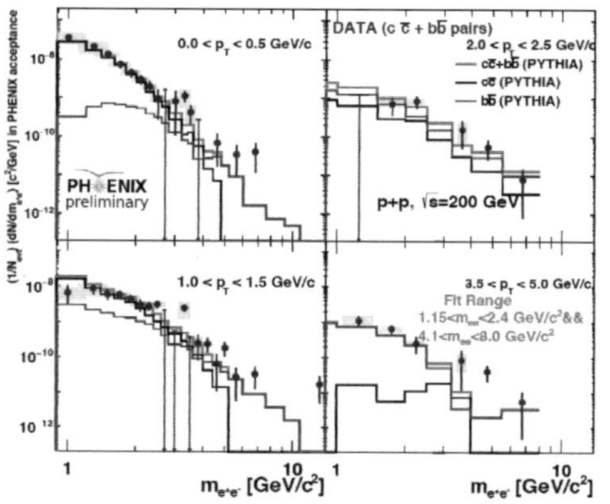

Figure 8. Invariant mass distributions of opposite sign electron pairs after subtraction of the yield from vector and pseudo-scalar mesons, and Drell-Yan in four p_T bins. Black: $c\bar{c}$ yield, blue: $b\bar{b}$ yield, red: the sum.

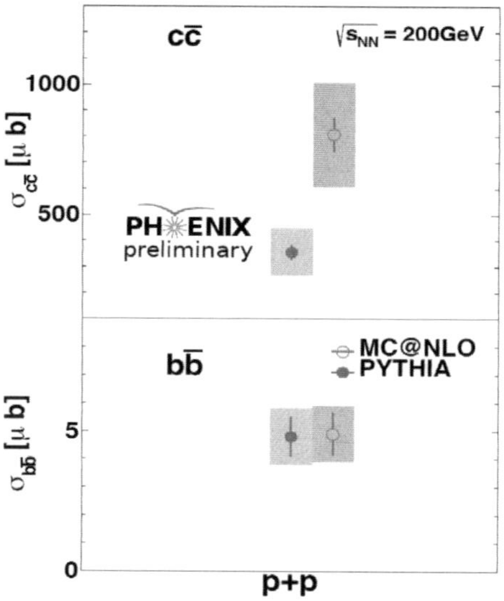

Figure 9. Integrated $c\bar{c}$ (top) and $b\bar{b}$ (bottom) cross-section obtained using PYTHIA and MC@NLO models.

This approach is model-dependent, since the decay electron spectra depend on original charm/beauty distributions. However, for heavier quarks (beauty) the effect of the original quark distribution gets smeared by the decay kinematics, and the results do not depend on the chosen model. This is not true in case of charm. We used two models, tuned PYTHIA [5] and MC@NLO in this study [6]. The integrated $c\bar{c}$ and $b\bar{b}$ cross-section obtained in this way is shown in Fig. 9. The $b\bar{b}$ cross-section shows, as expected, no model dependence, while $c\bar{c}$ cross-section is strongly dependent on rapidity shape of the original quark distribution, which is different in PYTHIA and MC@NLO.

3.2. Same sign dimuons at $\sqrt{s} = 510$ GeV
Another way of separating charm and beauty contributions to di-lepton spectra is to use same sign di-leptons (muons in this case). This approach is based on B-meson oscillations. The

Journal of Physics: Conference Series **736** (2016) 012010 doi:10.1088/1742-6596/736/1/012010

advantage of this method is that it has very low background. There is no contamination from Drell-Yan, quarkonia, or vector mesons.

The spectrum of same sign dimuons consistes of combinatorial background (calculated using mixed events method), and correlated pairs. The correlated same sign pairs contain, in addition to the signal, contribution from charm pairs (negligible in PHENIX acceptance), and the background from jets. This last contribution is the major source of the experimental uncertainty, and is calculated using hadronic simulations.

Fig. 10 shows invariant mass distribution of same sign dimuons measured in p+p collisions at $\sqrt{s} = 510$ GeV after combinatorial background subtraction at backward (top) and forward (bottom) rapidity. The difference between top and bottom plots is due to the fact that the amount of material in front of the PHENIX South and North Muon Arms is very different, resulting in different hadronic contamination.

Total $b\bar{b}$ integrated cross-section is shown in Fig. 11. The integration of cross-section to lower invariant masses is done using PYTHIA. This $b\bar{b}$ cross-section is slightly above the one predicted by NLO pQCD [10], but the two agree within experimental and theoretical uncertainties.

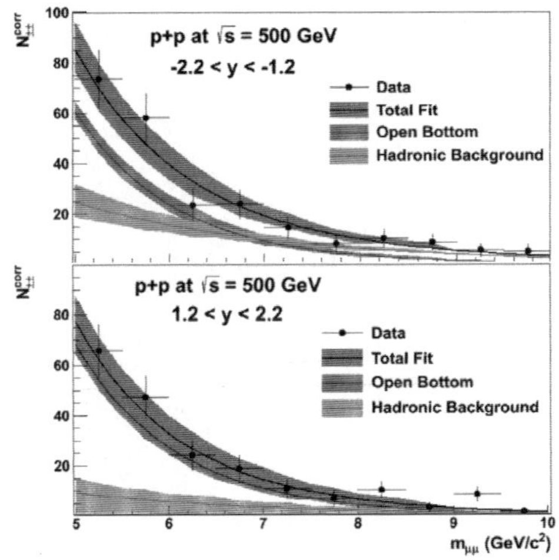

Figure 10. Invariant mass distributions of same sign muon pairs after subtraction of combinatorial background. The difference between the top and bottom plots is due to different amount of material in front of the PHENIX South and North muon arms.

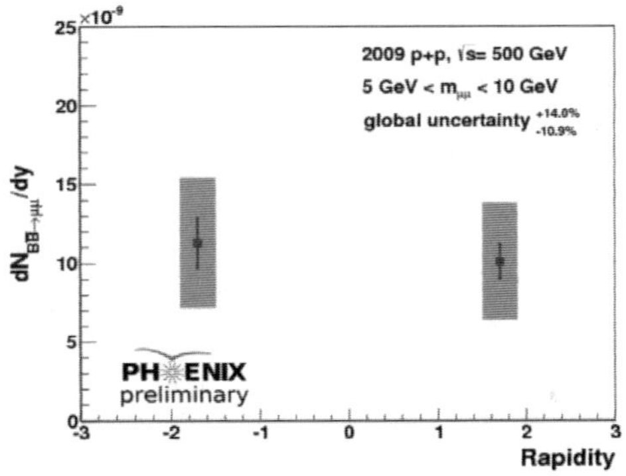

Figure 11. Total $b\bar{b}$ integrated cross-section as a function of rapidity.

4. Summary

We have reviewed recent quarkonia results from the PHENIX experiment at RHIC.

In proton-nucleus collisions $\psi(2S)$ suppression is larger than that of J/ψ in nucleus-going direction, and the same in proton-going direction. Qualitatively, this observation is unlike what was observed at LHC energies, but the difference is within experimental uncertainty. Very strong $\psi(2S)$ suppression is observed at low transverse momentum.

In U+U collisions, J/ψ suppression is weaker than in Au+Au collisions. This is consistent with dominance of coalescence over color screening. Centrality dependence of U+U/Au+Au ratio of nuclear modification factors is also consistent with the importance of coalescence.

Journal of Physics: Conference Series **736** (2016) 012010 doi:10.1088/1742-6596/736/1/012010

PHENIX measured $b\bar{b}$ production using opposite sign electrons at $\sqrt{s} = 200$ GeV, and same sign muons at $\sqrt{s} = 510$ GeV. Both results are somewhat higher than NLO pQCD calculation, but agree with it within uncertainty.

References

[1] A. Adare et al., (PHENIX Collaboration) Phys. Rev. Lett. **111**, 202301 (2013)
[2] A. Adare et al., (PHENIX Collaboration) Phys. Rev. Lett. **107**, 142301 (2013)
[3] X. Zhao and R.Rapp, Phys. Rev. C **82**, 064905 (2010)
[4] A. Adare et al., (PHENIX Collaboration) Phys. Rev. C **91**, 014907 (2015)
[5] T. Sjostrand, S. Mrenna, and P. Z. Skands, J. High Energy Phys. **5**, 026 (2006)
[6] S. Frixione and B. Webber, J. High Energy Phys. **6**, 029 (2002)
[7] H. Masui, B. Mohanty, and N. Xu, Phys. Lett. B **679**, 440 (2009)
[8] Q. Y. Show et al., Phys. Lett. B **749**, 215 (2015)
[9] B. Abelev et al., (ALICE Collaboration) J. High Energy Phys. (2014) 2014:73, doi:10,1007/JHEP12(2014)073
[10] R. Vogt, Eur.Phys.J.ST 155:213-222.2008, arXiv:0709.2531v1

Investigation of the properties of jets from p-Pb and Pb-Pb collisions with ALICE

B A Hess[1] **(for the ALICE Collaboration)**

[1] Physikalisches Institut, Eberhard Karls Universität Tübingen, Auf der Morgenstelle 14, 72076 Tübingen, Germany

E-mail: Benjamin-Andreas.Hess@Uni-Tuebingen.de

Abstract. Jets from hard scattering processes allow to study the properties of strongly interacting matter produced in ultra-relativistic heavy-ion collisions. The hot and dense medium created in such collisions is expected to cause energy loss of hard-scattered partons via elastic scattering and gluon radiation. Eventually, these processes modify the parton fragmentation. We report measurements of charged jets from lead-lead (Pb-Pb) and proton-lead (p-Pb) collisions at $\sqrt{s_{\mathrm{NN}}} = 2.76\,\mathrm{TeV}$ and $5.02\,\mathrm{TeV}$. To estimate cold nuclear matter effects, the jet production in p-Pb collisions is studied for different centrality classes and is compared to that in proton-proton (pp) collisions via the nuclear modification factor. In addition, we discuss the measurement of (charged) jets recoiling from a high-p_{T} trigger hadron, which allows to remove the contribution of combinatorial jets without introducing a bias on the jet population. Furthermore, we report about the measurement of strange hadrons (Λ, K^0_{S}) in association with charged jets from Pb-Pb and p-Pb collisions. The results are expected to clarify the role of the fragmentation process in the anomalous baryon-to-meson ratio observed at intermediate p_{T} in A-A collisions. In particular, the measurement allows disentangling the contributions from jet fragmentation and other hadronisation processes.

1. Introduction

Jets are phenomenological objects representing partons originating from hard scattering processes. They are defined and on an equal footing for both experimental measurement and theoretical models, thus allowing a direct comparison. Jets are a powerful tool to investigate the properties of the hot medium created in ultra-relativistic heavy-ion collisions. Since the hard scattering process occurs in the initial stage of a nucleus-nucleus collision, the hard-scattered partons interact with the medium and lose energy. The investigation of the jet-medium interaction is therefore a key tool to assess the properties of this hot medium.

Besides such final state interactions there can also be cold nuclear matter effects due to the presence of a nucleus in both initial and final state. They can be studied separately in, e.g., proton-lead collisions, where no hot medium is expected to be created.

2. Jet measurements with ALICE

The ALICE detector at the CERN LHC is described in detail in [1, 2]. Here, only a brief overview of the detectors relevant for the present analysis will be given. Charged particle tracking is performed via the six layers of silicon detectors in the Inner Tracking System (ITS) and the large Time Projection Chamber (TPC). Both detectors are located inside the L3 solenoid magnet,

Journal of Physics: Conference Series **736** (2016) 012011 doi:10.1088/1742-6596/736/1/012011

which provides a magnetic field of $B = 0.5\,\mathrm{T}$. This detector set covers the full azimuth and $|\eta_{\mathrm{lab}}| < 0.9$, where η_{lab} denotes the pseudo-rapidity in the ALICE laboratory frame.

The event and centrality selection for this analysis is based on the two scintillator detectors V0A and V0C, which cover $2.8 < \eta_{\mathrm{lab}} < 5.1$ and $-3.7 < \eta_{\mathrm{lab}} < -1.7$, respectively, and the Zero Degree Calorimeters (ZDCs). The latter consist of two sets of neutron (ZNA and ZNC) and proton calorimeters (ZPA and ZPC) located at a distance of $\pm 112.5\,\mathrm{m}$ from the interaction point.

In this analysis, charged jets have been studied, i.e. jets that are reconstructed from charged tracks only. The jet reconstruction is performed with the anti-k_{T} and k_{T} algorithms [3] with a resolution parameter R. The jet axis is required to lie within $|\eta_{\mathrm{jet}}| < 0.9 - R$, such that each jet is fully contained in the detector acceptance. In particular in lead-lead (Pb-Pb) and proton-lead (p-Pb) collisions, the jet reconstruction algorithm assigns tracks to the jet that stem from the so-called underlying event, i.e. bulk hadrons from soft interactions. This biases the reconstructed jet momentum $p^{\mathrm{raw}}_{\mathrm{T,ch\,jet}}$ towards higher values. The contribution from the underlying event can be subtracted as follows:

$$p_{\mathrm{T,ch\,jet}} = p^{\mathrm{raw}}_{\mathrm{T,ch\,jet}} - A_{\mathrm{ch\,jet}} \cdot \rho_{\mathrm{ch}}, \tag{1}$$

where $A_{\mathrm{ch\,jet}}$ denotes the jet area and ρ_{ch} is the event-by-event background density [4]. For p-Pb collisions, the estimation of ρ_{ch} needs an additional occupancy correction to take into account the sparse environment, see [5] for more details. On top of the average background, instrumental effects and background fluctuations can smear the reconstructed jet momentum. The impact of both sources on the jet spectrum is corrected via an unfolding technique [6].

In the following, the measurement of charged jets from p-Pb and Pb-Pb collisions at $\sqrt{s_{\mathrm{NN}}} = 5.02\,\mathrm{TeV}$ and $2.76\,\mathrm{TeV}$, respectively, will be reported.

3. Centrality dependence of charged jet production in p-Pb collisions

Proton-lead collisions can be used to study cold nuclear matter effects in the initial and final state, in particular to disentangle them from effects due to the hot medium created in the final state of Pb-Pb collisions. The nuclear modification is quantified by comparing the p_{T}-differential per-event yields in p-Pb collisions to those in proton-proton (pp) at the same centre-of-mass energy per nucleon pair and scaled by the number of binary collisions N_{coll}. Accordingly, the centrality-differential nuclear modification factor in p-Pb collisions, Q_{pPb}, is defined as [6]:

$$Q_{\mathrm{pPb}} = \frac{\mathrm{d}^2 N^c_{\mathrm{pPb}}/\mathrm{d}\eta \mathrm{d}p_{\mathrm{T}}}{\langle N^c_{\mathrm{coll}} \rangle \cdot \mathrm{d}^2 N_{\mathrm{pp}}/\mathrm{d}\eta \mathrm{d}p_{\mathrm{T}}},$$

where the superscript c indicates that the corresponding quantity is evaluated for a certain centrality class c.

The centrality selection uses a hybrid approach, which performs the centrality classification based on the zero-degree energy measured in the lead-going neutron detector ZNA and connects it with the charged particle multiplicity measurement in lead-going direction via the V0A detector. The hybrid approach is described in detail in [7].

The reconstructed jet momentum needs to be corrected for the underlying event contribution via Eq. 1. In case of p-Pb collisions, the background density ρ_{ch} reaches values not larger than $1.60\,\mathrm{GeV}/c$ in the $0 - 20\%$ ZNA centrality class and is $0.98\,\mathrm{GeV}/c$ for minimum bias collisions [6]. These values are small compared to the jet momentum of $p_{\mathrm{T,ch\,jet}} > 20\,\mathrm{GeV}/c$ used in the analysis.

In Fig. 1, the nuclear modification factors Q_{pPb} of charged jets are shown for various ZNA centrality classes. The jets have been reconstructed with the anti-k_{T} algorithm with resolution parameters $R = 0.2$ and 0.4. The Q_{pPb} factors are found to be consistent with unity for all centrality classes, which indicates the absence of a strong centrality-dependent nuclear

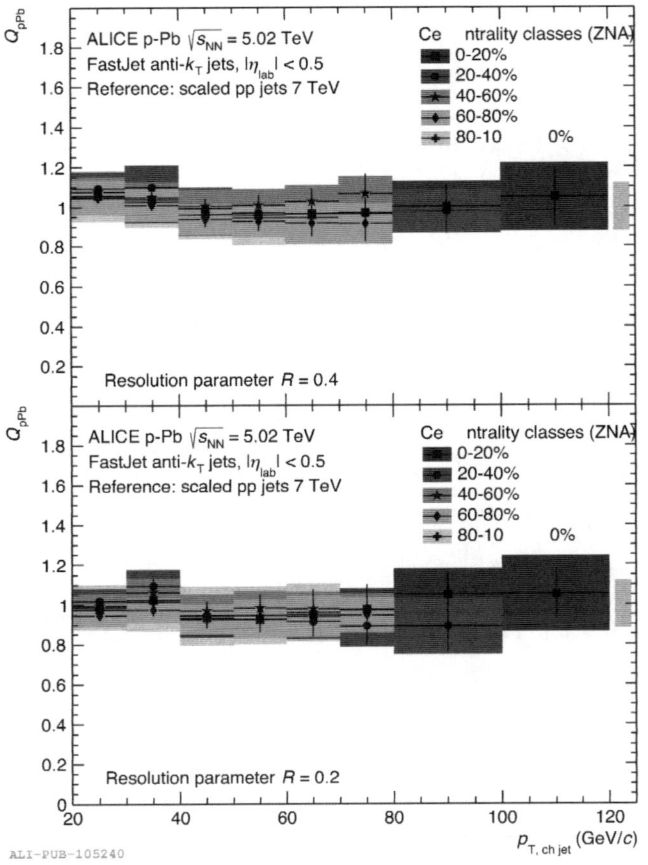

Figure 1. Nuclear modification factors Q_{pPb} of charged jets for several ZNA centrality classes (hybrid approach). The results are shown for jet resolution parameter R = 0.4 (top) and 0.2 (bottom). The global normalisation uncertainty is indicated by the box around unity. More details in [6].

modification of the jet yield in the probed kinematic regime. The result is consistent with single charged particle measurements [7].

The ratio of jet production cross-sections reconstructed with different resolution parameters contains information about the radial structure of the jet. In the following, the ratio between $R = 0.2$ and 0.4 in the common rapidity interval $|\eta_{\mathrm{lab}}| < 0.5$ is considered:

$$\mathscr{R}(0.2, 0.4) = \frac{\mathrm{d}\sigma_{\mathrm{pPb},R=0.2}/\mathrm{d}p_{\mathrm{T}}}{\mathrm{d}\sigma_{\mathrm{pPb},R=0.4}/\mathrm{d}p_{\mathrm{T}}}.$$

The more collimated a jet is, the larger the ratio $\mathscr{R}(0.2, 0.4)$. In the extreme case, when all particles are contained in the jet with smaller radius, the ratio is unity.

In Fig. 2, the ratio $\mathscr{R}(0.2, 0.4)$ is shown for several centrality classes and minimum bias events. As expected, the ratio rises with increasing jet momentum indicating the stronger collimation of jets with rising p_{T}. The ratio for all centrality classes is consistent with the ratio for minimum bias p-Pb collisions. The latter was found to be consistent with the ratio in pp collisions [5]. It is interesting to note that the cross-section ratio was found to be unchanged with respect to pp even in central Pb-Pb collisions [8], where a significant jet suppression was observed. Since we do not observe a strong nuclear modification in p-Pb collision, the unaltered jet production cross-section in p-Pb is in agreement with the expectation.

4. Hadron-jet coincidence measurements

In Pb-Pb collisions, jets are embedded in the huge background of the underlying event. The reconstructed jet spectrum is dominated by combinatorial jets at low p_{T}, where "low" can mean

Journal of Physics: Conference Series **736** (2016) 012011 doi:10.1088/1742-6596/736/1/012011

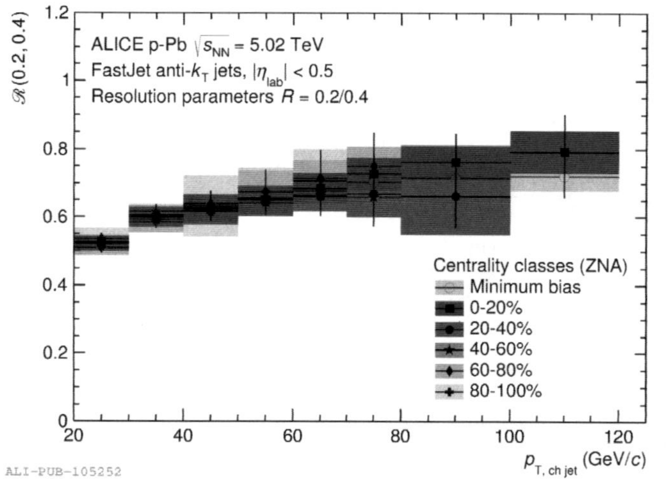

Figure 2. Charged jet production cross-section ratio between resolution parameters $R = 0.2$ and 0.4. The ratio is shown for several ZNA centrality classes and minimum bias. More details in [6].

several 10 GeV/c depending on the jet resolution parameter R, the lower track p_T cut, etc.. Combinatorial jets consist solely of hadrons generated by soft production processes. They can be suppressed by requiring a high-p_T track in each jet. This, however, introduces a fragmentation bias. A more elegant solution is to use hadron-jet coincidence measurements [9] that allow to suppress combinatorial jets without introducing fragmentation biases, i.e. the method is infrared and collinear safe. This enables measurements of jets with large R and/or low p_T.

The basic idea is to select events with a high-p_T Trigger Track (TT) and analyse jets on the away side (azimuthal distance $\Delta\varphi \sim \pi$). Combinatorial jets are independent of the trigger p_T. This is visible in Fig. 3, in which the semi-inclusive distributions of charged jets recoiling from a TT from two different p_T classes are shown. Jets have been reconstructed with the anti-k_T algorithm. The distributions are similar for low and negative p_T, where combinatorial jets dominate, but the yield of the higher TT class shows an excess towards higher p_T, due to the contribution of jets originating from hard-scattered partons. The p_T becomes negative if the background subtraction term in Eq. 1 becomes larger than the raw jet p_T.

To eliminate the contribution from combinatorial jets, the per-trigger-yield of the low TT class is subtracted from that of the high TT class:

$$\Delta_{\text{recoil}} = \left.\frac{1}{N_{\text{trig}}}\frac{\mathrm{d}^2N_{\text{jet}}}{\mathrm{d}p_{\text{T,jet}}^{\text{ch}}\mathrm{d}\eta}\right|_{\text{TT}\{20,50\}} - c_{\text{Ref}}\cdot\left.\frac{1}{N_{\text{trig}}}\frac{\mathrm{d}^2N_{\text{jet}}}{\mathrm{d}p_{\text{T,jet}}^{\text{ch}}\mathrm{d}\eta}\right|_{\text{TT}\{8,9\}}$$

where N_{trig} is the number of trigger tracks in a given TT class, with TT$\{20, 50\}$ being used as an abbreviation for $p_{\text{T,trig}} \in [20, 50]$ GeV/c and similarly for TT$\{8, 9\}$. Furthermore, a scale factor $c_{\text{Ref}} \approx 1$ has been introduced to normalise the spectra in Fig. 3 around the peaks, see [9] for more details. The resulting Δ_{recoil} spectrum for central Pb-Pb collisions is shown in Fig. 4 for jet resolution parameters $R = 0.2, 0.4$ and 0.5.

To investigate possible medium modifications of the spectra in Pb-Pb collisions, the Δ_{recoil} spectrum needs to be compared to the corresponding pp reference. At the time when this analysis was performed, no pp reference at the same centre-of-mass energy with sufficient statistics was available. Therefore, simulations with the PYTHIA Perugia tunes 2010 and 2011 [10] have been utilised instead. As was verified in [9], these tunes reproduce the Δ_{recoil} spectra measured in pp at $\sqrt{s} = 7$ TeV.

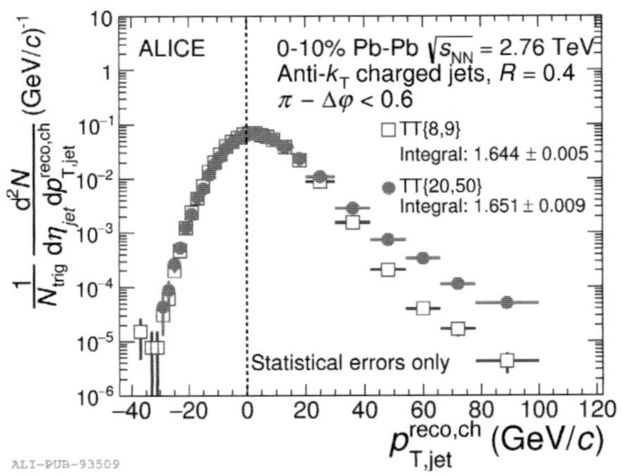

Figure 3. Semi-inclusive distributions of charged jets recoiling from a hadron trigger with $p_{T,trig} \in [8,9]$ GeV/c (blue boxes) and $p_{T,trig} \in [20,50]$ GeV/c (red dots) in central Pb-Pb collisions.

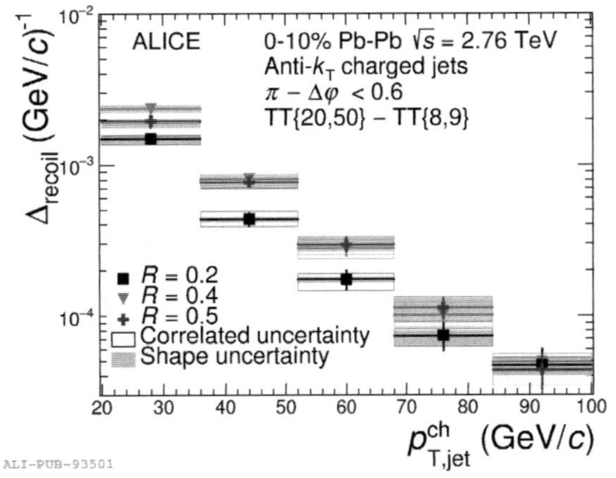

Figure 4. Δ_{recoil} distributions measured in 0–10% central Pb-Pb collisions for charged jets with $R = 0.2, 0.4$ and 0.5. The boxes indicate correlated and shape uncertainties as described in detail in [9].

The medium modification is quantified by the ratio ΔI_{AA} between the Δ_{recoil} spectra in (central) Pb-Pb collisions and the PYTHIA (pp) reference:

$$\Delta I_{AA} = \Delta_{recoil}^{PbPb} / \Delta_{recoil}^{PYTHIA}.$$

The resulting ΔI_{AA} is shown in Fig. 5 for charged jets with $R = 0.5$. The value of ΔI_{AA} is found to be around $0.5 - 0.6$ for all momenta, which indicates a strong suppression of the recoil jet yield for central Pb-Pb collisions. Similar results are found for jets with $R = 0.2$ and 0.4 [9].

This raises the question whether a radial redistribution of the energy inside the jet takes place. To investigate such an effect, the Δ_{recoil} ratios for different values of R are calculated for central Pb-Pb collisions and compared to predictions for pp obtained via the PYTHIA Perugia tunes 2010 and 2011. The resulting ratio between $R = 0.2$ and 0.5 is shown in Fig. 6. It is consistent with the PYTHIA (pp) prediction, i.e. there is no evidence for intra-jet broadening for a jet resolution parameter up to $R = 0.5$. Note that the observations are similar for $R = 0.2$ and 0.4 [9].

5. Neutral strange particle production in charged jets

The p_T-differential baryon/meson ratios (p/π, Λ/K^0_S) were found to exhibit a qualitatively similar evolution with event multiplicity in pp, p-Pb and Pb-Pb collisions, with an enhanced production of baryons at mid-p_T for high multiplicity [11]. This observation suggests similar production mechanism in all three collision systems. Since it cannot be explained solely by fragmentation, other hadronisation mechanisms like coalescence and parton recombination have been proposed [12, 13, 14, 15]. It has to be studied whether the hadronisation in jets exhibits similar features or it is purely a feature of the bulk.

For this the p_T-differential Λ/K^0_S ratio in jets can be compared to the inclusive ratio. The analysis is performed on data from p-Pb and Pb-Pb collisions. Jets have been reconstructed with the anti-k_T algorithm with resolution parameters $R = 0.2, 0.3$ and 0.4. In particular for Pb-Pb, the jet algorithm does not only cluster particles from real jets, that are produced by

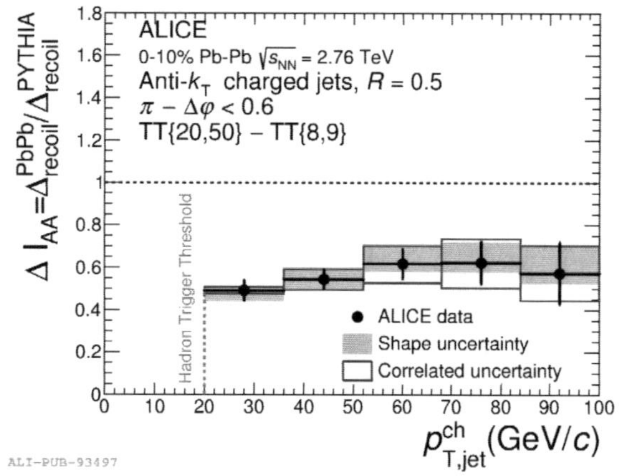

Figure 5. Δ_{recoil} ratio between Pb-Pb and (PYTHIA) pp collisions at $\sqrt{s_{NN}} = 2.76\,\text{TeV}$ for charged jets with $R = 0.5$.

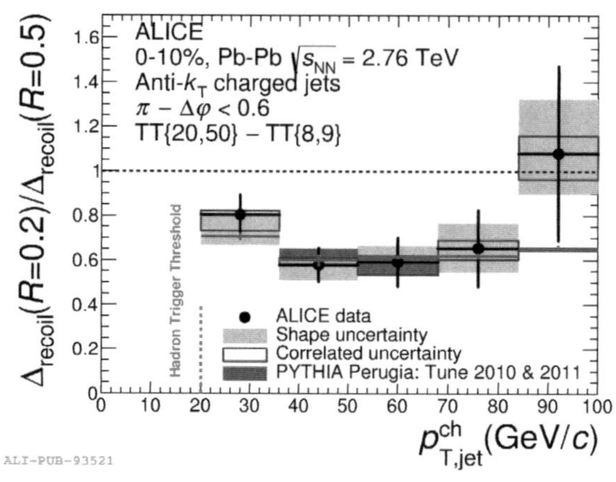

Figure 6. Ratio of Δ_{recoil} for $R = 0.2$ relative to $R = 0.5$ in central Pb-Pb collisions (black). For comparison, the corresponding ratio for pp collisions simulated with PYTHIA (red) is shown in addition. The width of the red band shows the difference between the Perugia 2010 and 2011 tunes.

fragmentation of hard-scattered partons, but also reconstructs combinatorial jets consisting of particles from the underlying event. To reduce the contribution of purely combinatorial jets, only jets containing a leading track with $p_T > 5\,\text{GeV}/c$ are accepted for analysis.

The K^0_S meson and Λ ($\overline{\Lambda}$) baryons do not have electric charge and decay into a pair of charged daughter particles by weak interaction. They are referred to as "V^0 particles". They are reconstructed and identified via the decay topology of their most frequent decay channels:

$$K^0_S \to \pi^+ + \pi^-,$$
$$\Lambda \to p + \pi^-,\quad \overline{\Lambda} \to \overline{p} + \pi^+.$$

Jets and V^0 particles are reconstructed independently of each other. The charged decay daughters of the V^0 particles are secondary tracks and, thus, excluded from the jet reconstruction procedure. To associate the V^0 candidates to a jet, the V^0 candidate is matched to a cone around the jet axis:

$$\sqrt{\Delta\varphi^2_{jet,V^0} + \Delta\eta^2_{jet,V^0}} < R,$$

where $\Delta\varphi^2_{jet,V^0}$ and $\Delta\eta^2_{jet,V^0}$ denote the azimuthal and pseudo-rapidity distance between the momentum of the V^0 candidate and the jet axis, and R is the resolution parameter for the jet reconstruction. Finally, the extracted V^0 yield in jets is corrected for reconstruction efficiency and the expected contribution from the underlying event is subtracted. The Λ baryon yields are corrected for the feed-down contribution from $\Xi^{0,-}$ baryons and correspondingly for $\overline{\Lambda}$.

The Λ/K^0_S ratio is shown in Figs. 7 and 8 for Pb-Pb and p-Pb collisions, respectively. In both cases, the results for $0-10\%$ most central collisions are shown, whereas for p-Pb collisions the $40-100\%$ centrality class is shown in addition. It is clearly visible that, while the inclusive ratio around $p_T \sim 3\,\text{GeV}/c$ is strongly enhanced, the ratio in jets is significantly lower in both collision systems. In the case of p-Pb, the analysis has been repeated with different jet resolution

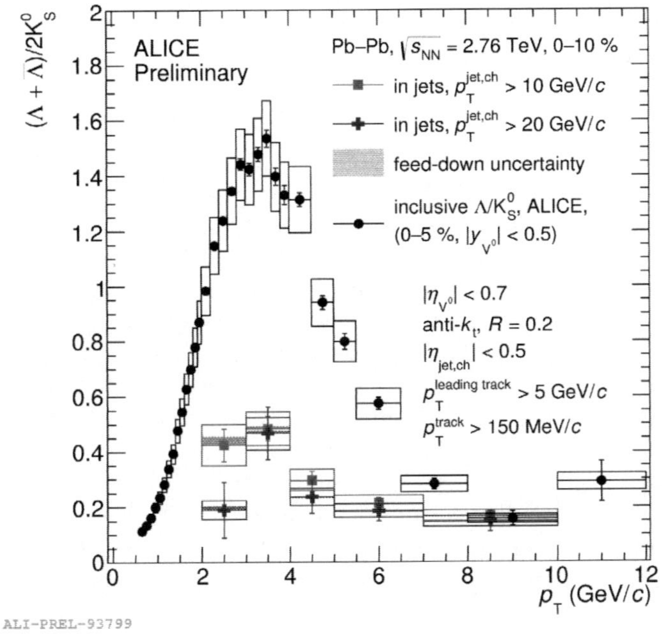

Figure 7. p_T-differential Λ/K^0_S ratio in charged jets with $R = 0.2$ and with transverse momentum larger than $10\,\text{GeV}/c$ (red boxes) and $20\,\text{GeV}/c$ (blue crosses), respectively, measured in $0-10\%$ most central Pb-Pb collisions at $\sqrt{s_{NN}} = 2.76\,\text{TeV}$. For comparison, the corresponding inclusive ratio is also shown (black dots).

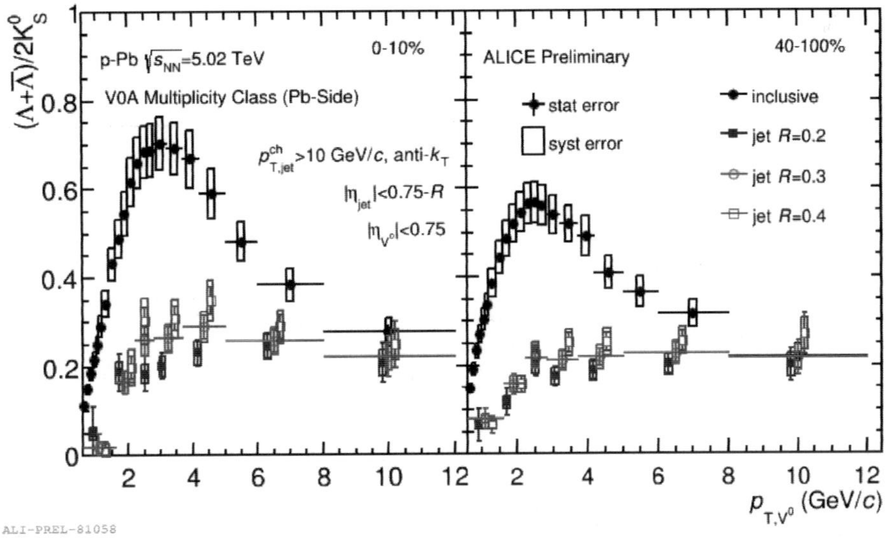

Figure 8. p_T-differential Λ/K^0_S ratio in charged jets with transverse momentum larger than $10\,\text{GeV}/c$ measured in p-Pb collisions at $\sqrt{s_{NN}} = 5.02\,\text{TeV}$ with a centrality of $0-10\%$ (left) and $40-100\%$ (right), respectively. The inclusive ratio (black dots) is compared with the ratio in jets reconstructed with resolution parameters $R = 0.2, 0.3$ and 0.4 (coloured boxes and discs).

parameters $R = 0.2, 0.3$ and 0.4. The Λ/K^0_S ratio in jets is found to be very similar for all R. Also, the ratios for centrality $0-10\%$ and $40-100\%$ are similar for jets, in contrast to the inclusive ratio. These observations suggest that the enhancement of the baryon/meson ratio is a feature of the bulk and the hadronisation in jets is not significantly modified with respect to the one observed in pp collisions.

6. Summary

We reported on measurements with ALICE of properties of jets from p-Pb and Pb-Pb collisions. Jets are an important probe to investigate the properties of the hot medium created in ultra-relativistic heavy-ion collisions. Their study can also help to disentangle the final state hot medium effects from cold nuclear matter effects in the initial and final state.

In our measurement of the charged jet production in p-Pb collisions, we do not observe a significant centrality-dependent nuclear modification. The jet cross-section ratio for $R = 0.2$ and 0.4 is observed to be independent of centrality, which indicates that the degree of collimation of jets at different centralities is not modified. Altogether, we do not observe strong cold nuclear matter effects in the probed kinematic regime.

Jet measurements in Pb-Pb collisions suffer from the huge underlying event leading to a significant contribution from combinatorial jets. The latter can be strongly suppressed by an analysis method based on the measurement of jets recoiling from a high-p_T trigger hadron and which does not introduce a fragmentation bias on quenched jets. The differential recoil jet yield in central Pb-Pb collisions is found to be suppressed with respect to pp by up to a factor two in charged jets with resolution parameter $0.2 \leq R \leq 0.5$. At the same time, we do not observe a significant change of the intra-jet energy profile for $R \leq 0.5$.

In order to clarify the role of the fragmentation process in the anomalous baryon-to-meson ratio at intermediate p_T that is observed in high-multiplicity Pb-Pb, p-Pb and pp collisions [11], we studied the p_T-differential Λ/K^0_S ratio in charged jets from Pb-Pb and p-Pb collisions. The comparison to the corresponding inclusive ratios indicates that the hadronisation in jets does not significantly contribute to the observed anomaly, but that it seems to emerge from processes with low momentum transfer.

Acknowledgments

The work was supported in part by BMBF under grant 05P12VTCAA.

References

[1] Aamodt K *et al.* (ALICE) 2008 *JINST* **3** S08002
[2] Abelev B B *et al.* (ALICE) 2014 *Int. J. Mod. Phys.* A **29** 1430044 (*Preprint* 1402.4476)
[3] Cacciari M, Salam G P and Soyez G 2012 *Eur. Phys. J.* C **72** 1896 (*Preprint* 1111.6097)
[4] Cacciari M and Salam G P 2008 *Phys. Lett.* B **659** 119–126 (*Preprint* 0707.1378)
[5] Adam J *et al.* (ALICE) 2015 *Phys. Lett.* B **749** 68–81 (*Preprint* 1503.00681)
[6] Adam J *et al.* (ALICE) 2016 Centrality dependence of charged jet production in p-Pb collisions at $\sqrt{s_{NN}} =$ 5.02 TeV (*Preprint* 1603.03402)
[7] Adam J *et al.* (ALICE) 2015 *Phys. Rev.* C **91** 064905 (*Preprint* 1412.6828)
[8] Abelev B *et al.* (ALICE) 2014 *JHEP* **03** 013 (*Preprint* 1311.0633)
[9] Adam J *et al.* (ALICE) 2015 *JHEP* **09** 170 (*Preprint* 1506.03984)
[10] Skands P Z 2010 *Phys. Rev.* D **82** 074018 (*Preprint* 1005.3457)
[11] Hess B A (for the ALICE collaboration) 2016 *Proc. 7th Int. Workshop on Multiple Partonic Interactions at the LHC (MPI@LHC 2015)* (Trieste, Italy) pp 203–206
[12] Minissale V, Scardina F and Greco V 2015 *Phys. Rev.* C **92** 054904 (*Preprint* 1502.06213)
[13] Greco V, Ko C M and Levai P 2003 *Phys. Rev.* C **68** 034904 (*Preprint* nucl-th/0305024)
[14] Fries R J, Muller B, Nonaka C and Bass S A 2003 *Phys. Rev.* C **68** 044902 (*Preprint* nucl-th/0306027)
[15] Fries R J, Muller B, Nonaka C and Bass S A 2003 *Phys. Rev. Lett.* **90** 202303 (*Preprint* nucl-th/0301087)

Journal of Physics: Conference Series **736** (2016) 012012 doi:10.1088/1742-6596/736/1/012012

Single electrons from semi-leptonic charm and bottom hadron decays in Au+Au collisions at PHENIX

Takashi Hachiya for the PHENIX collaboration

RIKEN BNL Research Center, Brookhaven National Laboratory, Upton, NY 11973-5000, USA

E-mail: hachiya@ribf.riken.jp

Abstract. Heavy quarks are clean probes to explore the nature of strongly coupled quark gluon plasma created in high energy heavy ion collisions. The strong suppression of single electrons from semi-leptonic decays of heavy flavor hadrons was observed. To further understand the heavy quark suppressions, PHENIX installed the silicon vertex detector (VTX) which allows us to measure the bottom and charm productions separately from measurement of displaced tracks. For the first time, we observed the electrons from bottom hadron decays are less suppressed than those from charms for $3 < p_T < 4$ GeV/c and are similarly strongly suppressed for higher p_T in minimum bias Au+Au collisions at $\sqrt{s_{NN}} = 200$ GeV. We present the results of separated bottom and charm productions using the 2011 dataset with the VTX.

1. Introduction

Heavy quarks are clean probes to explore the nature of strongly coupled quark gluon plasma (QGP) created in relativistic high energy heavy ion collisions. Heavy quarks, bottom and charm quarks, are mainly produced via initial hard scattering in the collisions because of their large masses ($M_b \approx 4.2$ GeV/c^2 and $M_c \approx 1.3$ GeV/c^2). Once produced, heavy quarks interact with QGP when they traverse it. Therefore, heavy quarks probe the property of QGP during its entire space-time evolution.

PHENIX previously measured a strong suppression of the heavy quark production at high transverse momentum (p_T) and a substantial flow in the measurement of single electrons from semi-leptonic decays of bottom and charm hadrons[1, 2]. The observed suppression of heavy quarks was found to be similar to that of light quarks. Before this discovery, it was expected that heavy quarks suffer less energy lose in QGP than light quarks and gluons via radiative energy loss due to "dead cone effect" [3], which provides a mass ordering of energy loss ($\Delta E_g > \Delta E_{u,d,s} > \Delta E_c > \Delta E_b$). Collisional energy loss is also considered to describe the suppressions in addition to radiative energy loss. To understand the phenomena of the heavy quark suppressions in detail, PHENIX installed a barrel silicon vertex detector (VTX) in 2011 to measure the bottom and charm components separately in the single electron measurement. VTX can disentangle the relative contributions between the radiative and collisional energy losses. In this article, the data analysis method are described and the results are discussed.

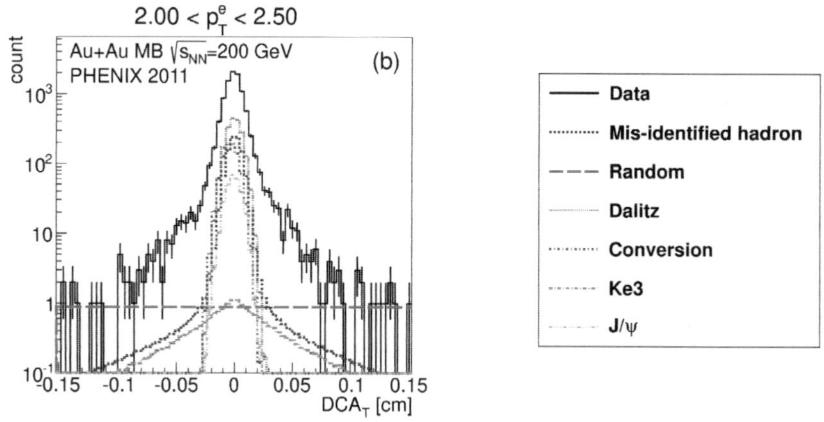

Figure 1. DCA_T distribution for electrons in $2 < p_T < 2.5$ GeV/c in minimum bias Au + Au collisions at $\sqrt{s_{NN}} = 200$ GeV. Also shown are the normalized components for the various background sources

2. Data analysis and bottom/charm separation

We measured single electrons using the data on Au+Au collisions at $\sqrt{s_{NN}} = 200$ GeV recorded in 2011[4]. Electron identification and tracking are performed by the PHENIX two central arm detectors. The analysis utilizes the difference of the decay lengths of B and D mesons ($c\tau_{B^0} \approx 455$ μm and $c\tau_{D^0} \approx 123$ μm). VTX measures the distance of closest approach (DCA) of the track from the collision vertex. The DCA is calculated in the transverse (DCA_T) and longitudinal (DCA_L) planes separately because VTX has better resolution in transverse plane by detector design. We use DCA_T for the analysis, where the DCA_T resolution is ≈ 60 μm for tracks with $p_T > 2$ GeV/c.

The data analysis is performed in three steps. First, we measure the DCA_T of inclusive electrons for $1.5 < p_T < 5.0$ GeV/c in five p_T bins. The electrons contain a large amount of backgrounds that originate from non-desired physics processes such as Dalitz decays of π^0 and η, photon conversions, K_{e3} decays, and $J/\psi \rightarrow e^+e^-$ decays as well as detector effects such as mis-identified hadrons and high multiplicity background. Most of them are rejected by vetoing pair-wise hits in VTX because very close pairs of hits are created by e^+e^- pairs from Dalitz decays and photon conversions, which are main background sources.

Second, we determined the background contributions remained in the DCA_T distributions. The background from mis-identified hadrons is evaluated by the event swap method and the high-multiplicity background is evaluated by embedding single simulated electrons into the real events. The backgrounds from Dalitz decays of π^0 and η, photon conversions, K_{e3}, and J/ψ decays are determined using a GEANT simulation of PHENIX detector with the inputs of measured spectra. DCA_T distributions from these backgrounds are normalized based on the background yields by cocktail calculations, which is similar with previous PHENIX single electron measurement in 2004[1, 2]. Figure 1 shows the DCA_T distribution for electrons in $2.0 < p_T < 2.5$ GeV/c. The normalized background components are also plotted in Fig. 1.

Third, to separate the bottom and charm components, we developed a new statistical separation method based on the Bayesian inference technique, which is called as an unfolding method. The unfolding method performs a simultaneous fitting to the DCA_T distributions from the 2014 data set and the published PHENIX invariant yield of inclusive heavy flavor electrons measured using the data recorded in 2004[2]. In the unfolding method, a set of trial parent bottom and charm hadron yields are chosen and converted to DCA_T distributions and

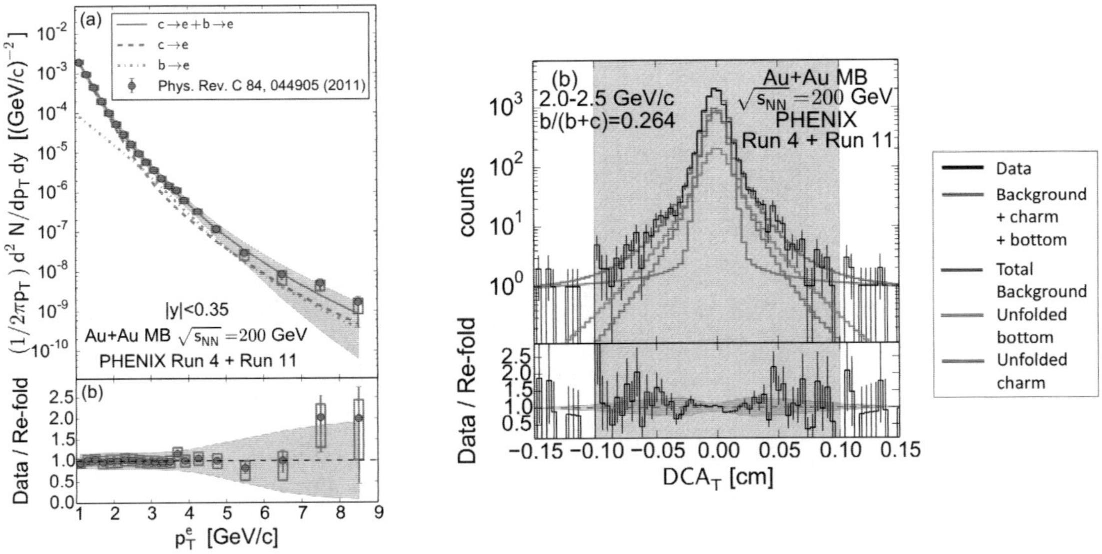

Figure 2. (Left) Invariant yield of inclusive heavy flavor electrons measured using the data recorded in 2004[2]. (Right) DCA_T distribution for single electrons measured in this analysis. The invariant yield and DCA_T distributions are compared with the calculated bottom and charm components from the unfolding method as well as their backgrounds. Bottom panels in left and right plots show the ratio between the data and the results from the unfolding method. The results are in good agreement with the measure data within the uncertainties.

invariant yield of electrons decayed from these parents using a decay matrix. The resulting distributions of electron DCA_T and invariant yield are compared with the measured data using a likelihood function. The benefit of the unfolding method is to constrain the bottom and charm components in both the DCA_T and the invariant yield spaces using the parent hadron yields because both the DCA_T shape and invariant yield of electrons are sensitive to the p_T distributions of parent bottom and charm hadrons. Figure 2 shows the comparisons between measured electrons and electrons calculated by the unfolding method in p_T (left) and DCA_T distribution (right), respectively. The bottom panels of Fig. 2 are the ratio between them. We found a good agreement in both invariant yield and DCA_T within the uncertainties.

3. Results

From the results of the unfolding method, the fraction of electrons from bottom decays to inclusive heavy flavors (both bottoms and charms) is calculated. The resulting bottom electron fraction is shown in the left panel of Fig. 3 and compared with FONLL perturbative QCD calculations for $p+p$ collisions at $\sqrt{s} = 200$ GeV[5]. We find that the measured bottom electron fraction in Au + Au collisions shows a steep rise in $2 < p_T < 4$ GeV/c with a possible peak at $p_T \sim 3$ GeV/c compared with the FONLL calculation. In Fig. 3, the bottom electron fraction in $p+p$ collisions at $\sqrt{s} = 200$ GeV [6, 7] is also plotted and found that they are in good agreement with the FONLL calculations within uncertainties.

We can calculate the nuclear modification factor (R_{AA}) for electrons from bottom and charm hadron decays separately using the bottom electron fraction in Au + Au and $p+p$, and R_{AA} of inclusive heavy flavor electrons. The resulting R_{AA} is shown in the right panel of Fig. 3. We find that electrons from bottom decays are less suppressed than that from charm decays for $p_T < 4$ GeV/c and similarly strongly suppressed for higher p_T. However, it is difficult to compare with

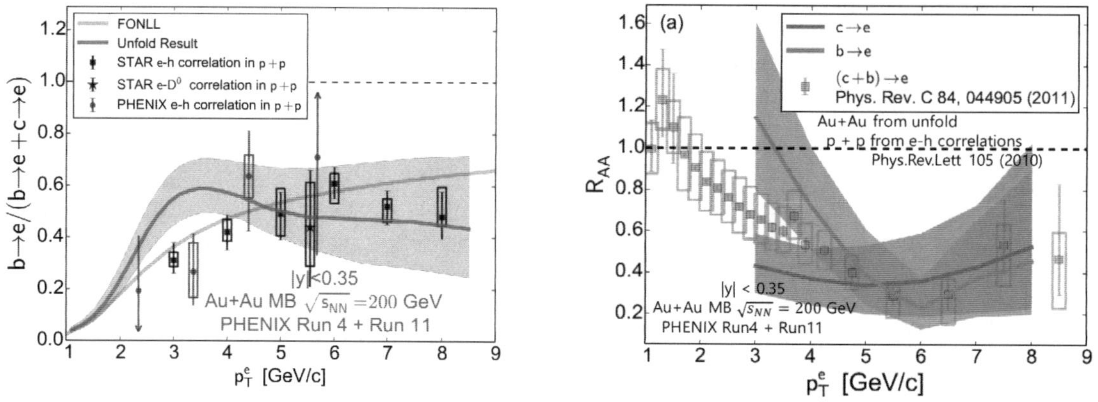

Figure 3. (Left) The unfolded result of bottom electron fraction in minimum bias Au + Au collisions at $\sqrt{s_{NN}} = 200$ GeV[4], which are compared with the FONLL calculation[5]. Also shown are the bottom electron fraction in $p+p$ collisions at $\sqrt{s} = 200$ GeV from PHENIX[6] and STAR[7]. (Right) R_{AA} for electrons from separated bottom and charm hadron decays as well as inclusive heavy flavor electrons

the theoretical calculations in detail because of the relatively large uncertainties.

4. Summary and Outlook

PHENIX has measured the electron yield from bottom and charm hadron decays separately. The VTX detector installed in 2011 enables us to separate the bottom and charm components in single electrons by measuring the DCA_T distributions. Using the DCA_T distributions and previous PHENIX results of invariant yield of heavy flavor electrons, the unfolding method successfully extracted the fraction of bottom electrons to inclusive heavy flavors as a function of p_T. From the comparison of the bottom electron fractions in Au + Au and $p+p$, we find a less suppression of electrons from bottom decays than that from charm decays for $p_T < 4$ GeV/c and a similar strong suppression for higher p_T.

This is the first measurement of the strong bottom suppression in Au + Au collisions at $\sqrt{s_{NN}} = 200$ GeV. However the uncertainties in the current measurement is relatively large so as to disentangle the energy loss mechanisms in detail. We collected approximately 10 times larger amount of Au + Au data in 2014 and large statistics $p+p$ data in 2015. These data sets should provide definitive measurements of separated bottom and charm electron suppressions.

References

[1] A. Adare *et al.* Energy Loss and Flow of Heavy Quarks in Au + Au Collisions at $\sqrt{s_{NN}} = 200$ GeV *Phys. Rev. Lett.* **98** 172301 (2007)
[2] A. Adare *et al.* Heavy-quark production in $p+p$ and energy loss and flow of heavy quarks in Au + Au collisions at $\sqrt{s_{NN}} = 200$ GeV *Phys. Rev.* C **84**, 044905 (2011)
[3] Y. L. Dokshitzer and D. E. Kharzeev *Phys. Lett.* B 519, 199 (2001)
[4] A. Adare *et al.* Single electron yields from semileptonic charm and bottom hadron decays in Au + Au collisions at $\sqrt{s_{NN}} = 200$ GeV *Phys. Rev.* C **93** 034904 (2016)
[5] M. Cacciari, P. Nason, and R.Vogt, QCD Predictions for Charm and Bottom Quark Production at RHIC *Phys. Rev. Lett.* **95** 122001 (2005)
[6] A. Adare *et al.* Measurement of Bottom versus Charm as a Function of Transverse Momentum with Electron-Hadron Correlations in $p+p$ Collisions at $\sqrt{s} = 200$ GeV *Phys. Rev. Lett.* 103 082002 (2009)
[7] M. Aggarwal *et al.* Measurement of the Bottom contribution to non-photonic electron production in $p+p$ collisions at $\sqrt{s} = 200$ GeV *Phys. Rev. Lett.* 105 202301 (2010)

Equilibration of hadrons in HICs via Hagedorn States

M Beitel, K Gallmeister and C Greiner

Institut für Theoretische Physik, Goethe-Universität Frankfurt am Main,
Max-von-Laue-Str. 1, 60438 Frankfurt am Main, Germany

E-mail: beitel@th.physik.uni-frankfurt.de

Abstract. Hagedorn states (HS) are a tool to model the hadronization process which occurs in the phase transition region between the quark gluon plasma (QGP) and the hadron resonance gas (HRG). These states are believed to appear near the Hagedorn temperature T_H which in our understanding equals the critical temperature T_c. A covariantly formulated bootstrap equation is solved to generate the zoo of these particles characterized baryon number B, strangeness S and electric charge Q. These hadron-like resonances are characterized by being very massive and by not being limited to quantum numbers of known hadrons. All hadronic properties like masses, spectral functions etc. are taken from the hadronic transport model Ultra Relativistic Quantum Molecular Dynamics (UrQMD). Decay chains of single Hagedorn states provide a well description of experimentally observed multiplicity ratios of strange and multi-strange particles. In addition, the final energy spectra of resulting hadrons show a thermal-like distribution with the characteristic Hagedorn temperature T_H. Box calculations including these Hagedorn states are performed. Indeed, the time scales leading to equilibration of the system are drastically reduced down to $2 \ldots 5\,\mathrm{fm}/c$.

1. Introduction

Back in the 60's of the last century and before the advent of quantum chromodynamics (QCD) as the theory of strong interactions, R. Hagedorn [1] proposed the existence of a whole zoo of massive, unobserved hadronic resonance states. The spectrum of these particles, known as Hagedorn spectrum, exhibits the specific feature of being exponential in the infinite mass limit with the slope given by the so called Hagedorn temperature T_H. This temperature denotes the limiting temperature for hadronic matter since any partition function of a HRG with Hagedorn-like mass spectrum diverges as long as $T > T_H$. Above the Hagedorn temperature a new state of matter, namely the QGP, shall be realized. Thus, Hagedorn states provide a tool to understand the phase transition from HRG to QGP and back. The Hagedorn states are created in multi-particle collisions most abundantly near T_H which in our understanding equals to the critical temperature T_c. Hagedorn states are color neutral objects which are allowed to have any quantum numbers as long as they are compatible with its mass. The appearance of Hagedorn states in multi-particle collisions and their role was already discussed in [2, 3, 4, 5, 6]. Their appearance near T_c can explain, as shown in [4, 5, 6], the fast chemical equilibration of (multi-) strange baryons B and their anti-particles $\bar{B}$ at Relativistic Heavy Ion Collider (RHIC) energies. The inclusion of Hagedorn states in a hadron resonance gas model provides also a lowering of the speed of sound, c_s and of the shear viscosity over entropy density ratio η/s at the phase transition and being in good agreement with lattice calculations [7, 8, 9, 10].

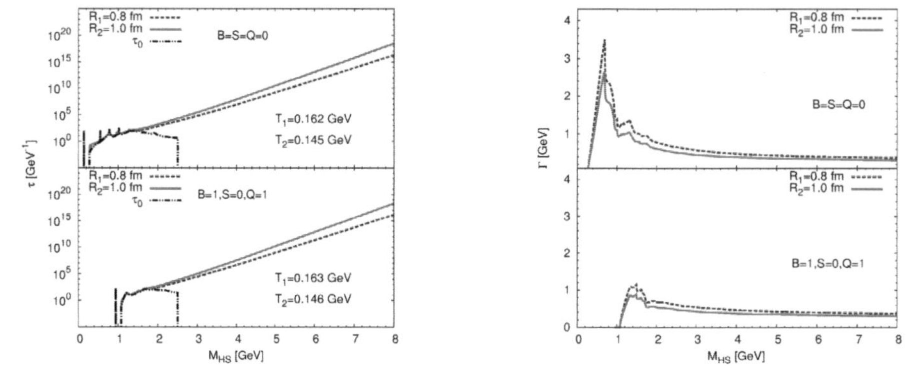

Figure 1. Hagedorn spectrum (left) and total decay width (right) for meson-like ($B = S = Q = 0$) (upper panel) and baryon-like ($B = 1, S = 0, Q = 1$) (lower panel) Hagedorn states.

The starting point of all calculations provided here is the postulate of the statistical bootstrap model (SBM) stating that fireballs consist of fireballs which in turn consist of fireballs etc. . As shown in [11] (cf. also [12]) the mathematical formulation of this postulate in the formulation of [13, 14] leads to the evolving bootstrap equation for the density of states,

$$\tau_{\vec{C}}(m) = \frac{R^3}{3\pi\, m} \sum_{\vec{C}_1,\vec{C}_2} \iint \mathrm{d}m_1 \mathrm{d}m_2\, m_1 \tau_{\vec{C}_1}(m_1)\, m_2 \tau_{\vec{C}_2}(m_2) \times\, p_{\mathrm{cm}}(m, m_1, m_2)\, \delta^{(3)}_{\vec{C},\vec{C}_1+\vec{C}_2}\,, \tag{1}$$

where the conservation of the quantum numbers baryon number, strangeness and electrical charge $\vec{C} = (B, S, Q)$ is guaranteed. A major aspect was to restrict on Hagedorn states, which just consist of two constituents in order to allow for generation processes as $2 \to 1$ and decay as $1 \to 2$. This is some necessary constraint in order to allow for a microscopical calculation in a transport model simulation based on cross sections. The above set of integral equations of Volterra type is solved numerically by disretizing the mass range and consecutively stepping to higher mass bins. The zeroth order input are the known hadrons according their spectral functions as implemented in the transport model UrQMD [15]. The radius R is the only free parameter of the model. Choosing reasonable values as $R = 0.8\,\mathrm{fm}$ ($1.0\,\mathrm{fm}$) yield Hagedorn temperatures, i.e. the slopes of the exponentially rising Hagedorn spectra $\tau_{\vec{C}}(m) \sim m^a \exp(m/T_H)$ as $T_H = 162\,\mathrm{MeV}$ ($145\,\mathrm{MeV}$) nearly independent of the quantum numbers, as indicated in fig. 1.

In order to treat the decay and the production of Hagedorn states, the decay width Γ and the production cross section $\sigma = \pi R^2$ are connected via detailed balance as

$$\Gamma_{\vec{C}}(m) = \frac{\sigma}{2\pi^2\, \tau_{\vec{C}}(m)} \sum_{\vec{C}_1,\vec{C}_2} \iint \mathrm{d}m_1 \mathrm{d}m_2\, \tau_{\vec{C}_1}(m_1)\, \tau_{\vec{C}_2}(m_2) \times\, p^2_{\mathrm{cm}}(m, m_1, m_2)\, \delta^{(3)}_{\vec{C},\vec{C}_1+\vec{C}_2}\,, \tag{2}$$

which represents the second major equation of the model. Corresponding Hagedorn spectra and their decay width are shown in fig. 1. One observes, that the slope decreases with increasing volume of the Hagedorn state. Correspondingly also the width decreases. Nevertheless, for larger masses, the width is nearly constant in the range of some hundreds of MeV, and also stays finite in the infinite mass limit.

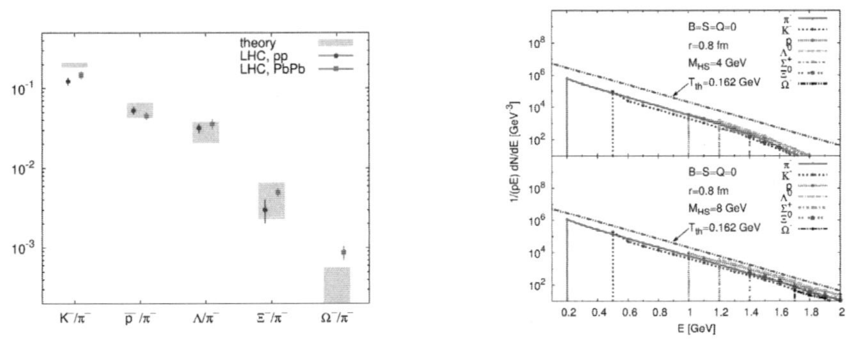

Figure 2. Left: Comparison of particle multiplicity ratios from theory vs. p-p at $\sqrt{s_{NN}} = 0.9\,\text{TeV}$ [16] and Pb-Pb at $\sqrt{s_{NN}} = 2.76\,\text{TeV}$ [17, 18, 19], both from ALICE at LHC. The bands show the range of calculations from cascade decay of a single charge neutral Hagedorn state with radius $R = 0.8\,\text{fm}$ and initial mass $M_{HS} = 4\,\text{GeV}$ and $M_{HS} = 8\,\text{GeV}$. Right: Energy spectra of hadrons stemming from cascade decay of charge neutral Hagedorn state with radius $R = 0.8\,\text{fm}$ and initial mass $M_{HS} = 4\,\text{GeV}$ and $M_{HS} = 8\,\text{GeV}$.

Having the total decay width and the corresponding composition, one is able to calculate the branching ratios for the decay into light Hagedorn states and/or hadrons. Performing such cascading decay simulations, the resulting hadron multiplicity coincide very well with experimentally observed values[11]. This is visualized in fig. 2, where the multiplicity ratios for strange and multi-strange particles are compared with experimental values measured at LHC. The most striking result is the fact, that the energy spectra of these decay hadrons *look* thermal, i.e. the functional behavior is exponential, as shown in fig. 2. The slope parameter coincides with the underlying Hagedorn temperature.

All these ingredients were implemented into UrQMD [15], which can handle the creation and the decay of Hagedorn states and their component dynamically. This enables us to perform box calculations and study the time evolution and the equilibration of the system. The model is able to handle different initialization scenarios, as e.g. the initialization with only nucleons (or pions) with a given particle and energy density. While this may stand for some 'bottom-up' scenario, alternatively one may start with a given Hagedorn state distribution and look into the resulting hadron distributions. This indicates a 'top-down' scenario, where the Hagedorn states may stem from some preceding deconfined phase.

Fig. 3 shows the time dependence of some stable hadrons with respect to strong interaction for box calculations, where only Hagedorn states according to different energy densities were initialized. The given multiplcities at some time are the resulting muliplicities, if all particles would decay down to the stable remnants. One indeed observes very short time scales for the equilibration: While the pions are almost equilibrated after $2\,\text{fm}/c$, the number of protons reaches values close to the maximum value already after $5\,\text{fm}/c$. For the hyperons Σ^- the chemical equilibration is on the same time scale as is for protons. The Ω^- baryon has a low production rate but still chemically equilibrates within the life time of a fireball being $t \approx 10\,\text{fm}/c$. The multiplicity ratios are almost stable for all energy densities, except for the smallest considered one, i.e. for $\varepsilon = 0.2\,\text{GeV}/\text{fm}^3$. In addition we find, that the energy spectra of the hadrons again follow an exponential shape, with slope parameters very close to the Hagedorn temperature $T_H = 162\,\text{MeV}$, see fig. 4.

We conclude, that Hagedorn states are a valuable tool for the understanding of the phase transition between a deconfined and a hadronic phase. While already the decay particles of one single Hagedorn state *look* thermal with the Hagedorn temperature as the only scale, a collective system of Hagedorn states leads to small equilibration times of hadrons in the order

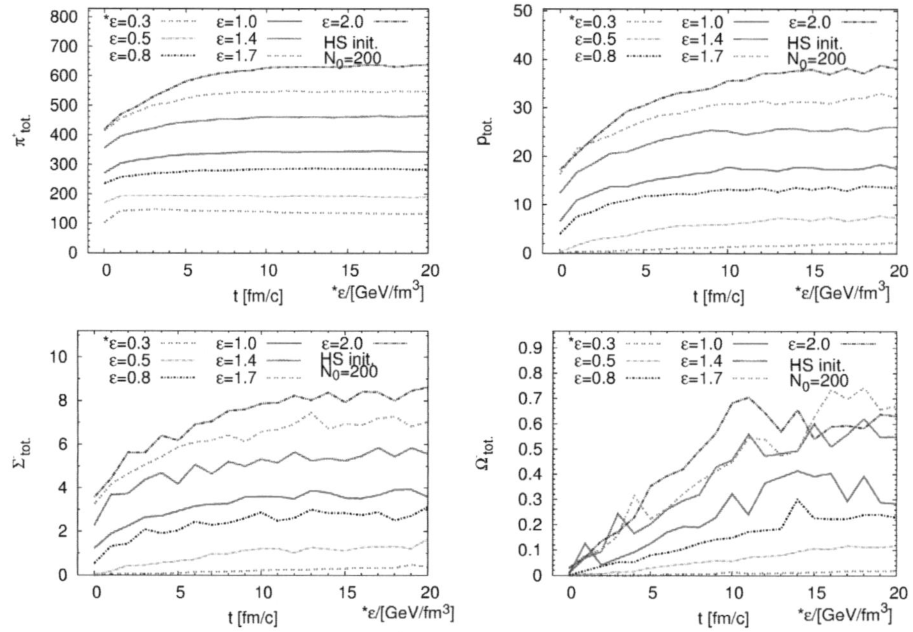

Figure 3. The time dependence of π^+ (up left) and p (up right) hadrons and Σ^- (bottom left) and Ω^- (bottom right) hyperons in a box calculation, initialized with Hagedorn states, for different initial energy densities as indicated in the plot.

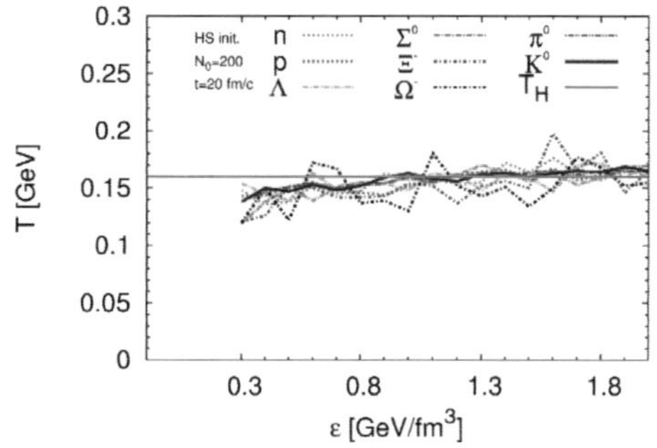

Figure 4. The slope parameter of the energy distribution of different particle species for different initialization energy densities. Red solid line denotes the Hagedorn temperature T_H.

of few fm/c. These remarkable features may also open the door for alternative descriptions of heavy ion collisions, where directly from a pure gluonic phase without quarks a transition to the confined phase is possible [20].

[1] Hagedorn R 1965 *Nuovo Cim. Suppl.* **3** 147–186
[2] Greiner C and Leupold S 2001 *J. Phys. G* **27** L95–L102
[3] Greiner C, Koch-Steinheimer P, Liu F M, Shovkovy I A and Stoecker H 2005 *J. Phys. G* **31** S725–S732
[4] Noronha-Hostler J, Greiner C and Shovkovy I 2008 *Phys. Rev. Lett.* **100** 252301
[5] Noronha-Hostler J, Greiner C and Shovkovy I 2010 *J. Phys. G* **37** 094017
[6] Noronha-Hostler J, Beitel M, Greiner C and Shovkovy I 2010 *Phys. Rev. C* **81** 054909
[7] Noronha-Hostler J, Noronha J and Greiner C 2009 *Phys. Rev. Lett.* **103** 172302
[8] Majumder A and Muller B 2010 *Phys. Rev. Lett.* **105** 252002

[9] Noronha-Hostler J, Noronha J and Greiner C 2012 *Phys. Rev. C* **86** 024913
[10] Jakovac A 2013 *Phys. Rev. D* **88** 065012
[11] Beitel M, Gallmeister K and Greiner C 2014 *Phys. Rev. C* **90** 045203
[12] Yellin J 1973 *Nucl. Phys. B* **52** 583–594
[13] Frautschi S C 1971 *Phys. Rev. D* **3** 2821–2834
[14] Hamer C and Frautschi S C 1971 *Phys. Rev. D* **4** 2125–2137
[15] Bass S A *et al.* 1998 *Prog. Part. Nucl. Phys.* **41** 255–369
[16] Aamodt K *et al.* (ALICE) 2011 *Eur. Phys. J. C* **71** 1594
[17] Abelev B *et al.* (ALICE Collaboration) 2013 *Phys. Rev. C* **88** 044910
[18] Abelev B B *et al.* (ALICE Collaboration) 2013 *Phys. Rev. Lett.* **111** 222301
[19] Abelev B B *et al.* (ALICE Collaboration) 2014 *Phys. Lett. B* **728** 216–227
[20] Stoecker H *et al.* 2016 *J. Phys.* **G43** 015105

Journal of Physics: Conference Series **736** (2016) 012014 doi:10.1088/1742-6596/736/1/012014

PHENIX results on collectivity in high-multiplicity p + Au, d + Au and ^{3}He + Au collisions

Qiao Xu, (for the PHENIX Collaboration)
Physics and Astronomy Department, Vanderbilt University, U.S.A
E-mail: qiao.xu@vanderbilt.edu

Abstract. We present recent PHENIX results for long-range two-particle correlation functions across a large pseudorapidity(η) gap in high-multiplicity p+Au, d+Au and ^{3}He+Au collisions. Enhanced correlations are observed for particles with small azimuthal separation. Measurements of the second order Fourier coefficients (v_2) in p/d/^{3}He+Au collisions, and the third order Fourier coefficient (v_3) in ^{3}He+Au collision are performed using the event plane method. Scaling of v_2 with the initial eccentricity (ε_2) is performed in each system to investigate the role of the initial geometry in the development of a final-state anisotropy in the particle emission. The v_2 coefficients for identified $\pi^{\pm}$, $K^{\pm}$ and (anti-)proton are measured as a function of transverse kinetic energy (KE_T) and their magnitudes are found to scale approximately with the number of constituent quarks in the hadron.

Introduction

Collective flow is one of the defining signatures of quark-gluon plasma (QGP) formation in high-energy nucleus-nucleus (AA) collisions. It has long been considered that a large volume, in which a thermal equilibrium can be achieved, is a prerequisite for QGP formation. However, a number of recent measurements from high-multiplicity collisions in small systems at RHIC [1, 2, 3] and LHC [4, 5, 6, 7] have found strong correlations between the produced particles reminiscent of those observed in the AA collisions. In AA collisions, the final-state particles anisotropies are well described by hydrodynamics models with small specific viscosity. In these models, the anisotropies are driven by the pressure gradients in the initial energy density. There are different initial state models, some of which, such as the IP-glasma model [8], involve gluon saturation. Alternatively, the system evolution can be described on a microscopic level with partonic and hadronic scatterings, such as in the AMPT model [9]. These models have different sensitivity to the initial geometry in the collisions [10]. To investigate if QGP can be formed in small system, and if the observed correlations can be attributed to collective hydrodynamic flow, the PHENIX experiment has conducted a series of geometry-controlled experiments using 0-5% central p/d/^{3}He + Au collision at $\sqrt{s_{NN}}$ = 200 GeV. The two-particle correlations and the Fourier coefficients in the single-particle azimuthal distributions are studied as a function of transverse momentum.

1. Experiment Details

In the PHENIX detector [11], charged particle trajectories are reconstructed with the two central arm using the drift chambers (DC) and multi-wire proportional chambers (PC) [11], covering

Journal of Physics: Conference Series **736** (2016) 012014 doi:10.1088/1742-6596/736/1/012014

$|\eta| < 0.35$ and $\pi/2$ in azimuthal angle each. The drift chamber tracks are projected to the third layer of the PC and matched to hits in this detector, to decrease the contribution from mis-reconstructed tracks. Time-of-Flight detectors (TOF) are used for hadron identification. TOF is comprised of two arms: east arm (TOFe) and west arms (TOEw). Identification of $\pi^{\pm}$, $K^{\pm}$ and (anti-)proton in a momentum range up to 3 GeV/c is achieved over 99% purity [12, 13].

The beam-beam counters (BBC) [11] consist two arrays of 64 photomultiplier tubes (PMTs) each, covering 2π in azimuth and $3.1 < |\eta| < 3.9$ in pseudorapidity. The Forward Silicon Vertex Detector (FVTX) consists of two identical end-cap assemblies, covering $1.0 < |\eta| < 3.0$. The event planes are measured in the Au-going direction using the charge deposited in the south BBC (BBC-S), as well as the reconstructed clusters in the south end-cap of FVTX (FVTX-S).

The p+Au, d+Au and ^{3}He+Au data sets were collected in year 2015, 2008, and 2014, respectively. In p+Au and ^{3}He+Au collisions, high multiplicity (HM) triggers were implemented to enrich the central events. The high multiplicity trigger is based on the minimum bias trigger, but imposes additional requirement of more than 48 photomultiplier tubes firing in the BBC-S in p+Au collisions, which corresponds approximately to the 5% most central events. A similar high multiplicity trigger was used on 2014 run for ^{3}He+Au collisions. The HM trigger increases the recorded central event samples, in comparison to the minimum bias sample, by 40 times in p+Au and 10 times in ^{3}He+Au collisions.

Centrality classes are determined as a percentile of the total multiplicity measured in the BBC-S, following the procedure documented in Ref. [14]. In this analysis we select the 0-5% most central events. The initial geometry for the corresponding centrality is computed and characterized by a standard Monte Carlo (MC) Glauber model [15] with a Gaussian smearing of the nucleon distributions with $\sigma = 0.4$ fm.

2. Analysis method

The two-particle azimuthal correlations are measured between tracks in the PHENIX central arm at a given transverse momentum(p_T) and charge measured in the PMTs in the BBC-S, which is on Au-going side. The correlation function is constructed as a function of relative azimuthal angle $\Delta\phi$ and p_T using pairs from the same event $S(\Delta\phi, p_T)$ and normalized to the correlations in mixed events $M(\Delta\phi, p_T)$, where the particles in the pair come from different events, following the same procedure as in [3]:

$$S(\Delta\phi, p_T) = \frac{d(w_{pmt} N_{sameevent}^{Track(p_T)-PMT})}{d\Delta\phi} \tag{1}$$

$$C(\Delta\phi, p_T) = \frac{S(\Delta\phi, p_T)}{M(\Delta\phi, p_T)} \frac{\int M(\Delta\phi', p_T) d\Delta\phi'}{\int S(\Delta\phi', p_T) d\Delta\phi'} \tag{2}$$

where the w_{pmt} is taken from the total charge from the BBC-S PMTs.

We measure the flow coefficient with the event plane method [16].

$$v_n = \frac{< \cos(n(\phi - \Psi_{n,BBC_s}) >}{Res(\Psi_{n,BBC_s})} \tag{3}$$

The event plane is determined by detectors at large pseudorapidity, and the standard re-centering and flattering procedures are applied to calibrate the event-plane distributions [16]. To determine the event plane resolution, the three sub-event method is used:

$$Res(\Psi_{n,BBC_s}) =< \cos(n(\Psi_{n,BBC_s} - \Psi_{RP})) >= \sqrt{\frac{< \cos(n(\Psi_{n,BBC_s} - \Psi_{n,CNT})) >< \cos(n(\Psi_{n,BBC_s} - \Psi_{n,FVTX_s})) >}{< \cos(n(\Psi_{n,CNT} - \Psi_{n,FVTX_s})) >}} \tag{4}$$

Here, $\Psi_{n,CNT}$ is the event plane angle determined by the central arm tracking detectors, Ψ_{n,BBC_s} is the event plane angle determined by the BBC-S PMTs. $\Psi_{n,FVTX_s}$ is the event plane angle determined by the south end-caps of the FVTX.

3. Results

3.1. Correlation function in p+Au, d+Au and ^{3}He+Au collisions

The correlation functions $C(\Delta\phi, p_T)$ for 0-5% central events in p+Au, d+Au and ^{3}He+Au collisions in selected p_T bins, , and compared to the correlation functions measured in minimum bias p+p collisions at the same center-of-mass energy. We analyze these shapes by fitting each correlation function to a four-term Fourier cosine expansion, $f(\Delta\phi) = 1 + \Sigma_{n=1}^{4} c_n(p_T)\cos(n\Delta\phi)$. The fitting results are plotted in Fig. 1 for each distribution. Central ^{3}He+Au and d+Au collisions show a clearly visible enhancement of near-side pairs; central p+Au collisions show a flat distribution in near-side area. In both cases, a sizable second order Fourier term is needed to describe the correlations. In contrast, the p+p collisions can be described almost completely by the dipole term, as expected from back-to-back jets contribution and transverse momentum conservation.

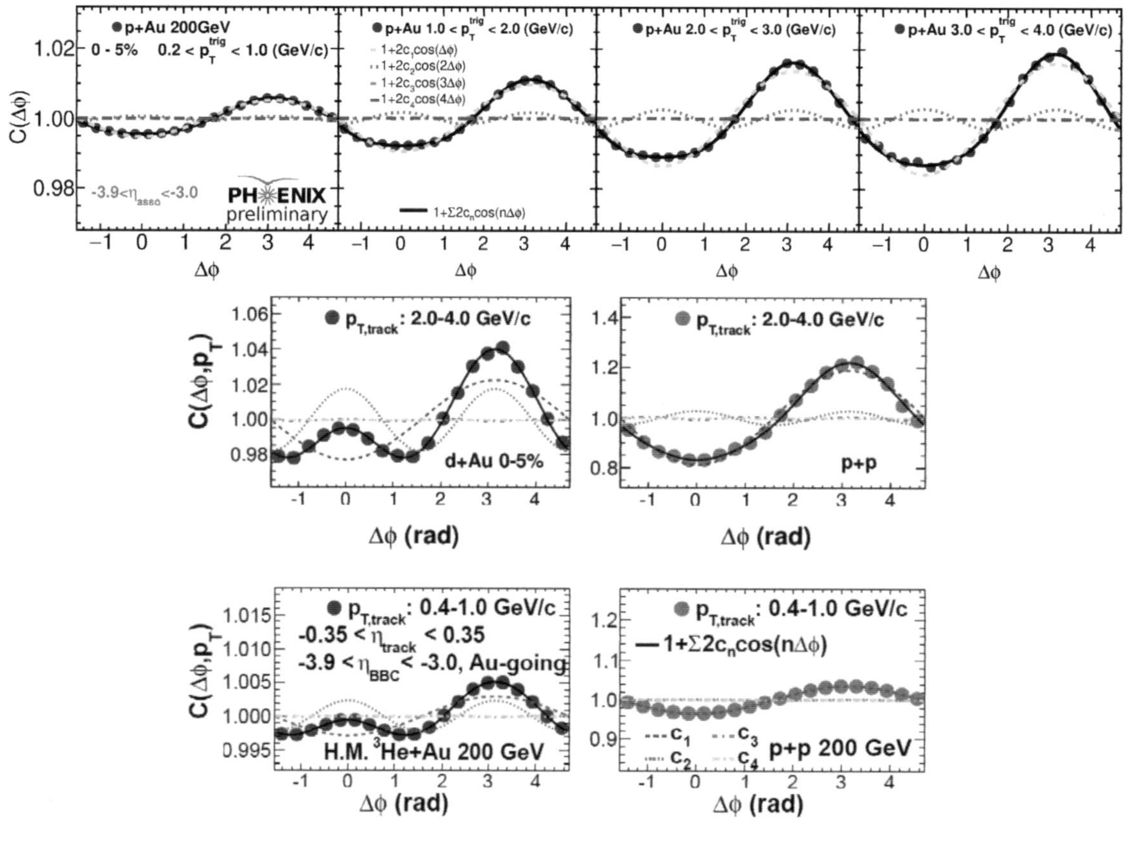

Figure 1. Two-particle correlation functions in high-multiplicity p+Au, d+Au and ^{3}He+Au collisions at $\sqrt{s_{NN}} = 200$ GeV, and comparisons to correlation functions in minimum bias p+p collisions at $\sqrt{s_{NN}} = 200$ GeV.

3.2. Charged hadron v_2 comparison in three systems

To better understand the origin of the near-side peaks and the azimuthal anisotropy in particle production in small system, we compare the v_2 vs p_T in the three systems for 0-5% central event, as shown in Fig. 2. The data clearly indicate that v_2 in p+Au collisions is significant, but smaller than that measured in d+Au [2] and ^{3}He+Au [3]. Given that from Glauber MC, the initial eccentricity ε_2 in central p+Au collisions is smaller than in d+Au and ^{3}He+Au collisions, it supports the hypothesis that the initial geometry plays a major role in determining these anisotropies. The data are compared to the predictions of the SONIC hydrodynamics model [17], which translates the initial state geometry to final-state momentum anisotropies. The SONIC model [17] can describe the data well in all the three systems.

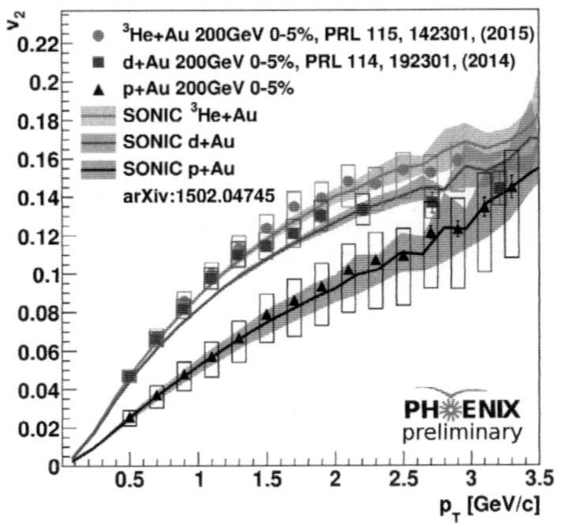

Figure 2. Measured v_2 from 0-5% p/d/^{3}He+Au $\sqrt{s_{NN}} = 200$ GeV collisions, in comparison to the predictions from the SONIC model [17].

3.3. Charged hadron v_2 and v_3 compared to models

Various models have predicted the anisotropy in small systems with different physics mechanism, including hydrodynamics models with [18] or without pre-flow [17], AMPT model [9], and gluon saturation [8]. In Fig. 3, we compared our results with the predictions of these models. SONIC model[17] and superSONIC [18] predict the v_2 values well in all three systems. Within the current systematics, we cannot tell one from another. AMPT model [9] can predict the three systems up to 1.5 GeV/c. IPGlasma+Hydrodynamic [8] model underpredicts the p+Au results but overpredicts the d/^{3}He+Au results. Clear v_3 signal is observed in central ^{3}He+Au events. The v_3 measured in ^{3}He+Au provides an additional constrain.

3.4. Eccentricity scaling

The quantity v_2/ε_2 is shown in the Fig. 4 in p/d/^{3}He+Au collisions at same energy. The quantity ε_2 of these three systems are from the standard Glauber model calculations[15]. The Glauber model predicts that the initial eccentricity in d+Au collisions is similar to that in ^{3}He+Au collisions but larger than in p+Au collisions. The measured v_2 follows the same order, however, the scaling of the values of v_2 with the corresponding eccentricities in these three system would not collapse to a common pattern. The larger v_2/ε_2 in p+Au collisions may indicate that the substructure of the proton also plays a role.

Journal of Physics: Conference Series **736** (2016) 012014 doi:10.1088/1742-6596/736/1/012014

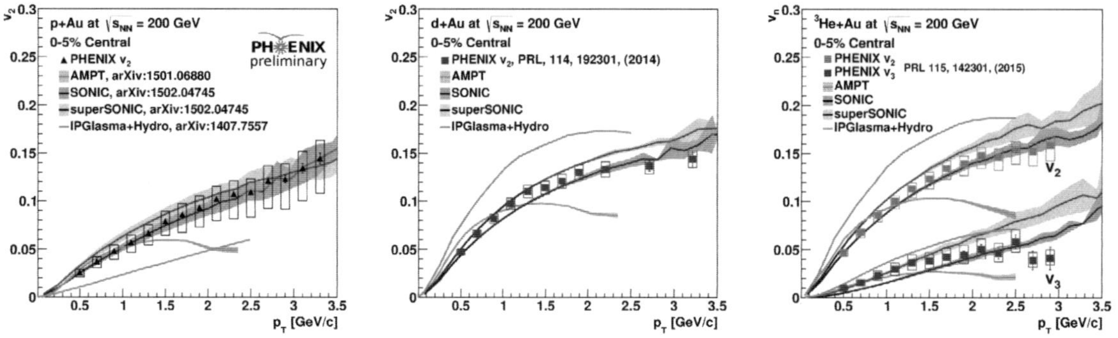

Figure 3. Measured v_2 from 0-5% p/d/^{3}He+Au $\sqrt{s_{NN}}$ = 200 GeV collisions. The data are compared with AMPT, SONIC/superSONIC and IPGlasma models predictions.

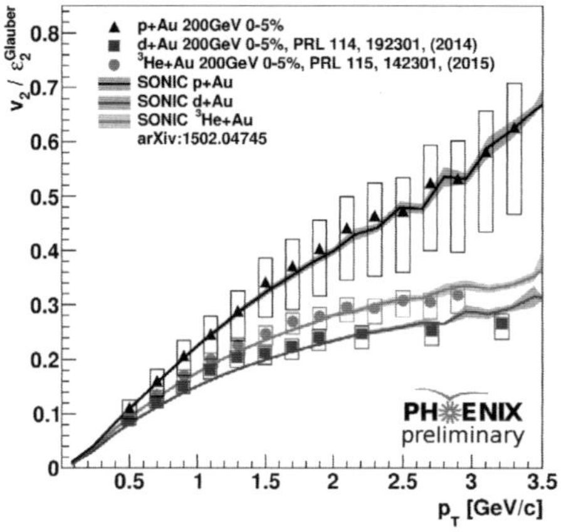

Figure 4. Measured v_2/ε_2 for mid-rapidity for charged tracks in 0-5% central p/d/^{3}He+Au collisions at $\sqrt{s_{NN}}$ = 200 GeV

3.5. Identified particle v_2 in d+Au and ^{3}He+Au collisions

The second order flow coefficients for identified $\pi^{\pm}$, $K^{\pm}$ and (anti-)proton are presented in the left panel of Fig. 5. Mass-ordering is observed in ^{3}He+Au collisions. The trend is consistent with hydrodynamic flow in which the heavier particles are boosted to higher p_T even with same expanding velocity of the medium. The right panel of the Fig. 5 shows that the v_2 values for identified particles can be described by a universal curve when scaled by number of quarks in the hadron. Such scaling previously observed in Au+Au collisions is also seen in the small ^{3}He+Au system, which is consistent with viscous hydro calculations or the quark coalescence models [19, 20].

Conclusions

We have presented measurements of long-range azimuthal correlations between tracks at mid-rapidity $|\eta| < 0.35$, and charges in BBC Au-going direction in p/d/^{3}He+Au collisions at $\sqrt{s_{NN}}$ = 200 GeV. Enhanced correlations with small azimuthal separation are observed for pairs across $|\Delta\eta| > 3.5$ in high multiplicity events. The anisotropic flow coefficients v_2 and v_3 are measured

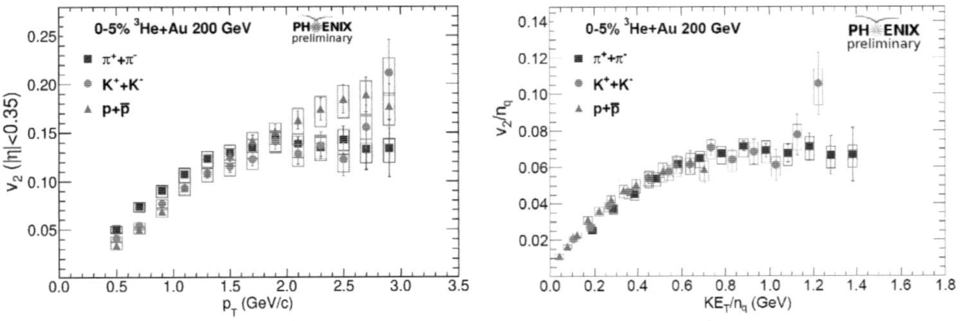

Figure 5. Left panel: Measured v_2 of identified $\pi^{\pm}$, $K^{\pm}$ and (anti-)proton from 0-5% ^{3}He+Au at $\sqrt{s_{NN}}$ = 200 GeV collisions. Right panel: Constituent quark-number (n_q) scaling of v_2 for $\pi^{\pm}$, $K^{\pm}$ and (anti-)proton as a function of transverse kinetic energy $KE_T = \sqrt{p_T^2 + m^2} - m$ in 0-5% central ^{3}He+Au collisions.

for mid-rapidity particles with the event plane method. The v_2 of 0-5% p+Au is sizable and found to be smaller than that of d/^{3}He+Au and indicates that the initial geometry plays an important role in describing system evolution. A sizable v_3 is seen in ^{3}He+Au collisions. Hydrodynamics models with initial conditions from the Glauber model describe v_2 and v_3 quite well. These values of η/s at or near the conjectured lower bound $1/4\pi$ [21] are similar to the ones needed to describe collective flow in large systems, such as Au+Au or Pb+Pb collisions. The measurements of v_2 for identified hadrons in 0-5% ^{3}He+Au collisions are found to scale with the number of quarks, as expected in systems where the flow develops at the partonic level.

References

[1] (PHENIX Collaboration) A. Adare, et al., Quadrupole anisotropy in dihadron azimuthal correlations in central d+Au collisions at $\sqrt{s_{NN}}$ = 200 GeV, Phys. Rev. Lett., 111:212301, Nov 2013.

[2] (PHENIX Collaboration) A. Adare et al., Measurement of long-range angular correlation and quadrupole anisotropy of pions and (anti)protons in central d+Au collisions at $\sqrt{s_{NN}}$ = 200 GeV, Phys. Rev. Lett., 114(19):192301, 2015.

[3] (PHENIX Collaboration) A. Adare et al., Measurements of Elliptic and Triangular Flow in High-Multiplicity ^{3}He+Au Collisions at $\sqrt{s_{NN}}$ = 200 GeV, Phys. Rev. Lett., 115:142301, 2015.

[4] (CMS Collaboration) S. Chatrchyan, et al., Observation of long-range, near-side angular correlations in pPb collisions at the LHC, Phys.Lett. B718 (2013) 795C814.

[5] (ATLAS Collaboration) G. Aad, et al., Observation of Associated Near-Side and Away-Side Long-Range Correlations in $\sqrt{s_{NN}}$ = 5.02 TeV Proton-Lead Collisions with the ATLAS Detector, Phys. Rev. Lett. 110 (2013) 182302.

[6] (ALICE Collaboration) B. Abelev, et al., Long-range angular correlations on the near and away side in p-Pb collisions at $\sqrt{s_{NN}}$ = 5.02 TeV, Phys.Lett. B719 (2013) 29C41.

[7] (CMS Collaboration) S. Chatrchyan, et al., Multiplicity and transverse momentum dependence of two- and four-particle correlations in pPb and PbPb collisions, Phys.Lett. B724 (2013) 213C240.

[8] Bjoern Schenke and Raju Venugopalan, Collective effects in light heavy ion collisions, Nucl. Phys., A931:1039C1044, 2014.

[9] Zi-Wei Lin, Che Ming Ko, Bao-An Li, Bin Zhang, and Subrata Pal. Multiphase transport model for relativistic heavy ion collisions, Phys. Rev. C, 72(6):064901, December 2005.

[10] J. L. Nagle, A. Adare, S. Beckman, T. Koblesky, J. O. Koop, D. McGlinchey, P. Romatschke, J. Carlson, J. E. Lynn, M McCumber, Exploiting Intrinsic Triangular Geometry in Relativistic 3He+Au Collisions to Disentangle Medium Properties, Phys. Rev. Lett. 113 (11) (2014) 112301. doi:10.1103/PhysRevLett.113.112301.

[11] (PHENIX Collaboration) K. Adcox, et al., PHENIX detector overview, Nucl. Instrum. Meth. A499 (2003) 469C479. doi:10.1016/S0168-9002(02)01950-2.

[12] (PHENIX collaboration)Adare, A. et al., Spectra and ratios of identified particles in Au+Au and d+Au collisions at $\sqrt{s_{NN}} = 200$ GeV, Phys. Rev. C 88 (2013) 024906.
[13] (PHENIX collaboration)Adare, A. et al., Deviation from quark number scaling of the anisotropy parameter v_2 of pions, kaons, and protons in Au+Au collisions at $\sqrt{s_{NN}} = 200$ GeV, Phys. Rev. C 85 (2012) 064914.
[14] (PHENIX Collaboration) A. Adare et al., Centrality categorization for Rp(d)+A in high-energy collisions, Phys. Rev. C 90, 034902.
[15] M. L. Miller, K. Reygers, S. J. Sanders, and P. Steinberg, Ann.Rev.Nucl.Part.Sci. 57, 205 (2007), nucl-ex/0701025.
[16] A. M. Poskanzer, S. A. Voloshin, Methods for analyzing anisotropic flow in relativistic nuclear collisions, Phys. Rev. C58 (1998) 1671C1678.
[17] M. Habich, J. L. Nagle, and P. Romatschke, Particle spectra and HBT radii for simulated central nuclear collisions of C + C, Al + Al, Cu + Cu, Au + Au, and Pb + Pb from s = 62.4 - 2760 GeV, Eur. Phys. J., C75(1):15, 2015.
[18] Paul Romatschke. Light-Heavy Ion Collisions: A window into pre-equilibrium QCD dynamics?, Eur. Phys. J., C75(7):305, 2015.
[19] R. C. Hwa, C. B. Yang, Scaling behavior at high p_T and the p/pi ratio, Phys. Rev. C67 (2003) 034902. doi:10.1103/PhysRevC.67.034902.
[20] R. J. Fries, B. Muller, C. Nonaka, S. A. Bass, Hadronization in heavy ion collisions: Recombination and fragmentation of partons, Phys. Rev. Lett. 90 (2003) 202303. doi:10.1103/PhysRevLett.90.202303.
[21] P. K. Kovtun, D. T. Son, and A. O., StarinetsViscosity in Strongly Interacting Quantum Field Theories from Black Hole Physics, Phys. Rev. Lett. 94, 111601.

Journal of Physics: Conference Series **736** (2016) 012015 doi:10.1088/1742-6596/736/1/012015

Latest results of charged hadron flow measurements in CuAu collisions at RHIC-PHENIX

Hiroshi Nakagomi for the PHENIX collaboration

University of Tsukuba, Tsukuba Tennodai 1-1-1 Ibaraki Japan

E-mail: hiro_n@rcf.rhic.bnl.gov

Abstract.
Measurements of azimuthal anisotropic flow v_n for inclusive charged hadrons and identified particles at mid rapidity in Cu+Au collisions at $\sqrt{s_{NN}}$ = 200GeV are presented. The data were recorded by the PHENIX experiment at Relativistic Heavy Ion Collider(RHIC). Directed, elliptic and triangular flow as a function of transverse momentum p_T are measured with respect to event planes. The inclusive charged hadron v_1 shows the negative value at high p_T. The v_2 and v_3 are compared to those in Au+Au and Cu+Cu collisions. We find the v_2 and v_3 follow an empirical scaling with $1/(\epsilon_n N_{part}^{1/3})$. We also compare the v_2 and v_3 to hydrodynamical predictions. The identified particles v_2 and v_3 show a mass ordering in low p_T region and baryon and meson splitting in high p_T region. However the identified hadron v_1 only shows mass ordering in mid p_T region.

1. Introduction

Azimuthal anisotropies of particle production in relativistic heavy ion collisions have been measured for the investigating the properties of the quark gluon plasma(QGP). The magnitude of these anisotropies can be evaluated by v_n which are the coefficients of the Fourier serious of the particle emission angle distributions with respect to the event planes Ψ_n[1],

$$\frac{dN}{d\phi} = 1 + \sum_{n=1} 2v_n \cos(n(\phi - \Psi_n)) \quad (1)$$

where n is the order of the harmonic, ϕ is the azimuthal angle of produced particle. The event plane Ψ_n is determined for each of the harmonics n on event by event basis. Until now the second harmonic coefficient v_2 has been studied well and concluded the QGP state is the nearly perfect fluid[2]. Initial spatial anisotropy of the nuclear overlap region, which is a rugby ball shape in A+A collisions, is considered to be the origin of the elliptic flow. The initial spatial anisotropy is converted to an anisotropy in momentum space.

In symmetric nuclear - nuclear collisions, when the nuclei are considered to be smooth spheres, odd harmonics have to vanish at mid-rapidity. If the participant nucleons fluctuate on event by event, the initial condition could make odd harmonic anisotropic particle production at mid rapidity. Indeed, sizeable odd harmonics have been observed at RHIC[3] and LHC[4]. In order for theory calculation to reproduce even and odd harmonics spontaneously, the experimental measurements of even and odd harmonics strongly constrain the initial condition and the value of η/s by comparing the theory predictions and the experimental observables[3].

Journal of Physics: Conference Series **736** (2016) 012015 doi:10.1088/1742-6596/736/1/012015

Although the many experimental observables and theoretical predictions exist, the uncertainties of the initial condition and the value of η/s remains. In 2012, the first asymmetric collisions, Cu+Au were operated at RHIC in Brook Haven National Laboratory to control the initial condition because the asymmetric collisions may lead to larger odd harmonics at mid rapidity.

2. Experimental details

In our analysis, drift chamber(DC), three layers of pad chamber(PC1,PC2 and PC3), the time of flight East and West(TOF.E, TOF.W) and the electromagnetic calorimeter(EMCal) are employed for Central arm particle tracking and identification[5]. The global detectors, the beam-beam counters(Bbc), the zero degree calorimeter and the shower max detectors(Smd) are employed for centrality and event plane determination[5],[6].

The event plane method[7] is used in the measurements of azimuthal anisotropies of emitted particle distributions in Cu+Au collisions. The event plane Ψ_n^{obs} is determined on event by event base for each harmonic order. The Fourier coefficients v_n are measured with respect to the $n-th$ order event plane and divided by the event plane resolution.

$$v_n = \frac{\left\langle cos(n[\phi - \Psi_n^{obs}]) \right\rangle}{Res\{\Psi_n^{obs}\}} \tag{2}$$

where ϕ is the azimuthal angle of the produced particle and Ψ_n^{obs} as defined by the spectator neutrons for the 1st harmonic and the produced particles for the 2nd and 3rd harmoincs. The first order event plane Ψ_1(directed plane) is determined by the Smd South(SmdS). The SmdS measures the center of shower profile given by Au spectator neutrons to determine the spectator plane. For the directed flow measurement, the spectator neutrons are preferred because they are insensitive to momentum conservation effect. The measurements of the charged hadron and identified hadron v_1 are with respect to the Au spectator event plane. Since the number of neutrons in Au spectator is larger than that in Cu spectator, the event plane resolution of the SmdS is greater than that of the SmdN. The combination of the Bbc South(BbcS) and the Bbc North(BbcN) are used to determine the second and third order event planes.

3. Results

3.1. Charged hadron v_n

Figure 1 shows the charged hadron v_1 at mid-rapidity as a function of p_T for 4 centrality bins in Cu+Au collisions at $\sqrt{s_{NN}} = 200$GeV. The red solid circles correspond to the v_1 data points and the shaded boxes are systematic uncertainties. The v_1 is measured with respect to the Au spectator event plane as described previous section. However, we flip the sign of v_1 to alight with previous conventions by taking smaller size (Cu in this case) spectator as forward going spectator. In all centrality classes, high p_T particles at mid rapidity are emitted towards the Au side as indicated by the negative v_1. Considering the momentum balance between the high and low p_T particles due to the momentum conservation effect, the v_1 of the low p_T charged particles could be slightly negative, however the data points are consistent with zero in the low pT region below 1GeV/c within large systematic uncertainties. At high p_T region, the magnitude of the v_1 decreases from central to peripheral collisions.

Figure 2 shows the charged hadron v_2 as a function of p_T in Cu+Au collisions compared with those in Au+Au and Cu+Cu collisions[8]. The v_2 component in Cu+Au collisions shows similar p_T and centrality dependence as seen in symmetric collision systems. We observed the v_2 in Cu+Au collisions are always between those in Au+Au and Cu+Cu. Except for 0-10% centrality bin, the v_2 in Au+Au collisions are higher than those in Cu+Au and Cu+Cu. This trend is not ordered according to system size dependence of second order initial spatial anisotropy. The v_2

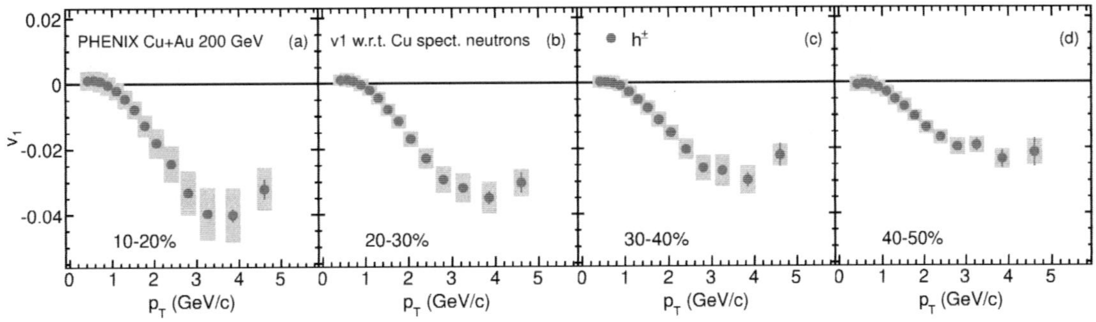

Figure 1. Charged hadron v_1 as a function of p_T for 4 centrality bins.

scaled with the initial spatial anisotropy ϵ_2 and $N_{part}^{1/3}$[8] for three collision systems are shown in Figure 3. Since the $N_{part}^{1/3}$ is proportional to the length of participant zone or expansion time, the $N_{part}^{1/3}$ could be responsible for the system size contribution to v_2. In Figure 3, it is shown that the empirical scaling works well also in Cu+Au collisions as well as in the other collision systems.

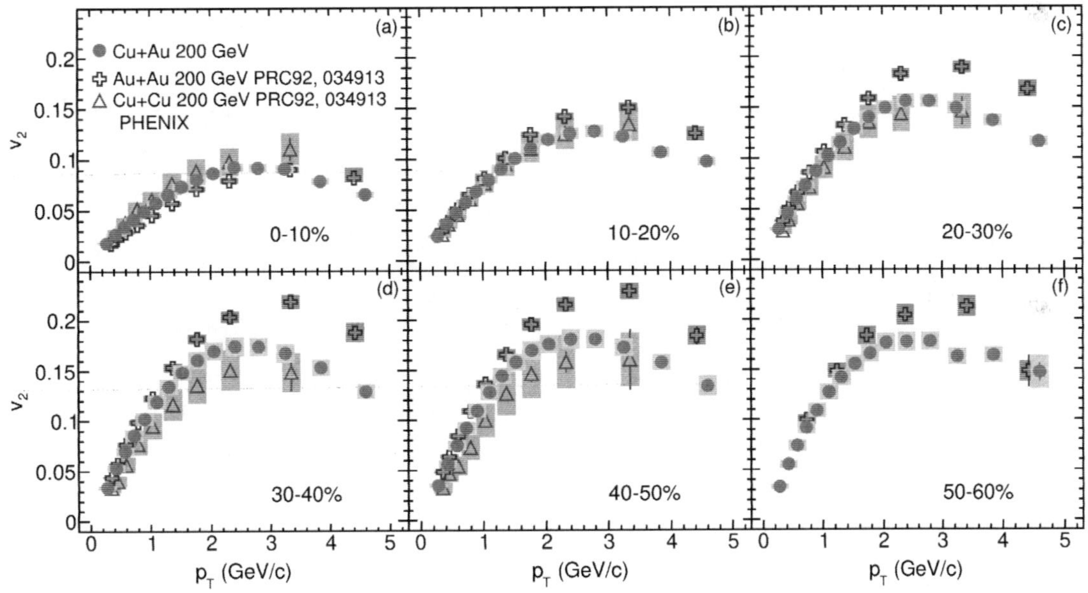

Figure 2. Charged hadron v_2 as a function of p_T in Cu+Au, Au+Au and Cu+Cu collisions for 6 centrality bins.

In Figure 4, the v_3 as a function of p_T at mid-rapidity in Cu+Au and Au+Au[3] collisions for three centrality bins are shown. The v_3 in Cu+Au collisions has similar centrality and p_T dependence as seen Au+Au collisions. Unlike the v_2, the trend of system size dependence of the v_3 are ordered according to the third order initial spatial anisotropy ϵ_3 dependence. The empirical scaling is perfomed to the v_3 using the ϵ_3 and $N_{part}^{1/3}$. The scaled v_3 are shown in Figure 5 which also hold a universal behavior like the scaled v_2. The empirical scaling is successfully extended to the third order azimuthal anisotropy.

Journal of Physics: Conference Series **736** (2016) 012015 doi:10.1088/1742-6596/736/1/012015

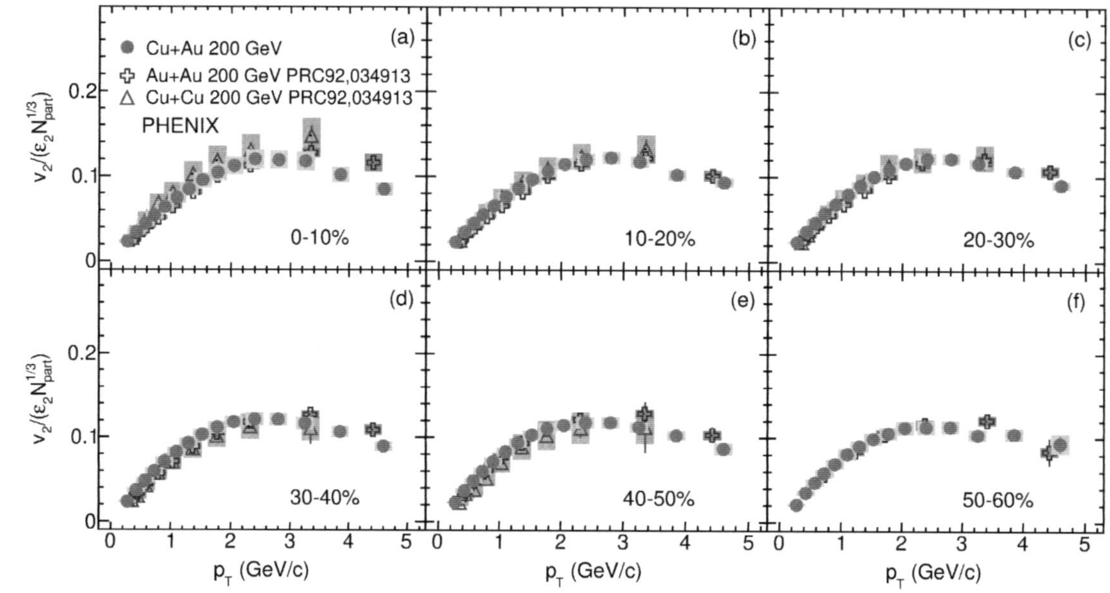

Figure 3. Scaled charged hadron v_2 as a function of p_T in Cu+Au, Au+Au,Cu+Cu for 6 centrality bins.

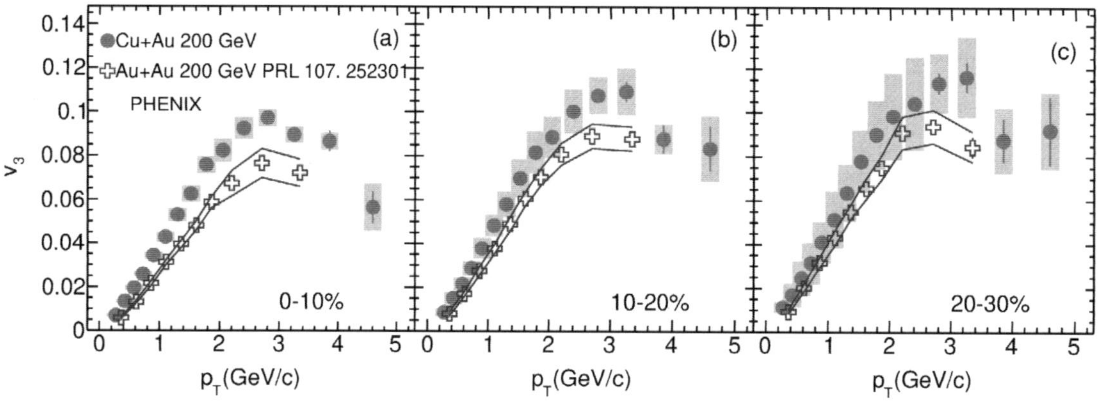

Figure 4. Charged hadron v_3 as a function of p_T in Cu+Au and Au+Au collisions for 3 centrality bins.

3.2. Identified particle v_n

Figure 6,7 and 8 show the identified particle v_2, v_3 and v_1 in Cu+Au collisions. The symbols are represented charge combined $\pi^{\pm}$,$K^{\pm}$, p and $\bar{p}$. For the v_1 and v_3 results, centrality bins are combined to improve statistical uncertainties. There are two trends for the v_2 and v_3 in Figure 6 and 7. In the low p_T region, the anisotropy becomes larger as hadron mass decreases. Hydrodynamics predicts similar mass dependence caused by radial flow effect. Above p_T >2GeV/c, this mass ordering becomes reversed. The anisotropy for baryons is larger than that for mesons. These particle species dependencies have been seen in symmetric collision systems[9]. Unlike the v_2 and v_3, the v_1 shown in Figure 8 have the mass ordering in the mid p_T region and don't show baryon and meson splitting at the higher p_T.

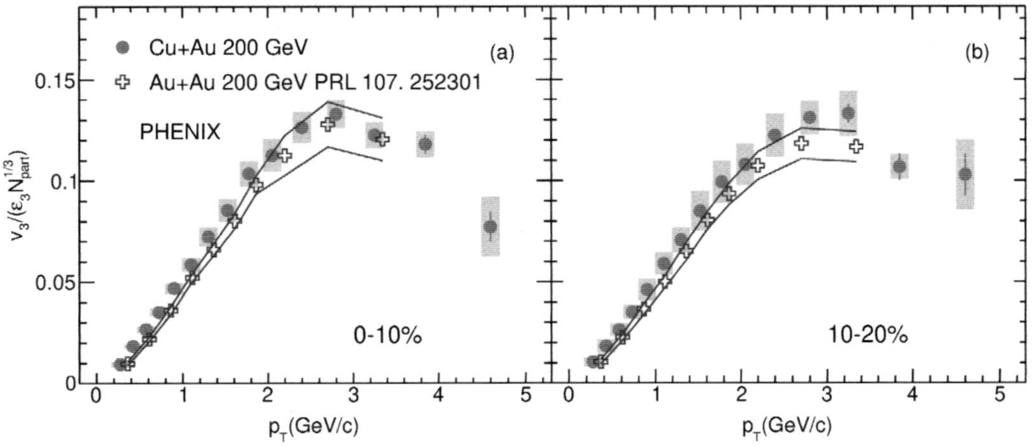

Figure 5. Scaled charged hadron v_2 as a function of p_T in Cu+Au and Au+Au for 2 centrality bins.

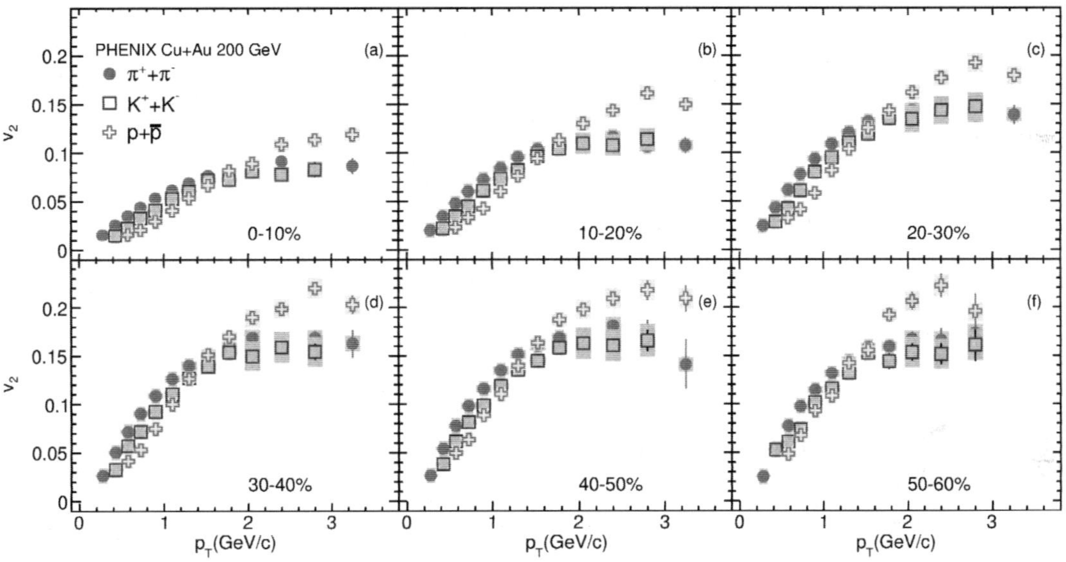

Figure 6. $\pi^\pm$,$K^\pm$, p v_2 as a function of p_T in Cu+Au collisions for 6 centrality bins

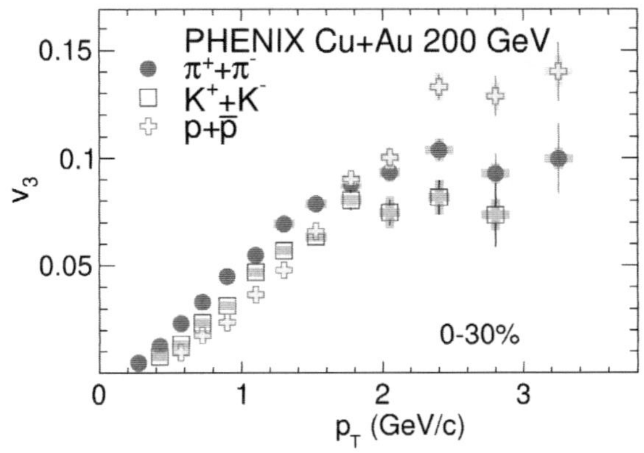

Figure 7. $\pi^\pm$,$K^\pm$, p v_3 as a function of p_T in Cu+Au collisions for 1 centrality bin

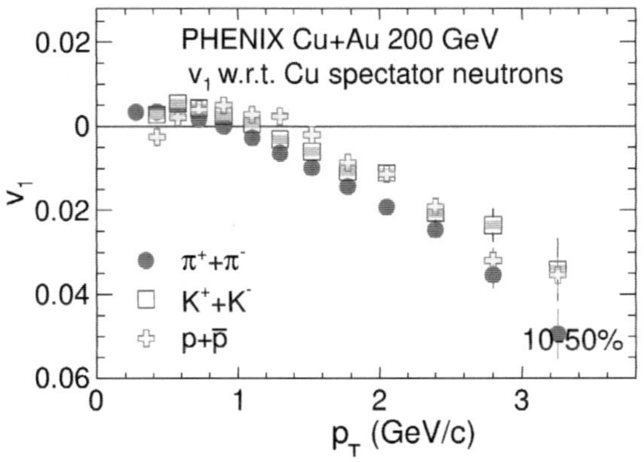

Figure 8. $\pi^\pm$,$K^\pm$, p v_1 as a function of p_T in Cu+Au collisions for 1 centrality bin

3.3. Theory comparisons

Event by event 3D+1 viscous hydrodynamic predictions are available[10]. In Figure 9 and 10, the hydrodynamic calculations with $\eta/s = 0.08$ and $\eta/s = 0.16$ for the v_2 and v_3 are compared to the PHENIX experimental data. In the 20-30% centrality bin, both η/s case reproduce the our measurements well. In the 0-5% centrality bin, $\eta/s = 0.08$ case has better agreement with the experimental data.

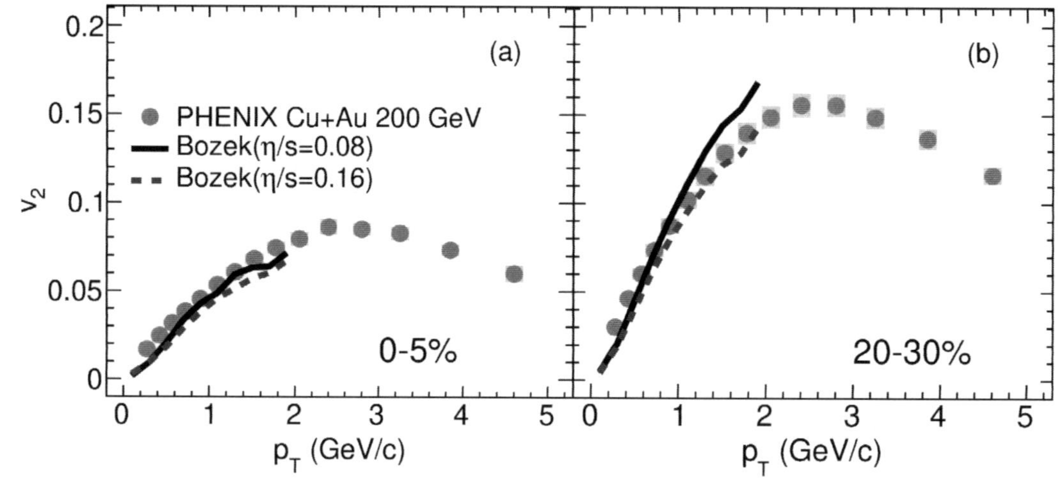

Figure 9. The charged hadron v_2 as a function of p_T in Cu+Au collisions in comparison to hydrodynamics calculations for two centrality bins

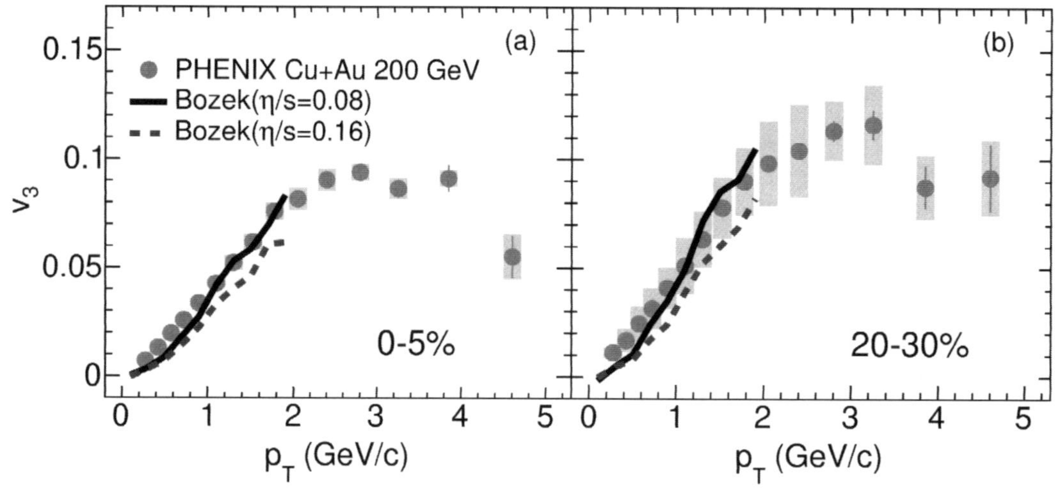

Figure 10. The charged hadron v_3 as a function of p_T in Cu+Au collisions in comparison to hydrodynamics calculations for two centrality bins

Journal of Physics: Conference Series **736** (2016) 012015 doi:10.1088/1742-6596/736/1/012015

4. Conclusion

We have presented directed, elliptic and triangular azimuthal anisotropies for inclusive charged hadrons and identified hadrons at mid-rapidity in Cu+Au collisions at $\sqrt{s_{NN}} = 200$GeV observed by PHENIX experiment at RHIC using event plane method.The v_1 coefficients for charged hadrons indicate high p_T particles are emitted to the Au side. For the system size dependence of the v_2 and v_3 measurements, the empirical scaling for the v_2 works well in Cu+Au, Au+Au and Cu+Cu and is extended for the v_3 measurements in Cu+Au and Au+Au.
The identified hadron v_2 and v_3 show the mass ordering in the low p_T region and baryon and meson splitting in the high p_T region. These trends have been seen in the previous symmetric collisions. The identified hadron v_1 shows the mass ordering in the mid p_T region and doesn't show baryon and meson splitting in the high p_T region.
The inclusive charged hadron v_2 and v_3 for the 0-5% and 20-30% centrality bins are compared to the event by event 3D+1 viscous hydrodynamics calculations. The hydrodynamics calculations with $\eta/s = 0.08 - 0.16$ reproduce the experimental data v_2 and v_3.

Reference

[1] S. Voloshin and Y. Zhang, Z. Phys. C 70, 665 (1996)
[2] K.Adcox et al.(PHENIX Collaboration) Nucl. Phys. A 757, 184 (2005)
[3] A. Adare et al. (PHENIX Collaboration), PhysRevLett.107.252301
[4] G. Aad et al. (ATLAS Collaboration), Phys. Rev. C 86 014907 (2012)
[5] K. Adcox et al.(PHENIX Collaboration) Nucl. Instrum. Methods Phys. Res. Sec. 499, 469(2003)
[6] C. Adler, A. Denisov, E. Garcia, M. Murray, H. Stro-bele, and S. White Nucl. Instrum. Methods Phys. Res. Sec. 499, 433(2003)
[7] A.M.Poskanzer, S.A. Voloshin PhysRevC.58.1671(1998)
[8] A. Adare el al.(PHENIX Collaboration) PhysRevC.92.034913
[9] A. Adare et al. (PHENIX Collaboration), arXiv:1412.1038
[10] Bozek Phys. Lett. B717, 287(2012)
[11] A. Adare et al (PHENIX Collaboration), arXiv:1509.07784

Journal of Physics: Conference Series **736** (2016) 012016 doi:10.1088/1742-6596/736/1/012016

Observation of Global Hyperon Polarization in Ultrarelativistic Heavy Ion Collisions

Isaac Upsal for the STAR Collaboration

The Ohio State University: 191 W Woodruff Ave. Columbus, Ohio, USA

E-mail: upsal.1@osu.edu

Abstract. Non-central heavy-ion collisions provide a system with non-zero total angular momentum which can be transferred, in part, to the fireball via baryon stopping. It has been predicted that a net spin of emitted particles aligned with the system angular momentum may emerge through coupling with the bulk material. Due to its parity violating decay the Λ baryon is self-analyzing, which allows us to associate the daughter proton decay direction with Λ particle spin. Ultimately this allows us to use them as a probe of net-particle spin. In preliminary STAR measurements of the net Λ baryon polarization from Au+Au collisions at 7.7, 11.5, 14.5, 19.6, 27, and 39 GeV we find that both Λ and $\overline{\Lambda}$ particles are polarized in the direction of the system angular momentum. Including previously published STAR results we see a polarization of about 2% for mid-central collisions when $\sqrt{s_{\mathrm{NN}}} < 100\mathrm{GeV}$.

Collisions of nuclei at ultra-relativistic energies create a system of deconfined colored quarks and gluons, called the Quark-Gluon Plasma (QGP)[1, 2, 3, 4]. The large angular momentum ($\sim 10^{(4-5)}\hbar$) present in non-central collisions may produce a "polarized QGP," in which quarks are polarized through spin-orbit coupling in QCD [7, 8, 9]. The polarization would be transmitted to hadrons in the final state and could be detectable through global hyperon polarization.

"Global" polarization refers to the phenomenon in which the spin of baryons emitted from the mid-rapidity fireball are correlated with the net angular momentum of the system. This is distinct from production plane polarization, an effect seen in $p+p$ and $p+A$ collisions, in which Λ baryons are polarized relative to the plane spanned by $\vec{p}_{\mathrm{beam}} \times \vec{p}_{\Lambda}$ [10, 11]. Production plane polarization is found in the beam fragmentation region (i.e. at forward rapidity) and vanishes by symmetry when averaged over rapidity or azimuthal angle.

Global polarization has been predicted in hydrodynamic calculations of non-central collisions [14, 15, 44, 17, 18]. In three-dimensional hydrodynamic calculations, the initial shear structure of the flow field results in a vortical structure. This vorticity is sensitive to the initial conditions [14, 15, 18], as well as the viscosity [14, 33, 18] and temperature [17] of the fluid as it evolves. Vortical structure of the velocity field has also been recently reported in microscopic transport simulations of heavy ion collisions at RHIC [19]. Global polarization arising from QCD spin-orbit coupling is expected to give a signal of similar size and magnitude for Λ and $\overline{\Lambda}$ baryons.

A global polarization signal may also arise due to a coupling of the hadronic magnetic dipole moment $\vec{\mu}_{\mathrm{H}}$ to the magnetic field $\vec{B}$ produced by the charged spectators in a non-central heavy ion collision. A purely magnetic polarization would give signals of similar magnitude but opposite sign for Λ and $\overline{\Lambda}$ baryons.

In this preliminary STAR analysis the first observation of global hyperon polarization in heavy ion collisions is seen. The measurements were performed by the STAR experiment at RHIC, in Au+Au

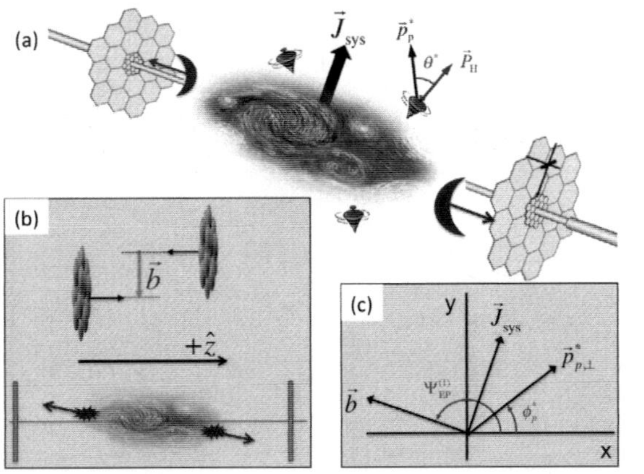

Figure 1. A sketch of the collision from several perspectives, illustrating the vectors and angles discussed in this study. Panel (a): The direction of the system angular momentum, $\hat{J}_{sys}$ is estimated by measuring the azimuthal angle of projectile fragments in forward detectors; the vorticity of the hot system at midrapidity and resulting polarized hyperons are schematically illustrated. Panel (b): At the energies under study, spectators "bounce off" [34]. The impact parameter vector $\vec{b}$ points from the center of the ion traveling in the $-\hat{z}$ direction, to the ion traveling in the $+\hat{z}$ direction; thus, $\hat{J}_{sys} = \hat{b} \times \hat{z}$. Panel (c): The angles used in equation 2 are shown; here, $\vec{p}^*_{\rm p}$ is the component of the daughter proton's momentum transverse to the beam, in the Λ particle's frame.

collisions at center-of-mass energies of $\sqrt{s_{NN}} = 7.7$, 11.5, 14.5, 19.6, 27, and 39 GeV. In this report, we focus on mid-central collisions, 20-50% of the total inelastic cross-section as estimated from the charged-particle yield at midrapidity [22] measured in the STAR TPC [21].

The reaction plane of an event is estimated by reconstructing the first-order event plane, $\Psi^{(1)}_{\rm EP}$, in two Beam-Beam Counters (BBCs) [23, 24] covering pseudorapidity range $3.3 < |\eta| < 5.0$, as described previously [25, 26]. Taking into account that the spectators "bounce-off" outward from the center of the system [34] the reaction plane estimate allows the determination of the angular momentum direction. The event plane resolution, $R^{(1)}_{\rm EP} = \left\langle \cos(\Psi^{(1)}_{\rm EP} - \Psi_{\rm RP}) \right\rangle$, is about 0.65 for midcentral events at $\sqrt{s_{\rm NN}} = 7.7$ GeV and decreases with energy to about 0.2 at $\sqrt{s_{\rm NN}} = 39$ GeV.

Hyperons were reconstructed in the TPC via the decay channel $\Lambda \to p + \pi^-$ ($\overline{\Lambda} \to \overline{p} + \pi^+$). The charged daughters were identified through their specific energy loss in the gas of the TPC for $|\vec{p}| \lesssim$ 0.6 GeV/c, and with the Time of Flight (TOF) detector for momenta up to 2.5 GeV/c. Hyperon candidates were identified according to topological properties of the decay vertex – projected distance of closest approach of the daughters to the primary vertex, of the parent hyperon to the primary vertex, and of the daughters to each other as well as apparent decay length. Our analysis is restricted to midrapidity hyperons, $|y_{\rm H}| < 1$.

We focus on Λ and $\overline{\Lambda}$ hyperons because their decay topology reveals their polarization. In particular, in the Λ ($\overline{\Lambda}$) particle's frame, the daughter proton (antiproton) tends to be emitted along (opposite) the parent's polarization direction. If θ^* is the angle between the daughter proton (antiproton) momentum and the hyperon polarization vector $\vec{P}_{\rm H}$, then

$$\frac{dN}{d\cos\theta^*} = \tfrac{1}{2}\left(1 + \alpha_{\rm H}|\vec{P}_{\rm H}|\cos\theta^*\right), \tag{1}$$

where $\alpha_\Lambda = 0.642$ and $\alpha_{\overline{\Lambda}} = -0.642$.

Symmetry demands that, averaged over all phasespace, $\vec{P}$ is parallel to $\hat{J}_{\rm sys}$, the unit vector in the

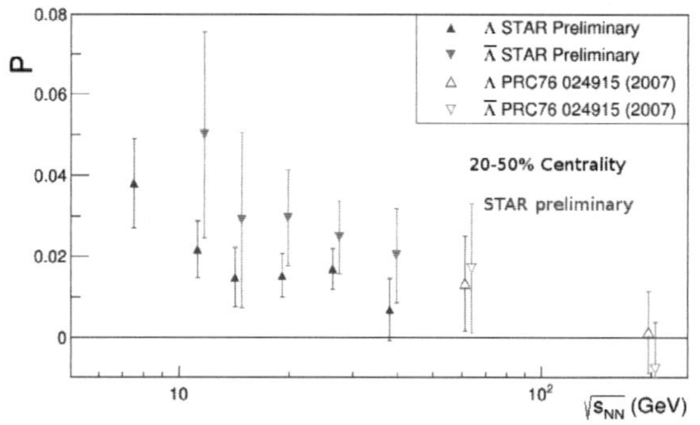

Figure 2. (Color online) The average polarization of Λ and $\overline{\Lambda}$ baryons from 20-50% central Au+Au collisions is plotted as a function of collision energy. The results of the present study ($\sqrt{s_{NN}} < 40$ GeV) are shown together with those reported earlier [6] for 62.4 and 200 GeV collisions. The results are corrected for feed-down coming only from Σ^0 decays. Only statistical errors are plotted.

direction of the angular momentum of the system. Both the magnitude ($0 \leq |\vec{P}| \leq 1$) and direction of $\vec{P}$ may depend strongly on transverse momentum, azimuthal angle, and rapidity. In this preliminary study the limited statistics of the datasets prohibits extensive exploration of these dependencies, and we extract only the average projection of the polarization on $\hat{J}_{sys}$. It may be shown [6] that

$$\overline{P}_{\rm H} \equiv \langle \vec{P}_{\rm H} \cdot \hat{J}_{sys} \rangle = \frac{8}{\pi \alpha_{\rm H}} \frac{\left\langle \sin\left(\Psi_{\rm EP}^{(1)} - \phi_p^*\right)\right\rangle}{R_{\rm EP}^{(1)}}, \tag{2}$$

where $\Psi_{\rm EP}^{(1)}$ is the angle of the first-order event plane, ϕ_p^* is the azimuthal angle of the daughter proton (antiproton) in the Lambda's rest frame, and $\langle \cdots \rangle$ indicates an average over the azimuthal angle of the hyperon momentum. Since $R_{\rm EP}^{(1)}$ depends on collision centrality, in the present analysis we calculate $\overline{P}_{\rm H}$ separately for events of centrality 20-30%, 30-40%, and 40-50%, and report the weighted (with particle yield) average.

The results are shown in Fig. 2. At collision energies below 39 GeV, a significant polarization is observed for both Λ and $\overline{\Lambda}$ baryons, at the level of 1.5-3.5 times statistical uncertainty. While including a broader centrality range into the analysis increases the hyperon statistics, the significantly poorer event plane resolution for more central or more peripheral collisions reduces the significance of the signal.

The results in Fig. 2 have been corrected to account for the combinatoric background of $p+\pi$ pairs that pass our topological criteria and fall within the selected invariant mass window. This corresponds to a $\sim 55\%$ ($\sim 35\%$) correction for Λ ($\overline{\Lambda}$) baryon polarization for most data points. For the $\sqrt{s_{NN}}$ =14.5 GeV data, the correction is larger (90% and 70%, respectively) due to the presence of additional material for runs at this energy. A very small residual signal, at the level of the statistical uncertainty, is observed in the combinatoric background for $p-\pi$ pairs with invariant mass slightly different from the m_Λ. This residual effect may be due to daughter protons of a real Λ particles which are paired with background pions to make a candidate Λ particle. Since the protons carry most of the momentum this may lead to a small polarization from such particles. This residual correlation is included in the combinatoric background correction. Not including it would lead to an upward shift on the order of 0.2% for most data points.

A significant number of Λ particles come from decays of others baryons. Since the Σ^0 baryon has a very short lifetime compared to the Λ baryon, a Λ baryon coming from a primary Σ^0 will look just like primary Λ so it is not possible to distinguish the two. It is clear from figure 2 that the vortical or

Journal of Physics: Conference Series **736** (2016) 012016 doi:10.1088/1742-6596/736/1/012016

spin-orbit contribution to the polarization is dominant since the polarization results for Λ and $\overline{\Lambda}$ baryons are the same sign and similar magnitude. If we assume that the polarization from this effect is the same for Σ^0 particles as it is for Λ particles then we can correct for the feed-down. It is known that the daughter Λ particle will carry, on average, $-\frac{1}{3}$ of the polarization of the Σ^0 particle. The contribution due to feed-down from multi-strange hadrons is not included. The number of Σ^0 baryons is estimated from a simple thermal scaling based on the mass difference of the particles – $N_{\Sigma^0}/N_\Lambda = e^{-\Delta M/T}$, where $T = 150$MeV. The results in figure 2 are corrected for Σ^0 feed-down.

Figure 2 also includes the polarization measurements STAR has published previously [6], for Au+Au collisions at $\sqrt{s_{NN}}$ =62.4 GeV and 200 GeV. In order to account for a sign error in the direction of $\hat{J}_{sys}$ in reference [6], the sign on these data have been flipped here.

Both Λ and $\overline{\Lambda}$ baryons exhibit a similar and positive polarization, consistent with a common spin-orbit or vortical origin. While the difference is of the order of the uncertainties, $\overline{P}_{\overline{\Lambda}}$ is slightly larger than $\overline{P}_\Lambda$ at all collision energies, suggesting the possibility of an additional magnetic contribution. This difference could also arise due to finite baryonic chemical potential [44, 42].

Early spin-orbit-based calculations [7] predicted quark polarization $P_q \sim 10-30\%$, though a more complete calculation [9] suggested $P_q \lesssim 0.4\%$ for collisions at $\sqrt{s_{NN}}$ =200 GeV. Hyperon polarizations were expected to be of the same order. While some hydrodynamic models predict a vanishing hyperon polarization [12, 13] in Au+Au collisions at RHIC, others [15] predict an average value of $P_Y \approx 1-2\%$ for collisions at $\sqrt{s_{NN}}$ =200 GeV, due to plasma vorticity. Spin-orbit-based calculations [9] suggest larger polarizations at LHC energies than at RHIC. Hydrodynamical calculations [17, 41, 14] suggest that P_Y will decrease with increasing collision energy as thermal fluctuations overcome vortical effects. Microscopic transport calculations [37, 19] find that 10-20% of the total angular momentum of the colliding system is retained by the plasma at midrapidity. Although this angular momentum increases with $\sqrt{s_{NN}}$, Jiang, *et al.* [19] report that the velocity-field vorticity *decreases* significantly with energy in the range $\sqrt{s_{NN}}$ =30-200 GeV. Meanwhile, Baznat, *et al.* [37] predict P_Y of several percent at forward rapidity for much lower-energy collisions ($\sqrt{s_{NN}}$ =5 GeV), but $\sim 10^{-4}$ at midrapidity.

Systematic, quantitative theoretical predictions for global polarization due to vorticity in collisions with $\sqrt{s_{NN}} < 200$ GeV would be valuable, to compare with these first measurements and explore the physical mechanisms driving the polarization. Similarly, a study of polarization due to magnetic coupling may shed light on the dynamics of the magnetic field generated in these collisions. The primary theoretical uncertainty in our understanding of B-field effects [29, 27, 28, 38, 39] concerns its time evolution, which depends strongly on the electric conductivity of the plasma. Calculations with diferent conductivities differ by a few orders of magnitude [36]; the calculations with conductivity determined in the lattice QCD, yields fields (at the time of a few fm/c) $eB \sim$ (a few) $\times 10^{-2} m_\pi^2$.

We have reported a preliminary first observation of global Λ and $\overline{\Lambda}$ baryon polarization in heavy ion collisions. The signal is on the order of a few percent and falls with increasing collision energy, though the $\sqrt{s_{NN}}$-dependence is small compared to experimental uncertainty. Theoretical calculations [9, 15, 17] for collisions at $\sqrt{s_{NN}}$ =200 GeV predict a polarization $\approx 0.5-2\%$, and others suggest [14, 19] that the signal may be larger at lower $\sqrt{s_{NN}}$. Quantitative predictions of the energy dependence may provide important constraints on the vortical structure of the quark-gluon plasma.

Specific theoretical predictions have been made on the dependence of $\overline{P}_\Lambda$ on p_T, rapidity, and azimuthal emission angle of the hyperon, as well as on the impact parameter. To test these predictions, and to quantify the level of magnetically-induced splitting, statistical uncertainties, depending on both event statistics and event plane resolution, should be reduced by at least a factor of 5. The second beam energy scan (BES-II) program at RHIC [1], planned for the end of the decade, will improve both statistics (through accelerator upgrades and increased beam time) and event plane resolution (through detector upgrades to the STAR experiment) and provide the opportunity to explore the systematics of global hyperon polarization more extensively.

[1] STAR Collaboration, STAR Note 0598, https://drupal.star.bnl.gov/STAR/starnotes/public/sn0598

The discovery of global hyperon polarization in heavy ion collisions may represent the first experimental access to the vortical structure of the quark gluon plasma, as well as providing constraints on magnetic effects in noncentral collisions. Further research on both the experimental and theoretical fronts is clearly required to understand this substructure fully.

References

[1] J. Adams *et al.* [STAR Collaboration], Nucl. Phys. A **757** (2005) 102 doi:10.1016/j.nuclphysa.2005.03.085 [nucl-ex/0501009].

[2] K. Adcox *et al.* [PHENIX Collaboration], Nucl. Phys. A **757** (2005) 184 doi:10.1016/j.nuclphysa.2005.03.086 [nucl-ex/0410003].

[3] B. B. Back *et al.*, Nucl. Phys. A **757** (2005) 28 doi:10.1016/j.nuclphysa.2005.03.084 [nucl-ex/0410022].

[4] I. Arsene *et al.* [BRAHMS Collaboration], Nucl. Phys. A **757** (2005) 1 doi:10.1016/j.nuclphysa.2005.02.130 [nucl-ex/0410020].

[5] U. Heinz and R. Snellings, Ann. Rev. Nucl. Part. Sci. **63** (2013) 123 doi:10.1146/annurev-nucl-102212-170540 [arXiv:1301.2826 [nucl-th]].

[6] B. I. Abelev *et al.* [STAR Collaboration], Phys. Rev. C **76** (2007) 024915 doi:10.1103/PhysRevC.76.024915 [arXiv:0705.1691 [nucl-ex]].

[7] Z. T. Liang and X. N. Wang, Phys. Rev. Lett. **94** (2005) 102301 Erratum: [Phys. Rev. Lett. **96** (2006) 039901] doi:10.1103/PhysRevLett.94.102301, 10.1103/PhysRevLett.96.039901 [nucl-th/0410079].

[8] S. A. Voloshin, nucl-th/0410089.

[9] J. H. Gao, S. W. Chen, W. t. Deng, Z. T. Liang, Q. Wang and X. N. Wang, Phys. Rev. C **77** (2008) 044902 doi:10.1103/PhysRevC.77.044902 [arXiv:0710.2943 [nucl-th]].

[10] G. Bunce *et al.*, Phys. Rev. Lett. **36** (1976) 1113. doi:10.1103/PhysRevLett.36.1113

[11] K. J. Heller *et al.*, Phys. Rev. Lett. **51** (1983) 2025. doi:10.1103/PhysRevLett.51.2025

[12] C. d. C. Barros, Jr. and Y. Hama, Phys. Lett. B **699** (2011) 74 doi:10.1016/j.physletb.2011.03.052 [arXiv:0712.3447 [hep-ph]].

[13] C. C. Barros, Jr., J. Phys. Conf. Ser. **509** (2014) 012056. doi:10.1088/1742-6596/509/1/012056

[14] B. Betz, M. Gyulassy and G. Torrieri, Phys. Rev. C **76** (2007) 044901 doi:10.1103/PhysRevC.76.044901 [arXiv:0708.0035 [nucl-th]].

[15] F. Becattini, L. Csernai and D. J. Wang, Phys. Rev. C **88** (2013) no.3, 034905 Erratum: [Phys. Rev. C **93** (2016) no.6, 069901] doi:10.1103/PhysRevC.93.069901, 10.1103/PhysRevC.88.034905 [arXiv:1304.4427 [nucl-th]].

[16] F. Becattini, V. Chandra, L. Del Zanna and E. Grossi, J. Phys. Conf. Ser. **509** (2014) 012055. doi:10.1088/1742-6596/509/1/012055

[17] L. P. Csernai, F. Becattini and D. J. Wang, J. Phys. Conf. Ser. **509** (2014) 012054. doi:10.1088/1742-6596/509/1/012054

[18] F. Becattini *et al.*, Eur. Phys. J. C **75** (2015) no.9, 406 doi:10.1140/epjc/s10052-015-3624-1 [arXiv:1501.04468 [nucl-th]].

[19] Y. Jiang, Z. W. Lin and J. Liao, arXiv:1602.06580 [hep-ph].

[20] S. A. Voloshin, A. M. Poskanzer and R. Snellings, arXiv:0809.2949 [nucl-ex].

[21] M. Anderson *et al.*, Nucl. Instrum. Meth. A **499** (2003) 659 doi:10.1016/S0168-9002(02)01964-2 [nucl-ex/0301015].

[22] L. Adamczyk *et al.* [STAR Collaboration], Phys. Rev. C **88** (2013) 014902 doi:10.1103/PhysRevC.88.014902 [arXiv:1301.2348 [nucl-ex]].

[23] C. A. Whitten [STAR Collaboration], AIP Conf. Proc. **980** (2008) 390. doi:10.1063/1.2888113

[24] F. S. Bieser *et al.*, Nucl. Instrum. Meth. A **499** (2003) 766. doi:10.1016/S0168-9002(02)01974-5

[25] G. Agakishiev *et al.* [STAR Collaboration], Phys. Rev. C **85** (2012) 014901 doi:10.1103/PhysRevC.85.014901 [arXiv:1109.5446 [nucl-ex]].

[26] L. Adamczyk *et al.* [STAR Collaboration], Phys. Rev. Lett. **112** (2014) no.16, 162301 doi:10.1103/PhysRevLett.112.162301(2014), 10.1103/PhysRevLett.112.162301 [arXiv:1401.3043 [nucl-ex]].

[27] V. Skokov, A. Y. Illarionov and V. Toneev, Int. J. Mod. Phys. A **24** (2009) 5925 doi:10.1142/S0217751X09047570 [arXiv:0907.1396 [nucl-th]].

[28] V. Voronyuk, V. D. Toneev, W. Cassing, E. L. Bratkovskaya, V. P. Konchakovski and S. A. Voloshin, Phys. Rev. C **83** (2011) 054911 doi:10.1103/PhysRevC.83.054911 [arXiv:1103.4239 [nucl-th]].

[29] D. E. Kharzeev, L. D. McLerran and H. J. Warringa, Nucl. Phys. A **803** (2008) 227 doi:10.1016/j.nuclphysa.2008.02.298 [arXiv:0711.0950 [hep-ph]].

[30] R. Armenteros *et al.*, Nucl. Phys. B **21** (1970) 15. doi:10.1016/0550-3213(70)90461-X

[31] S. A. Bass *et al.*, Prog. Part. Nucl. Phys. **41** (1998) 255 [Prog. Part. Nucl. Phys. **41** (1998) 225] doi:10.1016/S0146-6410(98)00058-1 [nucl-th/9803035].

[32] G. Van Buren [STAR Collaboration], J. Phys. G **31** (2005) S1127 doi:10.1088/0954-3899/31/6/072 [nucl-ex/0412034].

[33] L. P. Csernai, D. D. Strottman and C. Anderlik, Phys. Rev. C **85** (2012) 054901 doi:10.1103/PhysRevC.85.054901 [arXiv:1112.4287 [nucl-th]].

Journal of Physics: Conference Series **736** (2016) 012016 doi:10.1088/1742-6596/736/1/012016

[34] S. A. Voloshin and T. Niida, arXiv:1604.04597 [nucl-th].
[35] D. E. Kharzeev, J. Liao, S. A. Voloshin and G. Wang, Prog. Part. Nucl. Phys. **88** (2016) 1 doi:10.1016/j.ppnp.2016.01.001 [arXiv:1511.04050 [hep-ph]].
[36] L. McLerran and V. Skokov, Nucl. Phys. A **929** (2014) 184 doi:10.1016/j.nuclphysa.2014.05.008 [arXiv:1305.0774 [hep-ph]].
[37] M. I. Baznat, K. K. Gudima, A. S. Sorin and O. V. Teryaev, Phys. Rev. C **93** (2016) no.3, 031902 doi:10.1103/PhysRevC.93.031902 [arXiv:1507.04652 [nucl-th]].
[38] U. Gursoy, D. Kharzeev and K. Rajagopal, Phys. Rev. C **89** (2014) no.5, 054905 doi:10.1103/PhysRevC.89.054905 [arXiv:1401.3805 [hep-ph]].
[39] K. Tuchin, Int. J. Mod. Phys. E **23** (2014) 1430001. doi:10.1142/S021830131430001X
[40] A. Vilenkin, Phys. Rev. D **21** (1980) 2260. doi:10.1103/PhysRevD.21.2260
[41] L. P. Csernai, D. J. Wang, M. Bleicher and H. Stcker, Phys. Rev. C **90** (2014) no.2, 021904. doi:10.1103/PhysRevC.90.021904
[42] R. h. Fang, L. g. Pang, Q. Wang and X. n. Wang, arXiv:1604.04036 [nucl-th].
[43] F. Becattini and F. Piccinini, Annals Phys. **323** (2008) 2452 doi:10.1016/j.aop.2008.01.001 [arXiv:0710.5694 [nucl-th]].
[44] F. Becattini, V. Chandra, L. Del Zanna and E. Grossi, Annals Phys. **338** (2013) 32 doi:10.1016/j.aop.2013.07.004 [arXiv:1303.3431 [nucl-th]].
[45] M. Huang *et al.* [HyperCP Collaboration], Phys. Rev. Lett. **93** (2004) 011802. doi:10.1103/PhysRevLett.93.011802
[46] S. Wheaton and M. Hauer, Phys. Part. Nucl. Lett. **8** (2011) 869. doi:10.1134/S1547477111080152
[47] S. Wheaton and J. Cleymans, Comput. Phys. Commun. **180** (2009) 84 doi:10.1016/j.cpc.2008.08.001 [hep-ph/0407174].

Journal of Physics: Conference Series **736** (2016) 012017 doi:10.1088/1742-6596/736/1/012017

The First Transverse Single Spin Measurement in High Energy Polarized Proton-Nucleus Collision at the PHENIX experiment at RHIC

I. Nakagawa for the PHENIX Collaboration

RIKEN, 2-1 Hirosawa, Wako, Saitama 351-0198, Japan

E-mail: itaru@riken.jp

Abstract. Large single spin asymmetries in very forward neutron production seen using the PHENIX zero-degree calorimeters are a long established feature of transversely polarized proton-proton collisions at RHIC. Neutron production near zero degrees is well described by the one-pion exchange framework. The absorptive correction to the OPE generates the asymmetry as a consequence of a phase shift between the spin flip and non-spin flip amplitudes. However, the amplitude predicted by the OPE is too small to explain the large observed asymmetries. A model introducing interference of pion and a_1-Reggeon exchanges has been successful in reproducing the experimental data. During the RHIC experiment in year 2015, RHIC delivered polarized proton collisions with Au and Al nuclei for the first time, enabling the exploration of the mechanism of transverse single-spin asymmetries with nuclear collisions. The observed asymmetries showed surprisingly strong A-dependence in the inclusive forward neutron production, while the existing framework which was successfull in p+p only predicts moderate A-dependence. Thus the observed data are absolutely unexpected and unpredicted. In this report, experimental and theoretical efforts are discussed to disentangle the observed A-dependence using somewhat semi-inclusive type measurements and Monte-Carlo study, respectively.

1. Single Spin Asymmetry of Forward Neutron Production in p+p

The single spin asymmetry of very forward (almost zero degree) was first discovered[1] at RHIC in transversely polarized proton+proton collision at $\sqrt{s} = 200$ GeV. Neutrons were measured by a zero-degree calorimeter (ZDC) [2] with a position-sensitive shower-maximum detector (SMD). One ZDC module is composed of Cu-W alloy absorbers with PMMA-based communication grade optical fibers and corresponds to 1.7 nuclear interaction lengths. A single photomultiplier collects Cherenkov light via optical fibers. Three ZDC modules are located in series (5.1 nuclear interaction lengths) at 1800 cm away from the collision point, covering 10 cm in the transverse plane.

The detector covers the psudo-rapidity range of $6.8 \leq \eta \leq 8.8$ where the neutron emission angle with respect to the incident proton direction is limited to less than 2.2 mrad. Relatively large energy neutrons are selected in the analysis, i.e. the neutron energy cut was applied at the energy fraction of incident proton $x_{\rm F} = E_{\rm n}/E_{\rm p} \geq 0.4$. The production mechanism of the such a neutron is driven by soft process and of which cross section is well described by one pion exchange (OPE) model. The corresponding four momentum transfer can be defined $0.02 \leq -t \leq 0.5$ (GeV/c)^2 at under the assumption of OPE.

The single spin asymmetries for such a forward neutron turned out to be large as shown in Fig. 1 plotted as a function of transverse momentum p_T of detected neutron. The data points are measured at three different collision energies of 64, 200, and 500GeV, respectively. The data demonstrate the absolute amplitude of the asymmetries grow as a function of p_T. Although the OPE model was quite successful to reproduce the cross section data, the predicted asymmetry within the OPE framework appeared to be very tiny and far underestimates the data[3]. The forward neutron asymmetry is formulated as

$$A_N = \phi_{\rm flip}\phi_{\rm non-flip} \sin \delta \tag{1}$$

where $\phi_{\rm flip}$ ($\phi_{\rm non-flip}$) is spin flip (spin non-flip) amplitude between incident proton and outgoing neutron, and δ is the relative phase between these two amplitudes. Although the OPE can contribute to both spin flip and non-flip amplitudes, resulting A_N is small due to the small relative phase. The decent amplitude can be generated only by introducing the interference between spin flip π exchange and spin non-flip a_1-Reggeon exchange which has large phase shift in between[3]. So calculated A_N are compared with data points in Fig. 1 and are in good agreement with data.

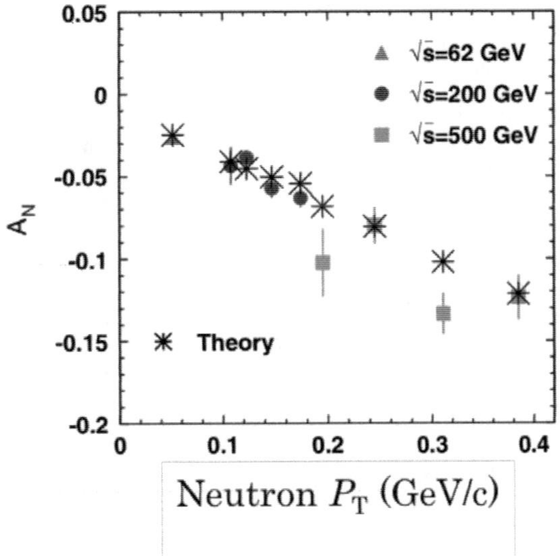

Figure 1. The forward neutron asymmetries observed p+p collision system at three different collision energies 64, 200, and 500GeVplotted as a function of transverse momentum p_T. The data points are well reproduced by the model calculations assumes the interference between π and a_1-Reggeon[3].

2. New Forward Neutron Asymmetries in p+A

In year 2015, the first attempt was made to collide the transversely polarized proton and nucleus in RHIC at $\sqrt{s} = 200$ GeV. In order to explore the evolving origin of the asymmetry as a function of atomic mass number (A), the forward (p-going side) neutron asymmetries were measured for p+Au and p+Al collisions. Since the proton beam is unpolarized at LHC, the measurement is the high energy frontier of the polarized proton and nucleus collision. The existing polarized proton+nucleus experiments are carried out in one order of magnitude lower $\sqrt{s}$, for instance, at RHIC[5] and Fermi Lab.[4] using polarized proton beam and fixed nuclear targets.

Journal of Physics: Conference Series **736** (2016) 012017 doi:10.1088/1742-6596/736/1/012017

The asymmetry analysis basically followed the same procedures and conditions as previous p+p measurements. The measurement in p+p collision system was repeated in year 2015 and resulting asymmetry was consistent with previous measurements. The ZDC acceptance was segmented into eight azimuthal bins and evaluated left-right raw asymmetries ϵ_N^{exp} defined by the Eqn. (2) in the each bin independently.

$$\epsilon_N^{raw}(\phi) = \frac{N^+(\phi) - N^-(\phi)}{N^+(\phi) + N^-(\phi)} \tag{2}$$

where $N^+(\phi)$ and $N^-(\phi)$ are number of neutrons observed in the given azimuthal ϕ bin when protons are polarized up and down, respectively.

The observed ϕ dependent asymmetries normalized by the proton polarization P are plotted in Fig. 3 as a function of ϕ. The $\phi = 0$ is set by the polarization direction of the proton. The analyzing power $A_{\rm N}$ is evaluated by fitting the data with a sine function:

$$\frac{\epsilon_N^{exp}(\phi)}{P} = A_{\rm N}\sin(\phi - \phi_0) \tag{3}$$

where ϕ_0 allows a deviation of the maximum asymmetry axis from vertical.

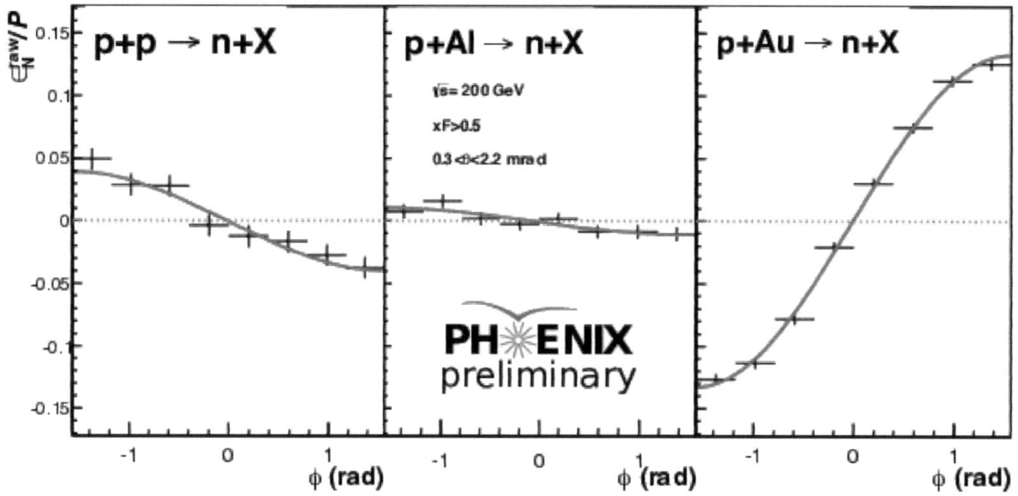

Figure 2. (Observed azimuthal angular modulations of the inclusive forward neutron production in p+p (left), p+Al (middle) and p+Au (right) collisions in year 2015.

Fig.3 shows preliminary results of forward neutron $A_{\rm N}$ measurements in 2015. The red points are the results of forward neutron $A_{\rm N}$ inclusive measurements. Surprisingly, they show unexpectedly strong mass number (A) dependence. The asymmetry even flips the sign from p+p, p+Al to p+Au. The p+Au data point shows magnificently large $A_{\rm N}$ of about 0.18 which is three times larger than that of p+p in absolute amplitude. The existing π and a_1-Reggeon interference framework predicts only moderate evolution as growing A and does not have any mechanism to flip the sign of $A_{\rm N}$ in any p+A collision systems.

More interestingly, another drastic dependence of $A_{\rm N}$ was observed in correlation measurements in addition to the inclusive neutron. In these measurements, another out-going

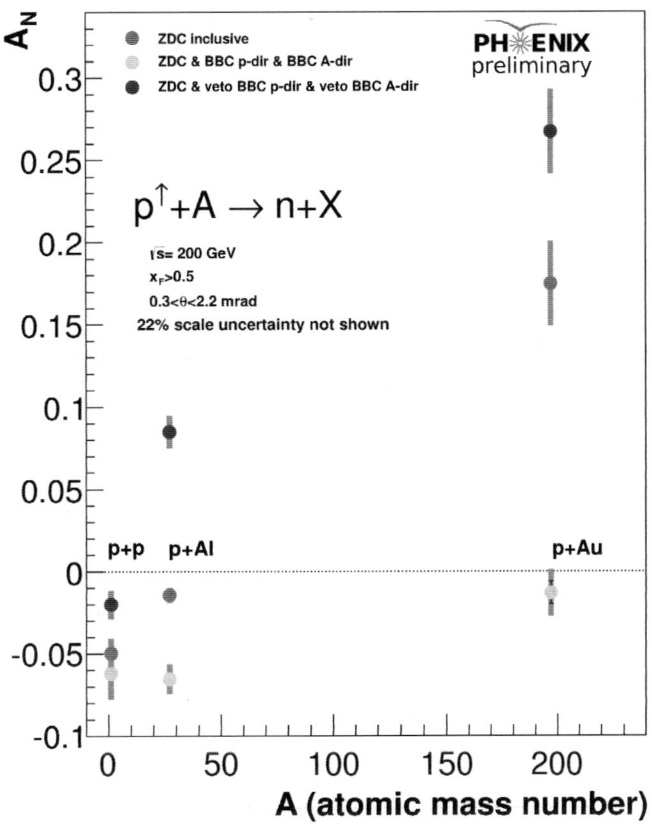

Figure 3. (Observed forward neutron A_{N} in transversely polarized proton-nucleus collisions. Data points are A=1, A=27, and A=197 are results of p+p, p+Al, and p+Au, respectively. Red, Blue and Green data points are neutron inclusive, neutron + BBC veto, and BBC tagged events, respectively

charged particle was either tagged or vetoed within the acceptance of the beam-beam counter (BBC) in both North and South arms which covers $3.1 \leq |\eta| \leq 3.9$. The BBCs cover such a limited acceptance, but the resulting asymmetries behaved remarkably contradicts. Once BBC hits (BBC tagging) are required in both arms (green data points), the drastic behavior of inclusive A_{N} is vanished and no flipping sign was observed between p+p and p+Au. On the contrary, the asymmetries are pushed even more positive for p+Al and p+Au data points once no hits in BBC are required (BBC vetoed) as represented by blue data points.

3. Ultra-Peripheral Collision Effects

It was pointed out the importance of the electro-magnetic (EM) interaction effect in the existing fixed target experiments [4] and [5], which were carried out in small t, one may need to consider the effect here, although it was ignored in p+p data. Due to the smallness of the four momentum transfers of the present kinematics, i.e. $-t \leq 0.5$ (GeV/c)^2, the EM interaction may play a role which becomes increasingly important in large atomic number nucleus. The EM field of the nucleus becomes rich source of exchanging photons between the polarized proton. This is known as the ultra-peripheral collision (UPC) in heavy ion collider experiments. In the UPC process, there is no charge exchange at the collision vertex unlike π or a_1 meson exchange. The forward neutron in the final state then can be produced through the diagrams shown in Fig. 4.

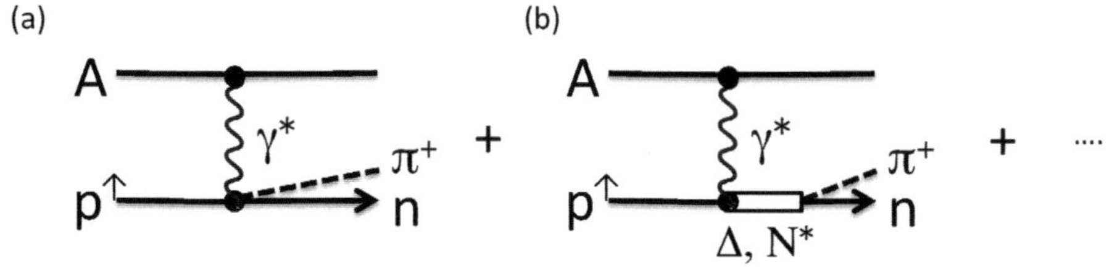

Figure 4. Feynman diagrams of UPC process to leave the high-z neutron in the final state to the proton going directions. (a) threshold photo pion production (b) Δ or N^* excitation and its decay into n+π^+ channel.

The description of $A_{\rm N}$ is thus extended from Eqn. (1) to Eqn. 5, which includes not only hadronic but also EM amplitudes:

$$\begin{aligned} A_{\rm N} &= \phi^{\rm had}_{\rm flip}\phi^{\rm had}_{\rm non-flip}\sin\delta_1 + \phi^{\rm EM}_{\rm flip}\phi^{\rm had}_{\rm non-flip}\sin\delta_2 \\ &+ \phi^{\rm had}_{\rm flip}\phi^{\rm EM}_{\rm non-flip}\sin\delta_3 + \phi^{\rm EM}_{\rm flip}\phi^{\rm EM}_{\rm non-flip}\sin\delta_4 \end{aligned} \quad (4)$$

where 'EM' and 'had' stand for electromagnetic and hadronic interactions, and $\delta_1 \sim \delta_4$ are relative phases, respectively. The second and the third terms are known as Coulomb nuclear interference (CNI), which is observed to cause $< 5\%$ asymmetry of elastic scattering in p + p, and p + C processes [5].

According to MC study [6], the neutron and its counter part π^+ via UPC process is substantially boosted towards the proton beam direction and therefore the fragmenting π^+ are mostly emitted in even higher rapidity region than BBC. Only small fraction of π^+ are detected by BBC. Thus EM processes are suppressed in the BBC tagging events while enhanced in the BBC vetoed events. As a consequence, one may draw a hypothesis that the moderate $A_{\rm N}$ evolution of the BBC tagging data are consistent with the prediction based on π and a_1-Reggeon interference scenario because the data are dominated by the hadronic amplitude. Although it is not trivial to predicts the sign of the $A_{\rm N}$ caused by EM amplitude, the observed growing feature of $A_{\rm N}$ for BBC vetoed events makes sense if EM amplitude guides to opposite sign of hadronic interactions at least qualitatively. Nevertheless it is not known at all at this moment why EM should have opposite sign nor should have such a large amplitude or large relative phases between hadronic/EM spin-flip and spin-nonflip amplitudes to produce such the large asymmetries. It remains mystery.

4. Summary

We have observed surprisingly strong A-dependence in the forward neutron asymmetry in transversely polarized proton-nucleus collision at $\sqrt{s}$=200GeV in PHENIX using ZDC detector. Such a drastic A-dependence in the forward neutron asymmetry $A_{\rm N}$ is absolutely unexpected within the current framework although it was successful in reproducing the $A_{\rm N}$ data in p+p. Another drastic dependence has been observed in the semi-inclusive measurements by requiring both North and South BBCs $3.1 < |\eta| < 3.9$ to be fired or not-fired in addition to forward neutron. The EM effect may play key role to disentangle these drastic behavior because 1) its amplitude glows as a function of atomic number Z, and 2) the fraction of EM process events are suppressed in BBC fired events while enhanced in BBC vetoed events. Further experimental dependency study is now underway by segmenting $p_{\rm T}$ into multiple bins.

Journal of Physics: Conference Series **736** (2016) 012017 doi:10.1088/1742-6596/736/1/012017

References

[1] Y. Fukao et al., Phys. Lett. B 650, 325 (2007), A. Adare et al.: Phys. Rev. D 88, 032006 (2013).
[2] C. Adler, A. Denisov, E. Garcia, M. J. Murray, H. Strobele, and S. N. White, Nucl. Instrum. Methods Phys. Res., A 470, 488 (2001).
[3] B. Z. Kopeliovich, I. K. Potashnikova, and Ivan Schmidt: Phys. Rev. Lett. 64, 357 (1990).
[4] D. C. Carey et al.: Phys. Rev. D 79, 094014 (2009).
[5] I. G. Alekseev et al.: Phys. Rev. D 79, 094014 (2009).
[6] G. Mitsuka: Eur. Phys. J. C 75:614 (2015).

Journal of Physics: Conference Series **736** (2016) 012018 doi:10.1088/1742-6596/736/1/012018

Sequential hadronization and the opportunities it presents

R. Bellwied[1]

[1] Department of Physics, University of Houston, USA

E-mail: bellwied@uh.edu

Abstract. Continuum extrapolated lattice QCD calculations of quantum number specific susceptibilities and the most recent RHIC and LHC data on produced particle yields, as well as their higher moment fluctuations, can be interpreted using a scenario of sequential flavor dependent hadronization during the QCD crossover transition. I will present the latest data from lattice QCD and experiment and confront the question whether the separation of strangeness and light quark chemical freeze-out could have consequences beyond a simple strangeness enhancement, firstly in the production of exotic strange states and secondly in the flavor dependent evolution of dynamic quantities such as in-medium energy loss and anisotropic flow, which are generated predominantly during the collective partonic phase.

1. Indications of sequential hadronization from the lattice

Lattice QCD data have started to reveal fine differences between conserved quantum numbers when continuum extrapolated data of susceptibilities became available a few years ago. As an example I show in Fig.1a the difference between light and strange quark second order susceptibilities in the QCD crossover region [1] . The inflection point of the $\chi_{2,s}$is clearly shifted to higher temperatures, which on the lattice can be attributed to a simple quark mass effect. In fact recent calculations have shown that the effect is even more prominent for the charm quark susceptibilities [2]. Although the simple χ_2 behavior does not allow us to determine a pseudo-critical temperature, the ratio of χ_4/χ_2 was proposed as a powerful tool to determine the chemical freeze-out temperature in the thermalized system from first principles [3], rather than relying on statistical fits to the particle yields or ratios [4]. In this context our original paper on flavor hierarchy [1] showed that the continuum extrapolated susceptibility ratio reveals indeed maxima at different temperatures for light and strange quarks. The final proof of a sequential hadronization scenario has to come from the comparison to experimental fluctuation data, though, which can be mapped to the susceptibility ratios on the lattice. It is by no means proven that the maxima obtained in the lattice ratios indeed correspond to the chemical freeze-out data although comparisons to Hadron Resonance Gas (HRG) model calculations, which describe the hadronic phase quite reliably, show that the maxima coincide with the deviations of the HRG curves from the lattice data, see Fig.1b.

Furthermore comparisons of lattice calculations [5, 6] to fluctuation data from STAR on net-protons and net-charge [7, 8], see Fig.2a, seem to indicate lower freeze-out temperatures than the ones obtained from statistical hadronization fits to the yields of all produced particles [4], see Fig2b. Two important points should be noted:

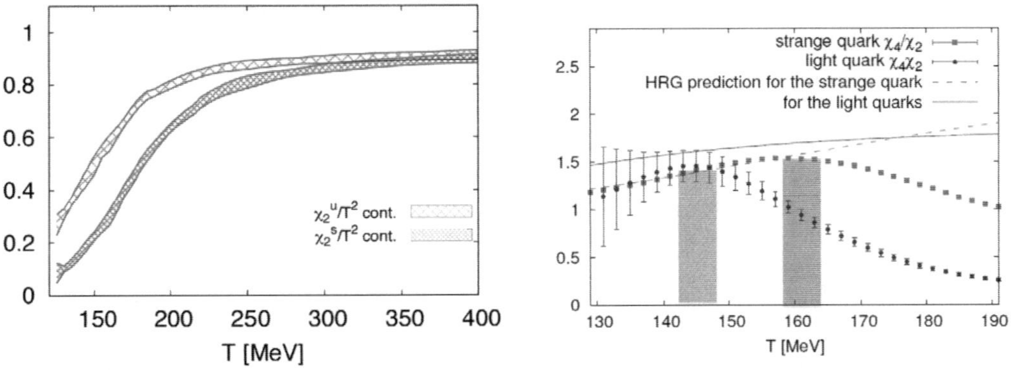

Figure 1. Results from continuum extrapolated lattice caclulations on flavor specific susceptibilities [1]. (a) χ_2 for up and strange quarks, (b)χ_4/χ_2 for lgiht and strange quarks.

1.) for the comparison to be valid one needs to be able to equate the efficiency corrected net-proton measurements in a finite acceptance bin to the total net-baryon number calculations on the lattice. Many independent studies have been performed to ensure this equivalence [9, 10, 11, 12, 13, 14] and one is now reasonably confident that the comparison can indeed be made to the level accuracy shown in Fig.2. Furthermore HRG results based on a parallel analysis of both net-number measurements show excellent agreement with the lattice results and push them out to even larger chemical potential, as shown in Fig.2b [15].

2.) the reason that the freeze-out temperatures deduced from the fluctuations of net-protons and net-charges are lower than the ones from the fits to all particle yields can again be attributed to the potential flavor hierarchy. Both, net-proton and net-charge multiplicities are dominated by light particles, protons in the one case, pions in the other. If light quarks indeed prefer to freeze-out at a lower temperature then this would explain the measured differences between total yield and specific fluctuation fits. The final proof will lie in a comparison between lattice data and a net-strangeness measurement, which will be discussed in the following sections.

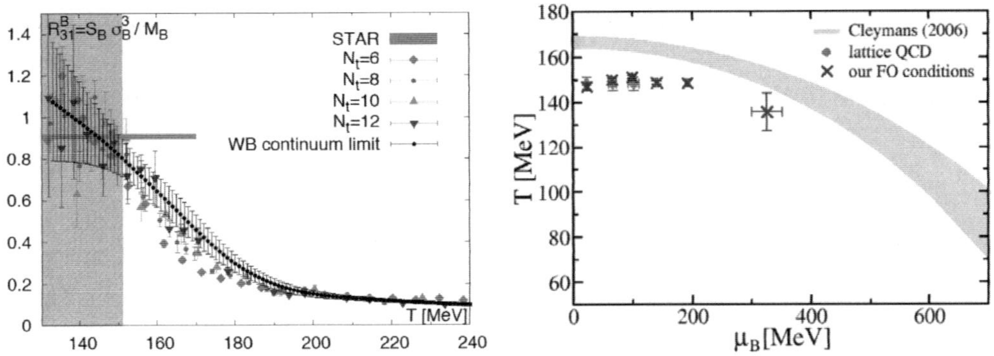

Figure 2. (a) Comparison of continuum extrapolated lattice QCD results for baryon-number χ_3/χ_1 to the equivalent ratio of STAR net-proton moments. An upper freeze-out temperature of 151 $\pm$ 4 meV can be deduced [6]. (b) the lattice QCD temperature (in red) compared to a HRG model analysis based on the STAR net-charge and net-proton data [15] and to a chemical freeze-out curve determined from fits to yields with Statistical Hadronization models [4].

Journal of Physics: Conference Series **736** (2016) 012018 doi:10.1088/1742-6596/736/1/012018

2. Indications of sequential hadronization from experimental data

The first indication of slight deviations in measured particle yields from a fit based on a common chemical freeze-out temperature came from the high precision multiplicity measurements of identified particles in the ALICE experiment, in particular the protons. In contrast to the RHIC data, the excellent position resolution of the Inner Tracking System (ITS) in ALICE enabled the collaboration to perform an in-depth feed-down correction for the protons from the weak decays of strange baryons. After subtracting this contribution the preliminary data shown at SQM in 2014 revealed a difference between the protons and strange and multistrange baryons chemical freeze-out temperature of almost 20 MeV (148 MeV for protons vs. 164 MeV for multi-strange baryons) [16]. A fit with a common freeze-out temperature of 156 MeV was presented at QM 2014 [17], but the fit still shows a deviation of almost 4σ between the protons and the multi-strange baryons as shown in Fig.3a. Several alternative explanations for the low proton yield were proposed following the release of the data [18, 19, 20, 21], some also proposing two freeze-out temperatures, but for different reasons [22]. The suggestion of enhanced hadronic re-interaction in the dense hadronic medium shortly after the chemical freeze-out [21, 23] garnered the most interest since it also tries to address the potential shortcoming of any HRG calculation, which assumes a non-interacting hadron gas as the hadronic phase. Although the idea of enhanced baryon annihilation in the dense cooling system seems appealing, the near perfect agreement between the freeze-out temperature from net-charge fluctuations, which are independent of annihilation, and net-proton fluctuations [6] seems to indicate that the effect of hadronic re-interactions is small. Nevertheless a detailed study of the hadronic interactions using a realistic transport approach is underway and has shown intriguing effects on bringing the chemical freeze-out temperature closer to the QCD crossover temperature deduced from the lattice, in particular at larger chemical potential, i.e. smaller collision energies (RHIC beam energy scan)[23].

The final missing experimental input into the question of flavor separation during the crossover will be a high precision measurement of net-strangeness fluctuations. Unfortunately, in order to fully determine net-strangeness the experiment needs to measure the fluctuations of net-yields of K, Λ, Ξ, and Ω (incl. Σ's). Since the measurement of multi-strange baryons is significantly statistics limited, even at the highest collision energies and data acquisition rates at the LHC, the question arose whether a subset of strange particles might be sufficient to determine the strange chemical freeze-out. Our recent study showed that just the kaons alone follow the trend of the strange freeze-out, but an in-depth study will likely require the Λ and Ξ as well.

Progress has been achieved experimentally by the STAR collaboration, which has shown preliminary net-kaon moments at QM 2015 [24] and by Ji Xu at this conference, see Fig.3b. Data from ALICE on net-charge, net-proton, and net-kaon are forthcoming, and one can look forward to an interesting interpretation of these results, since the lattice QCD data at μ_B=0 are the most reliable.

3. The latest results of comparisons between data and models

The analysis of the STAR kaon fluctuation data shown in Fig.3b using the lattice QCD and HRG methods employed in [6] and [15], respectively, is hampered by the large systematic error on the lower moment ratio, in particular at the relevant highest RHIC energies. Still, the fit procedure in the framework of the HRG formalism [15] leads to a rather clear trend towards a higher freeze-out temperature for the Kaons as shown in Fig.4a. Although the result is still in agreement with the combined fit to the net-proton and net-charge results, a reduction of the systematic to 10% will lead to a significant result as shown by the yellow points in Fig.4a.

Even more importantly the lattice approach, which is independent of the HRG result, has been recently refined to allow a direct comparison between lattice data and a particular particle species, in this case the kaons [25]. Until now lattice QCD only addressed the conserved quantum

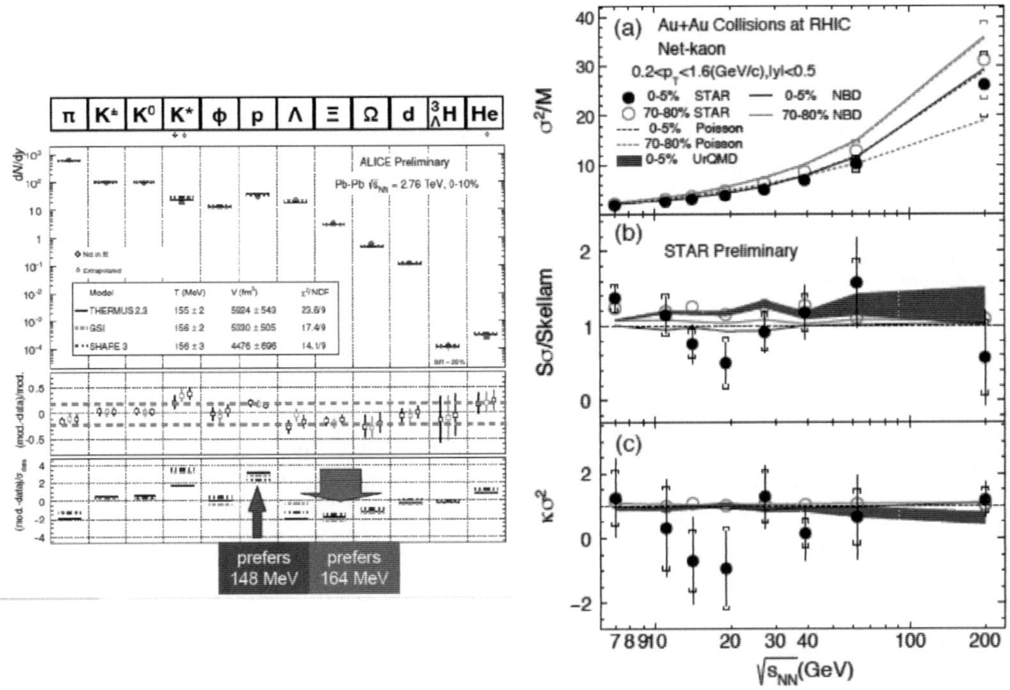

Figure 3. (a) Statistical Hadronization model fits to all measured particle yields in ALICE [17]. Although all three models converge on a common temperature, the main deviations seem to be flavor dependent. (b) First preliminary measurements of higher moment fluctuations of the net-kaon distribution in STAR [24] as function of collision energy.

number as a whole, not its individual particle species contributions. The new method is based on showing the equivalence between the lattice and a Boltzmann approximation to the partial pressure, which then allows to break out individual baryonic and mesonic contributions to the pressure on the lattice on the basis of this equivalence. Based on these calculations the data can now be confronted with lattice QCD calculations, which is shown in Fig.4b [25]. Again, the presently large experimental systematic error only allows to set a lower limit for the strange freeze-out temperature, which is still in line with the freeze-out temperatures deduced for the non-strange particles. The figure shows nicely, though, how any improvement on the kaon fluctuation uncertainty will further constrain the temperature.

4. Opportunities for the formation of new states

The concept of a sequential freeze-out of quark flavors poses significant questions for the hadronization process. How does a multi-flavored hadron freeze-out and when does it freeze-out ? There seem to be indications from the lattice and from the measurement of strange baryon multiplicities that the light quarks dominate the hadronization process inasmuch as any light quark dominated hadron seems to freeze-out rather late at a lower temperature. This has been shown on the lattice by the HotQCD group studying the D-mesons [26] in contrast to the studies of pure charm states by a subset of the same authors [27]. It also can be seen by an in-depth study of the freeze-out temperatures of strange baryons in ALICE, the Λ seems to prefer a lower freeze-out temperature than the multi-strange baryons. Therefore in the strangeness sector the ideal case to study will be an in-depth comparison of the proton to the Ω baryon. Unfortunately the Ω measurements are still statistics limited, although the upgraded ALICE detector promises

Journal of Physics: Conference Series **736** (2016) 012018 doi:10.1088/1742-6596/736/1/012018

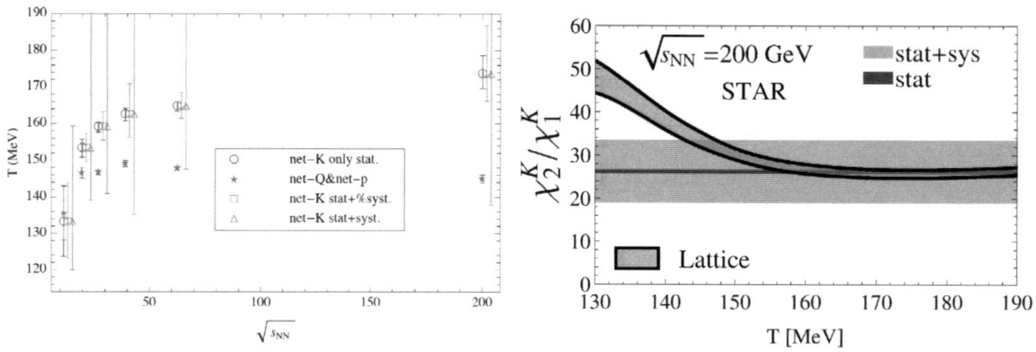

Figure 4. (a) A HRG fit to the preliminary STAR net-kaon χ_3/χ_1 ratio for central events at a collision energy of 200 GeV shown in Fig.3. The fit, shown in green follows the same procedure than the combined fit to net-charge and net-proton [15]shown in red here. The blue and yellow curve show the uncertainty on the net-Kaon fit assuming statistical errors only or an additional systematic error of 10%, respectively. (b) A direct comparison between the same experimental net-kaon ratio and a continuum extrapolated lattice QCD calculation that breaks out the kaon fluctuations based on partial pressure contributions [25].

an improvement of at least one order of magnitude.

The easiest way to remedy the apparent strangeness enhancement is to assume that the number of measured strange hadronic states is incomplete and thus requires us to postulate higher mass resonant states that will absorb the abundant strange quarks [28]. This concept is not unrealistic since the Particle Data Group has included many more experimentally confirmed heavy strange resonances in its booklet between 2008 and 2014 [29, 30]. The ultimate population of hadronic states can be estimated via the Quark Model [31]. If these states are included in the hadronic calculations, they seemingly can absorb the extra strangeness released by the sequential freeze-out. A new program to find more of these postulated states has been initiated by JLab.

On the other hand the apparent discrepancy of strange hadronic states and a partonic medium still domianted by quasi-free light quarks should lead to strangeness clustering and thus to the possibility of formation of multi-quark strange states, along the lines of the recently discovered charm tetraquark and pentaquark [32, 33]. ALICE has investigated the production of H-Dibaryons and the originally claimed strange pentaquark [34]. Unfortunately both of these searches turned out negative [35, 36] with limits well within the range of production estimated by thermal models. The reason could again be the dominance of light quarks in these states, and I suggest to search for multi-strange multi-quark states such as the ΞΞ-dibaryon [38].

5. Opportunities for a new interpretation of dynamic quantities in the partonic phase

Finally I am trying to address whether dynamic quantities governed by the evolution of the collective partonic phase could be affected by a sequential freeze-out scenario. The two main parameters in question are the shear viscosity as obtained by anisotropic flow v_n and the partonic energy loss as obtained by the nuclear modification factor R_{AA}. In both cases the modeling of the dynamic evolution of the partonic system very much impacts the final results and the correlation between these two effects. In other words, is it possible that a differing hadronization temperature for different particle species could lead to more or less energy loss and/pr more less anisotropic flow. This question arises mostly on the basis of particle species dependent R_{AA} at RHIC and the LHC. Fig.5(a) shows the published data on R_{AA} from PHENIX [39]. Special emphasis should be given to the apparent difference between the light mesons (π and η) and the strange mesons (K and ϕ). Below 8 GeV, i.e. in the region where collectivity plays a role and the particles are likely thermal, the strange mesons clearly exhibit less of an energy loss than the light mesons. Fig.5(b) shows preliminary data from ALICE for the energy loss measurements of D-meson vs. D_s-mesons [40]. Although the error bars are still large one sees a similar trend of the strange D-meson being less suppressed than the light D-meson.

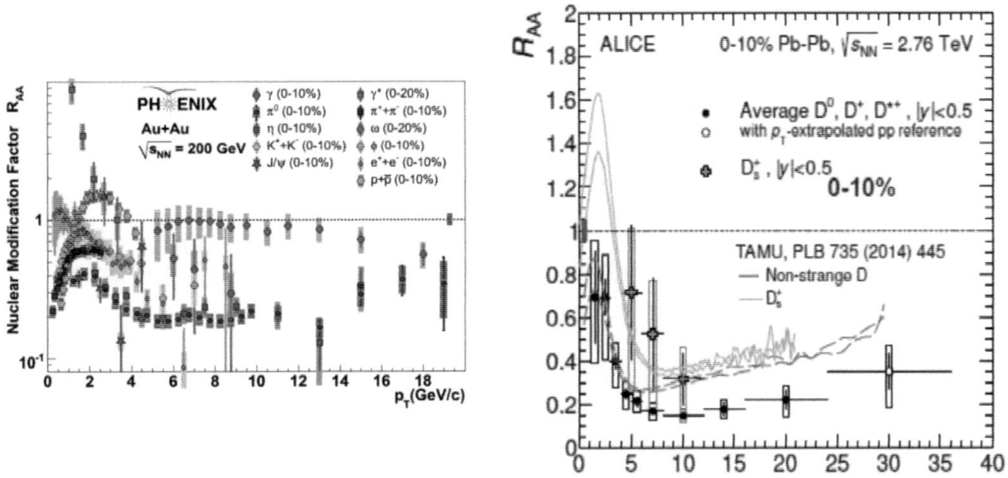

Figure 5. (a)The complete set of nuclear suppression factors measured by PHENIX. Special emphasis should be given to the π and η points compared to the K and ϕ points [39]. (b) The nuclear supreesion factors measured by ALICE for D- and D_s-mesons [40].

Both effects can potentially be attributed to canonical suppression, i.e. the suppression of strangeness formation in the small proton-proton reference system. But in our picture a canonical suppression is not necessary since all of the strangeness enhancement stems from the increased freeze-out temperature of the thermal deconfined strange quark system. These thermal quarks could potentially lead to an increased recombination cross section with charm quarks in the partonic phase [41] or alternately the strange charmed mesons lose less energy in the partonic system. The only requirement here is that the energy loss is predominantly generated very close to hadronization, i.e. if the system hadronizes at a higher temperature the partonic energy loss ceases earlier and thus generates less of an R_{AA} for the strange final states. This concept of late energy loss has been proposed and tested by Greco et al. [37] and they found that a temperature dependent energy loss close to freeze-out paired with a temperature-dependent coupling constant indeed leads to an excellent correlation between the experimental R_{AA} and v_n results. This agreement eludes most of the other models on the market. One should note, though, that this flavor effect on R_{AA} should also impact v_n in the same p_T range. Hydrodynamical calculations with differing switching temperature from the partonic to the hadronic phase point in that direction [42]. The preliminary ALICE data [43] show much less of an effect for the flow harmonics. One should also recognize that the effect seen, in particular by PHENIX (Fig.4(a)) cannot be explained by recombination or hydrodynamics since all particles that are compared are mesons and have very similar mass. I propose an in-depth study of energy loss in the p_T-range where thermalization is likely for multi-strange baryons in comparison to protons, in particular using the high statistics soon available from ALICE Run-II. Besides the aforementioned dynamic coefficients it might also be interesting to study bulk viscosity in a more flavor and particle species dependent way [44].

Journal of Physics: Conference Series **736** (2016) 012018 doi:10.1088/1742-6596/736/1/012018

6. Conclusions

I have shown that experimental data on particle yields and fluctuations and lattice QCD calculations of conserved quantum number susceptibilities can be reconciled if one assumes a flavor dependent approach to chemical freeze-out. This concept will enable a more detailed experimental investigation of the process of hadronization and will thus lead to a deeper understanding of the non-perturbative part of QCD.

In addition to the ongoing fluctuation measurements of net-charges (baryon number, electric charge and strangeness) I propose several new measurements that would be possible as a consequence of sequential hadronization. In the particle production sector we suggest to search for higher mass hadronic resonances, in particular in the strange sector, and for evidence of strangeness clustering, either through angle-dependent yield and fluctuation measurements and/or through the discovery of strangeness dominated multi-quark states.

Finally, I propose that sequential hadronization might impact dynamic state variables of the collective partonic system, such as R_{AA} and v_n if it can be shown that these quantities develop predominantly close to the QCD crossover temperature.

Acknowledgments

I would like to thank Paolo Alba, Claudia Ratti and Vincenzo Greco for useful discussion. Paolo also generously provided the calculations for Fig.3a. This work was supported by the U.S. Department of Energy under Grant DE-FG02-07ER41521. Part of this research used computer time that was provided by the DOE-INCITE program at the Argonne Leadership Computing Facility, which is a DOE Office of Science User Facility supported under Contract DE-AC02-06CH11357.

References

[1] R. Bellwied, S. Borsanyi, Z. Fodor, S. D. Katz and C. Ratti, Phys. Rev. Lett. **111**, 202302 (2013)
[2] R. Bellwied, S. Borsanyi, Z. Fodor, S. D. Katz, A. Pasztor, C. Ratti and K. K. Szabo, Phys. Rev. D **92**, no. 11, 114505 (2015)
[3] F. Karsch, Central Eur. J. Phys. **10**, 1234 (2012)
[4] J. Cleymans, H. Oeschler, K. Redlich and S. Wheaton, Phys. Rev. C **73**, 034905 (2006)
[5] S. Borsanyi, Z. Fodor, S. D. Katz, S. Krieg, C. Ratti and K. K. Szabo, Phys. Rev. Lett. **111**, 062005 (2013)
[6] S. Borsanyi, Z. Fodor, S. D. Katz, S. Krieg, C. Ratti and K. K. Szabo, Phys. Rev. Lett. **113**, 052301 (2014)
[7] L. Adamczyk *et al.* [STAR Collaboration], Phys. Rev. Lett. **112**, 032302 (2014)
[8] L. Adamczyk *et al.* [STAR Collaboration], Phys. Rev. Lett. **113**, 092301 (2014)
[9] V. Skokov, B. Friman and K. Redlich, Phys. Rev. C **88**, 034911 (2013)
[10] A. Bzdak, V. Koch and V. Skokov, Phys. Rev. C **87**, no. 1, 014901 (2013)
[11] A. Bzdak and V. Koch, Phys. Rev. C **86**, 044904 (2012)
[12] A. Bzdak and V. Koch, Phys. Rev. C **91**, no. 2, 027901 (2015)
[13] M. Kitazawa and M. Asakawa, Phys. Rev. C **86**, 024904 (2012) Erratum: [Phys. Rev. C **86**, 069902 (2012)]
[14] M. Nahrgang, M. Bluhm, P. Alba, R. Bellwied and C. Ratti, Eur. Phys. J. C **75**, no. 12, 573 (2015)
[15] P. Alba, W. Alberico, R. Bellwied, M. Bluhm, V. Mantovani Sarti, M. Nahrgang and C. Ratti, Phys. Lett. B **738**, 305 (2014)
[16] R. Preghenella [ALICE Collaboration], Acta Phys. Polon. B **43**, 555 (2012)
[17] M. Floris, Nucl. Phys. A **931**, 103 (2014)
[18] J. Noronha-Hostler and C. Greiner, Nucl. Phys. A **931**, 1108 (2014)
[19] M. Petrn, J. Letessier, V. Petrek and J. Rafelski, Phys. Rev. C **88**, no. 3, 034907 (2013)
[20] K. A. Bugaev, D. R. Oliinychenko, J. Cleymans, A. I. Ivanytskyi, I. N. Mishustin, E. G. Nikonov and V. V. Sagun, Europhys. Lett. **104**, 22002 (2013)
[21] F. Becattini, M. Bleicher, T. Kollegger, T. Schuster, J. Steinheimer and R. Stock, Phys. Rev. Lett. **111**, 082302 (2013)
[22] S. Chatterjee and B. Mohanty, Phys. Rev. C **90**, no. 3, 034908 (2014)
[23] F. Becattini, E. Grossi, M. Bleicher, J. Steinheimer and R. Stock, Phys. Rev. C **90**, no. 5, 054907 (2014)
[24] J. Thder [STAR Collaboration], arXiv:1601.00951 [nucl-ex].

[25] J. Noronha-Hostler, R. Bellwied, J. Gunther, P. Parotto, A. Pasztor, I. P. Vazquez and C. Ratti, arXiv:1607.02527 [hep-ph]
[26] A. Bazavov *et al.*, Phys. Lett. B **737**, 210 (2014)
[27] S. Mukherjee, P. Petreczky and S. Sharma, Phys. Rev. D **93**, no. 1, 014502 (2016)
[28] A. Bazavov *et al.*, Phys. Rev. Lett. **113**, no. 7, 072001 (2014)
[29] C. Amsler *et al.* [Particle Data Group Collaboration], Phys. Lett. B **667**, 1 (2008).
[30] K. A. Olive *et al.* [Particle Data Group Collaboration], Chin. Phys. C **38**, 090001 (2014).
[31] S. Capstick and N. Isgur, Phys. Rev. D **34**, 2809 (1986)
[32] R. Aaij *et al.* [LHCb Collaboration], Phys. Rev. Lett. **115**, 072001 (2015)
[33] R. Aaij *et al.* [LHCb Collaboration], arXiv:1606.07895 [hep-ex].
[34] C. Alt *et al.* [NA49 Collaboration], Phys. Rev. Lett. **92**, 042003 (2004)
[35] J. Adam *et al.* [ALICE Collaboration], Phys. Lett. B **752**, 267 (2016)
[36] B. B. Abelev *et al.* [ALICE Collaboration], Eur. Phys. J. C **75**, no. 1, 1 (2015)
[37] F. Scardina, M. Di Toro and V. Greco, Phys. Rev. C **82**, 054901 (2010).
[38] T. Yamazaki *et al.* [PACS-CS Collaboration], Phys. Rev. D **84**, 054506 (2011)
[39] A. Adare *et al.* [PHENIX Collaboration], Phys. Rev. C **83**, 024909 (2011)
[40] J. Adam *et al.* [ALICE Collaboration], JHEP **1603**, 082 (2016)
[41] M. He, R. J. Fries and R. Rapp, Phys. Lett. B **735**, 445 (2014)
[42] J. Noronha-Hostler, J. Noronha, G. S. Denicol, R. P. G. Andrade, F. Grassi and C. Greiner, Phys. Rev. C **89**, no. 5, 054904 (2014)
[43] A. Dobrin [ALICE Collaboration], Nucl. Phys. A **931**, 910 (2014).
[44] J. Noronha-Hostler, G. S. Denicol, J. Noronha, R. P. G. Andrade and F. Grassi, Phys. Rev. C **88**, 044916 (2013)

Solving the $R_{AA} \otimes v_2$ puzzle

Jacquelyn Noronha-Hostler
Department of Physics, University of Houston, Houston TX 77204, USA
E-mail: jakinoronhahostler@gmail.com

Abstract. For the past ten years $R_{AA}(p_T)$, the nuclear modification factor that encodes the suppression of high p_T particles due to energy loss within the medium was fairly well described by many theoretical models. However, the same models systematically underpredicted the high p_T elliptic flow, v_2, which is experimentally measured as the correlation between soft and hard hadrons. All previous calculations neglected the effect of event-by-event fluctuations of an expanding viscous hydrodynamical background as well as the soft-hard flow harmonic correlations in the experimentally measured v_2. In this talk I show how event-by-event viscous hydrodynamics (computed using the v-USPhydro code) coupled to an energy loss model (BBMG) is able to simultaneously describe soft physics observables as well as the high-p_T R_{AA} and v_2. Suggestions for future more differential calculations at the LHC run2 are made to explore soft-hard flow correlations.

1. Introduction

Two key signatures of the Quark Gluon Plasma- collective flow and jet quenching- have been thoroughly measured experimentally over the years at the Relativistic Heavy Ion Collider RHIC at BNL and the Large Hadron Collider LHC at CERN. Initially, it was thought that the soft physics regime (low transverse momentum p_T where collective flow is measured) and the hard physics regime (high p_T where jet physics is relevant) could be tackled separately. The early measurements of high p_T elliptic flow [1] demonstrated flow was possible even within the hard sector though theoretical calculations [2, 3, 4] taking into account only energy loss physics and simplified assumptions for the medium evolution were not able to reproduce the large measured v_2 and, in fact, the difficulties in simultaneously describing $R_{AA} \otimes v_2$ remained for years to come (see the discussion and references in [5, 6]).

Meanwhile, comparisons of relativistic hydrodynamical models to collective flow measurements had a breakthrough circa 2010 [7] with the inclusion of event-by-event calculations [8, 9, 10, 11, 12] to make comparisons to nonzero triangular flow measurements [7]. Event-by-event fluctuating initial conditions imply that within the same centrality class (i.e. holding the density constant) there is a wide range of initial shapes (eccentricities ε_n) that relativistic hydrodynamics map into the final flow harmonics [13, 14]. In Fig. 1 the distribution of ε_2 are shown for MCGlauber and MCKLN [15] (smoothing out initial energy density fluctuations to $\lambda = 0.3$ fm and $\lambda = 1$ fm- see [16] for more details). Regardless of the initial conditions, a wide range of initial elliptical shapes are produced (and more complex shapes such as "triangles", "squares" etc are also produced), which experimentally are proven via v_n distributions [17].

Event-by-event viscous hydrodynamical models have been enormously successful at describing soft physics observables such as the cumulants of flow harmonics (as shown in Fig. 2) as well as

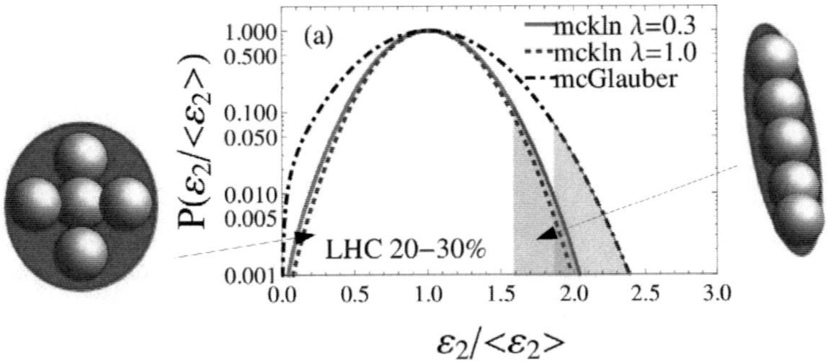

Figure 1. Distributions of ε_2 at LHC for MCGlauber and MCKLN (with two different smoothing scales) produce initial conditions that are at one end very circular and at the other end elliptical.

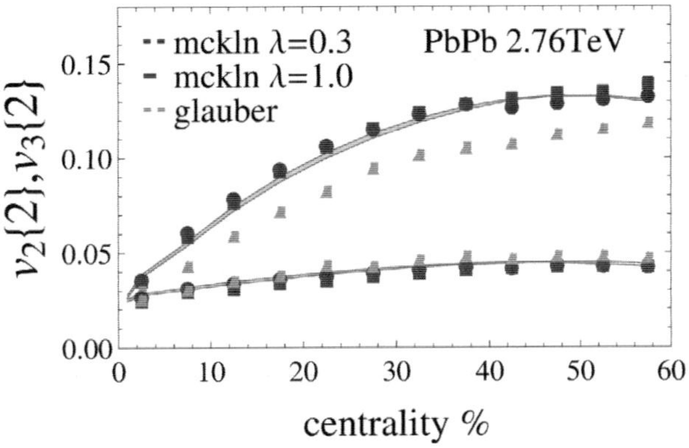

Figure 2. $v_2\{2\}$ and $v_3\{2\}$ for MCGlauber and MCKLN (with two different smoothing scales) at LHC.

even making predictions for the highest LHC energies [18]. While there has been some progress made in combining the knowledge of hydrodynamical backgrounds [19, 20] and fluctuating initial conditions [21] with energy loss models, it was not until [22] that full event-by-event viscous hydrodynamical backgrounds (v-USPhydro [23, 24]) were combined with an energy loss model (BBMG [5, 25, 26]). The effects of event-by-event fluctuations as well as the correlation between soft and hard flow harmonics in [22] provided the key step needed in order to solve the longstanding $R_{AA} \otimes v_2$ puzzle.

2. Event-by-event viscous hydrodynamics+energy loss

In [22] there were two crucial differences between all previous calculations of $v_2(p_T > 10GeV)$:

(i) Full event-by-event viscous hydrodynamical backgrounds were fed into the energy loss model to obtain v_2^{hard}.

(ii) $v_2^{exp}(p_T)$ from Eq. (1) was calculated correlating the soft integrated v_2^{soft} and hard $v_2^{hard}(p_T)$ to simulate the two particle (one soft, one hard) correlations used to measure elliptical flow experimentally.

The soft v_2^{soft} was taken using the best fitting parameters in the soft sector, see [22] for more details. The v_2^{hard} (the 2nd Fourier coefficient of $R_{AA}(p_T, \phi)$) is computed using the BBMG code [5, 25, 26] in the pQCD-like scenario. In order to make comparisons to experimental data then the soft and hard flow harmonics are correlated via

$$v_n^{exp}(p_T) = \frac{\langle v_n^{soft}\, v_n^{hard}(p_T) \cos\left[n\left(\psi_n^{soft} - \psi_n^{hard}(p_T)\right)\right]\rangle}{\sqrt{\left\langle \left(v_n^{soft}\right)^2 \right\rangle}}, \tag{1}$$

where $\langle \ldots \rangle$ denote event averages and ψ_n^{soft} and $\psi_n^{hard}(p_T)$ are the soft event plane angle and hard event plane angle, respectively, as was discussed in [27]. Note that if one assumes a smoothed, averaged hydrodynamical background then there is only one "event" and Eq. (1) reduces down to $v_n^{exp}(p_T)|_{(ic=avg)} = v_n^{hard}(p_T)$.

The enhancement of v_2 due to the soft-hard v_n correlation in Eq. (1) becomes clear if one considers linear response (a good approximation as long as one is not considering peripheral collisions [28]). In the soft sector we can assume $v_2^{soft} \sim c\,\varepsilon_2$ and for the hard sector since we have a p_T dependent quantity one takes $v_2^{hard} \sim \chi_2(p_T)\varepsilon_2$. Substituting them into Eq. (1), one finds

$$v_2^{exp}(p_T) \sim \chi_2(p_T)\sqrt{\langle \varepsilon_2^2 \rangle}. \tag{2}$$

Since $\langle \varepsilon_2 \rangle < \sqrt{\langle \varepsilon_2^2 \rangle}$ one can see that event-by-event fluctuations will always increase the flow harmonics (soft v_2 is also proportional to the root mean squared not the mean [29]). Note that the use of Eq. (1) is not the only source of enhancement for v_2, rather it is the combination of full event-by-event hydrodynamically expanding backgrounds with the correct calculation of Eq. (1) that leads to allows for a simultaneous description and solution of the $R_{AA} \otimes v_2$ puzzle.

3. Results

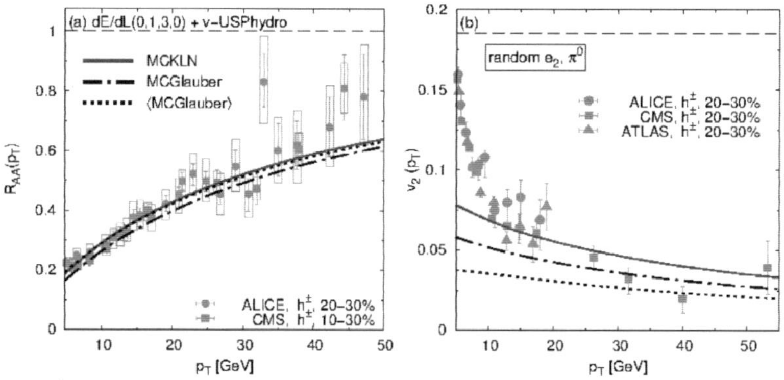

Figure 3. Comparison of event-by-event calculations in mid-central $\sqrt{s} = 2.76$ TeV Pb+Pb collisions of R_{AA} and v_2 from v-USPhydro+BBMG [22] to experimental data at LHC from ALICE [30, 32], CMS [31, 33], and ATLAS [34].

In Fig. 3 the result of the v-USPhydro+BBMG calculations are shown for R_{AA} and v_2 compared to ALICE [30, 32], CMS [31, 33], and ATLAS [34] data at the LHC. It is clear that the inclusion of event-by-event fluctuations with soft-hard v_2 correlation was a needed ingredient in order to bring up the elliptical flow to the experimental data. In fact, one can see in Fig. 3 that while R_{AA} is relatively insensitive to the effect of initial conditions and event-by-event fluctuations, v_2 has a clear splitting between MCGlauber and MCKLN initial conditions

Journal of Physics: Conference Series **736** (2016) 012019 doi:10.1088/1742-6596/736/1/012019

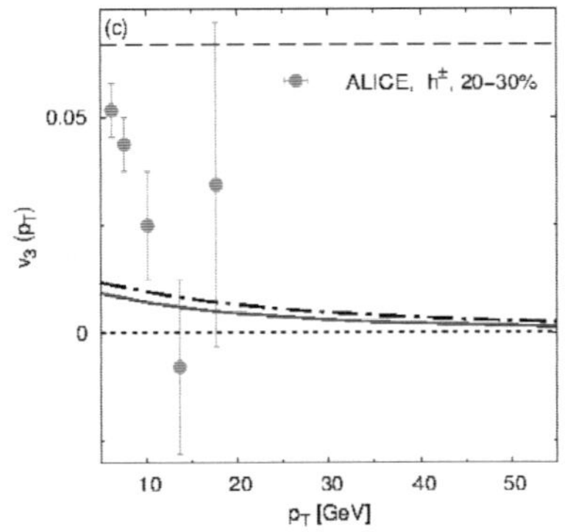

Figure 4. Comparison of event-by-event calculations in mid-central $\sqrt{s} = 2.76$ TeV Pb+Pb collisions of v_3 from v-USPhydro+BBMG [22] to experimental data at LHC from ALICE [32].

exactly as in the soft sector (see that in Fig. 2 the v_2 from MCKLN is always larger than from MCGlauber due to the larger eccentricities of MCKLN).

In fact, the similarities in the soft sector also hold true for v_3 at high p_T as seen in Fig. 4. MCKLN produces smaller ε_3 than MCGlauber, which correlates to a smaller v_3 both in the soft sector Fig. 2 and in the hard sector Fig. 4. Furthermore, the simple existence of a non-zero v_3 at high p_T demonstrates that event-by-event fluctuations are needed. Finally, an interesting avenue to explore for the future is the decorrelation between the soft and hard the event-planes for triangular flow (and above) [35].

4. Conclusions and Outlook

In this talk, it was shown that that event-by-event fluctuations combined with the proper calculation of v_2^{exp} at high p_T provided a natural solution to the $R_{AA} \otimes v_2$ puzzle. Event-by-event fluctuations have been clearly measured already at both RHIC and LHC and they have been shown to be necessary to describe v_3 [7], v_n distributions [17], event shape engineering [36], $SC(n,m)$ correlations [37, 38], and higher order $v_n\{m\}$ cumulants [39, 40, 41]. Thus, a proper comparison to high p_T v_2 experimental data necessarily requires the inclusion of event-by-event hydrodynamic fluctuations as well as the calculation of the soft-hard correlation in Eq. (1).

Once one accepts that event-by-event fluctuations also play a key role in the hard sector, it opens up an entire new field of possible calculations. In [22] it was suggested that event-shape engineering could be preformed also in the hard sector. Fig. 5 illustrates the underlying concept. For the same centrality class, e.g. density, one has a wide range of eccentricities ε_n's. It is already known that the shape of the event determines the path length seen by the jet, which is why centrality class selection is so important. However, using centrality class alone as an estimate for the path length is inherently flawed because within a single centrality class there are multiple possible path lengths depending on the eccentricities of the event. Thus, one possibility is to select on events with the same soft p_T flow harmonic in order to hold the path length constant within the centrality class. Additionally, comparisons can be made while holding the density fixed and changing only the path length by selecting on a variety of soft p_T flow harmonics to study the effect of energy loss for a very specific path length. Qualitatively, one can see from Fig. 5 that highly eccentric events have a shorter path length (since statistically its much more likely for a jet to cross the thin part of the event vs. the long part) whereas circular events have the longest average path length.

Experimental studies at ATLAS [42] and more recently at CMS [43] have already begun comparing soft-hard flow harmonics. Both demonstrate a very clear linear correlation between

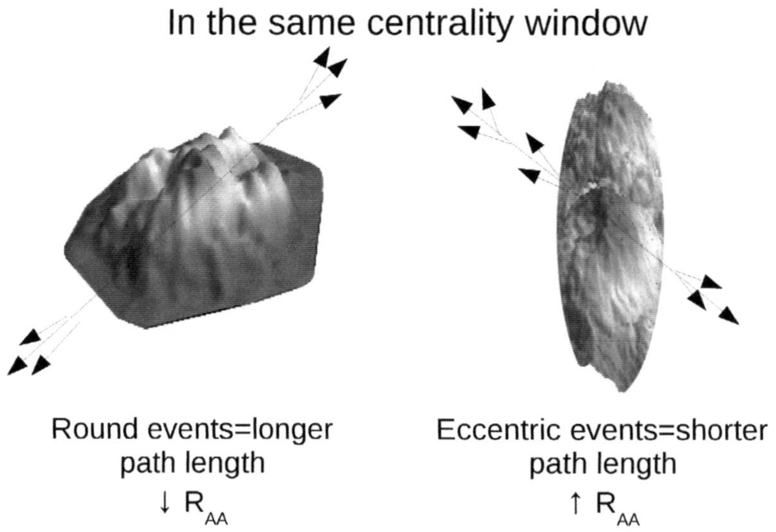

Figure 5. Within the same centrality class the path length can vary strongly on an event-by-event basis producing different R_{AA}'s.

the soft and hard v_2 (as shown in [22]) further solidifying the connection between v_2 and $\sqrt{\langle\varepsilon_2^2\rangle}$ in Eq. (2). Additionally, CMS [43] found a convergence of the cumulants of flow harmonics that is a good indication that there is collective flow even at high p_T.

The success of v-USPhydro+BBMG in fitting $v_2(p_T > 10GeV)$ makes one think whether the hard flow harmonics are also just as sensitive to the usual hydrodynamic parameters such as the initial time τ_0, the switching temperature T_{sw}, the transport coefficients etc. While the heavy flavor shows a strong dependence on the heavy quark drag coefficient [44], it also shows a very strong dependence on a temperature dependent shear viscosity to entropy density ratio $\eta/s(T)$ [45]. It is natural to wonder how a temperature dependent $\eta/s(T)$ would affect high p_T flow given its connection with the jet transport coefficient $\hat{q}$ [46]. Furthermore, using jet physics one can even extract a minimum at T_c [6] analogous to what is found in the hadronic phase [47, 48]. Since there is still a wide variation in $\eta/s(T)$ calculations [49] any additional constraint on the values of transport coefficients is sorely needed.

The most obvious step towards the future of combining soft and hard physics is to include hard scattering effects within relativistic hydrodynamics in order to study the intermediate p_T range of flow harmonics that are not possible with only relativistic hydrodynamics or energy loss models on their own. Initial work has been done in this direction [20, 50, 51] with the hope that this will be explored in more depth in the near future.

References

[1] A. Adare *et al.* [PHENIX Collaboration], Phys. Rev. Lett. **105**, 142301 (2010) doi:10.1103/PhysRevLett.105.142301 [arXiv:1006.3740 [nucl-ex]].
[2] X. N. Wang, Phys. Rev. C **63**, 054902 (2001) doi:10.1103/PhysRevC.63.054902 [nucl-th/0009019].
[3] M. Gyulassy, I. Vitev and X. N. Wang, Phys. Rev. Lett. **86**, 2537 (2001) doi:10.1103/PhysRevLett.86.2537 [nucl-th/0012092].
[4] E. V. Shuryak, Phys. Rev. C **66**, 027902 (2002) doi:10.1103/PhysRevC.66.027902 [nucl-th/0112042].
[5] B. Betz and M. Gyulassy, JHEP **1408**, 090 (2014) [JHEP **1410**, 043 (2014)] doi:10.1007/JHEP10(2014)043, 10.1007/JHEP08(2014)090 [arXiv:1404.6378 [hep-ph]].
[6] J. Xu, J. Liao and M. Gyulassy, Chin. Phys. Lett. **32**, no. 9, 092501 (2015) doi:10.1088/0256-307X/32/9/092501 [arXiv:1411.3673 [hep-ph]].

[7] B. Alver and G. Roland, Phys. Rev. C **81**, 054905 (2010) [Phys. Rev. C **82**, 039903 (2010)] doi:10.1103/PhysRevC.82.039903, 10.1103/PhysRevC.81.054905 [arXiv:1003.0194 [nucl-th]].
[8] S. Paiva, Y. Hama and T. Kodama, Phys. Rev. C **55**, 1455 (1997). doi:10.1103/PhysRevC.55.1455
[9] M. Gyulassy, D. H. Rischke and B. Zhang, Nucl. Phys. A **613**, 397 (1997) doi:10.1016/S0375-9474(96)00416-2 [nucl-th/9609030].
[10] C. E. Aguiar, Y. Hama, T. Kodama and T. Osada, Nucl. Phys. A **698**, 639 (2002) doi:10.1016/S0375-9474(01)01447-6 [hep-ph/0106266].
[11] O. Socolowski, Jr., F. Grassi, Y. Hama and T. Kodama, Phys. Rev. Lett. **93**, 182301 (2004) doi:10.1103/PhysRevLett.93.182301 [hep-ph/0405181].
[12] R. Andrade, F. Grassi, Y. Hama, T. Kodama and O. Socolowski, Jr., Phys. Rev. Lett. **97**, 202302 (2006) doi:10.1103/PhysRevLett.97.202302 [nucl-th/0608067].
[13] F. G. Gardim, F. Grassi, M. Luzum and J. Y. Ollitrault, Phys. Rev. C **85**, 024908 (2012) [arXiv:1111.6538 [nucl-th]].
[14] F. G. Gardim, J. Noronha-Hostler, M. Luzum and F. Grassi, Phys. Rev. C **91**, no. 3, 034902 (2015) doi:10.1103/PhysRevC.91.034902 [arXiv:1411.2574 [nucl-th]].
[15] H.-J. Drescher and Y. Nara, Phys. Rev. C **75**, 034905 (2007) doi:10.1103/PhysRevC.75.034905 [nucl-th/0611017].
[16] J. Noronha-Hostler, J. Noronha and M. Gyulassy, Phys. Rev. C **93**, no. 2, 024909 (2016) doi:10.1103/PhysRevC.93.024909 [arXiv:1508.02455 [nucl-th]].
[17] G. Aad *et al.* [ATLAS Collaboration], JHEP **1311**, 183 (2013) doi:10.1007/JHEP11(2013)183 [arXiv:1305.2942 [hep-ex]].
[18] J. Noronha-Hostler, M. Luzum and J. Y. Ollitrault, Phys. Rev. C **93**, no. 3, 034912 (2016) doi:10.1103/PhysRevC.93.034912 [arXiv:1511.06289 [nucl-th]].
[19] B. Betz, F. Senzel, C. Greiner and M. Gyulassy, arXiv:1512.07443 [hep-ph].
[20] J. Crkovska *et al.*, arXiv:1603.09621 [hep-ph].
[21] X. Zhang and J. Liao, Phys. Rev. C **87**, 044910 (2013) doi:10.1103/PhysRevC.87.044910 [arXiv:1210.1245 [nucl-th]].
[22] J. Noronha-Hostler, B. Betz, J. Noronha and M. Gyulassy, arXiv:1602.03788 [nucl-th].
[23] J. Noronha-Hostler, G. S. Denicol, J. Noronha, R. P. G. Andrade and F. Grassi, Phys. Rev. C **88**, 044916 (2013) [arXiv:1305.1981 [nucl-th]].
[24] J. Noronha-Hostler, J. Noronha and F. Grassi, Phys. Rev. C **90**, no. 3, 034907 (2014) [arXiv:1406.3333 [nucl-th]].
[25] B. Betz, M. Gyulassy and G. Torrieri, Phys. Rev. C **84**, 024913 (2011) doi:10.1103/PhysRevC.84.024913 [arXiv:1102.5416 [nucl-th]].
[26] B. Betz and M. Gyulassy, Phys. Rev. C **86**, 024903 (2012) doi:10.1103/PhysRevC.86.024903 [arXiv:1201.0281 [nucl-th]].
[27] M. Luzum and H. Petersen, J. Phys. G **41**, 063102 (2014) doi:10.1088/0954-3899/41/6/063102 [arXiv:1312.5503 [nucl-th]].
[28] J. Noronha-Hostler, L. Yan, F. G. Gardim and J. Y. Ollitrault, Phys. Rev. C **93**, no. 1, 014909 (2016) doi:10.1103/PhysRevC.93.014909 [arXiv:1511.03896 [nucl-th]].
[29] J. Y. Ollitrault, A. M. Poskanzer and S. A. Voloshin, Phys. Rev. C **80**, 014904 (2009) doi:10.1103/PhysRevC.80.014904 [arXiv:0904.2315 [nucl-ex]].
[30] B. Abelev *et al.* [ALICE Collaboration], Phys. Lett. B **720**, 52 (2013).
[31] S. Chatrchyan *et al.* [CMS Collaboration], Eur. Phys. J. C **72**, 1945 (2012).
[32] B. Abelev *et al.* [ALICE Collaboration], Phys. Lett. B **719**, 18 (2013).
[33] S. Chatrchyan *et al.* [CMS Collaboration], Phys. Rev. Lett. **109**, 022301 (2012).
[34] G. Aad *et al.* [ATLAS Collaboration], Phys. Lett. B **707**, 330 (2012).
[35] J. Jia, Phys. Rev. C **87**, no. 6, 061901 (2013) doi:10.1103/PhysRevC.87.061901 [arXiv:1203.3265 [nucl-th]].
[36] J. Adam *et al.* [ALICE Collaboration], Phys. Rev. C **93**, no. 3, 034916 (2016).
[37] J. Adam *et al.* [ALICE Collaboration], arXiv:1604.07663 [nucl-ex].
[38] G. Giacalone, L. Yan, J. Noronha-Hostler and J. Y. Ollitrault, arXiv:1605.08303 [nucl-th].
[39] S. Chatrchyan *et al.* [CMS Collaboration], Phys. Rev. C **87**, no. 1, 014902 (2013) doi:10.1103/PhysRevC.87.014902 [arXiv:1204.1409 [nucl-ex]].
[40] B. B. Abelev *et al.* [ALICE Collaboration], Phys. Rev. C **90**, no. 5, 054901 (2014) doi:10.1103/PhysRevC.90.054901 [arXiv:1406.2474 [nucl-ex]].
[41] G. Aad *et al.* [ATLAS Collaboration], Eur. Phys. J. C **74**, no. 11, 3157 (2014) doi:10.1140/epjc/s10052-014-3157-z [arXiv:1408.4342 [hep-ex]].
[42] G. Aad *et al.* [ATLAS Collaboration], Phys. Rev. C **92**, no. 3, 034903 (2015) doi:10.1103/PhysRevC.92.034903 [arXiv:1504.01289 [hep-ex]].

[43] CMS Collaboration [CMS Collaboration], CMS-PAS-HIN-15-014.
[44] S. K. Das, F. Scardina, S. Plumari and V. Greco, Phys. Lett. B **747**, 260 (2015) doi:10.1016/j.physletb.2015.06.003 [arXiv:1502.03757 [nucl-th]].
[45] R. Esha, M. Nasim and H. Z. Huang, arXiv:1603.02700 [nucl-th].
[46] A. Majumder, B. Muller and X. N. Wang, Phys. Rev. Lett. **99**, 192301 (2007) doi:10.1103/PhysRevLett.99.192301 [hep-ph/0703082].
[47] J. Noronha-Hostler, J. Noronha and C. Greiner, Phys. Rev. Lett. **103**, 172302 (2009) doi:10.1103/PhysRevLett.103.172302 [arXiv:0811.1571 [nucl-th]].
[48] J. Noronha-Hostler, J. Noronha and C. Greiner, Phys. Rev. C **86**, 024913 (2012) doi:10.1103/PhysRevC.86.024913 [arXiv:1206.5138 [nucl-th]].
[49] J. Noronha-Hostler, arXiv:1512.06315 [nucl-th].
[50] R. P. G. Andrade, J. Noronha and G. S. Denicol, Phys. Rev. C **90**, no. 2, 024914 (2014) doi:10.1103/PhysRevC.90.024914 [arXiv:1403.1789 [nucl-th]].
[51] L. Pang, Q. Wang and X. N. Wang, Phys. Rev. C **86**, 024911 (2012) doi:10.1103/PhysRevC.86.024911 [arXiv:1205.5019 [nucl-th]].

Journal of Physics: Conference Series **736** (2016) 012020 doi:10.1088/1742-6596/736/1/012020

Rapidity Correlation Structures from Causal Hydrodynamics

Sean Gavin[1], **George Moschelli**[2], **and Christopher Zin**[1]

[1]Department of Physics and Astronomy, Wayne State University, Detroit, MI, 48202, USA
[2]Lawrence Technological University, 21000 West Ten Mile Road, Southfield, MI 48075, USA

E-mail: [1]sean.gavin@wayne.edu, [2]gmoschell@ltu.edu, [1]aj3980@wayne.edu

Abstract. Viscous diffusion can broaden the rapidity dependence of two-particle transverse momentum fluctuations. Surprisingly, measurements at RHIC by the STAR collaboration demonstrate that this broadening is accompanied by the appearance of unanticipated structure in the rapidity distribution of these fluctuations in the most central collisions. Although a first order classical Navier-Stokes theory can roughly explain the rapidity broadening, it cannot explain the additional structure. We propose that the rapidity structure can be explained using the second order causal Israel-Stewart hydrodynamics with stochastic noise.

1. Introduction

Many experimental measurements demonstrate the profound effect that initial state fluctuations of nuclear collisions have on the subsequent dynamics of the system, but there is growing awareness that fluctuations and their dissipation also occur throughout the evolution. In earlier work we suggested that viscous diffusion broadens the rapidity dependence of two-particle transverse momentum correlations [1]. That work stimulated an experimental analysis by STAR [2] and the discovery that the rapidity structure of these correlations not only broadens but also develops a distinctively non-Gaussian structure not predicted by ref. [1].

We employ second order hydrodynamics with stochastic noise to develop an evolution equation for two particle transverse momentum correlations [3]. In this work we will compare rapidity correlation structures evolving from first and second order diffusion. In sec. 2 we briefly derive evolution equations for the transverse momentum correlation function as well as describe how this correlation function relates to experimental observables. In the interest of brevity, we leave the rigorous derivations to [3]. Instead we focus on the results in sec. 3. We find that second order hydrodynamics is required to explain the measured rapidity correlation structure and consequently that these measurements can constrain a transport coefficient that is sensitive to the thermalization process.

2. Fluctuations and Correlations

Nuclear collisions produce a high energy density fluid that flows outward with an average transverse velocity v_r. Small deviations of the flow occur in each event. These deviations perturb the transverse momentum current of the fluid by an amount $g_t = T_{0r} - \langle T_{0r}\rangle$, where $T^{\mu\nu}$ is the stress energy tensor. Viscous friction arises as neighboring fluid elements flow past one another. This friction reduces velocity fluctuations, driving the velocity toward v_r and the momentum deviations g_t to zero. The final size of the fluctuations depends on the magnitude of the viscosity and the lifetime of the fluid.

We consider fluctuations of a fluid at rest with energy density e and pressure p. Small fluctuations produce a small velocity $\mathbf{v}$ corresponding to a momentum current $\mathbf{M} \approx (e+p)\mathbf{v}$. To linear order in the fluctuations, we write the conservation form of the relativistic Navier-Stokes equation:

$$\frac{\partial}{\partial t}\mathbf{M} + \boldsymbol{\nabla} p = \frac{\zeta + \frac{1}{3}\eta}{w}\boldsymbol{\nabla}(\nabla \cdot \mathbf{M}) + \frac{\eta}{w}\nabla^2\mathbf{M} \tag{1}$$

where η and ζ are the shear and bulk viscosity coefficients and $w = e+p$ is the enthalpy density. The momentum fluctuation density can be written in terms of curl free longitudinal modes $\mathbf{g}_l$ and divergence free shear modes $\mathbf{g}$ so we have $\mathbf{M} = \mathbf{g}_l + \mathbf{g}$, where $\nabla \times \mathbf{g}_l = 0$ and $\nabla \cdot \mathbf{g} = 0$. By taking the curl of (1) we find the shear modes satisfy

$$\frac{\partial}{\partial t}\mathbf{g} = \nu\nabla^2\mathbf{g}, \tag{2}$$

where $\nu = \eta/w$ is the kinematic viscosity. Equation (2) shows that Fick's Law holds for the density of transverse momentum fluctuations from local equilibrium.

We look for the diffusion of *correlations* of the fluctuations $\mathbf{g}$ at different points in the fluid $r_g = \langle \mathbf{g}_1\mathbf{g}_2\rangle - \langle \mathbf{g}_1\rangle\langle \mathbf{g}_2\rangle$. In equilibrium, thermodynamic noise will cause such correlations to occur so $r_g^{le} \neq 0$ and the interesting correlations are those that differ from the equilibrium value $\Delta r_g = r_g - r_g^{le}$. It turns out that for two particle momentum correlations (2) becomes $\left[\frac{\partial}{\partial t} - \nu\left(\nabla_1^2 + \nabla_2^2\right)\right]\Delta r_g = 0$. For a detailed derivation please see ref. [3].

We assume the event-averaged flow velocity has the Bjorken form, $u^\mu = (t/\tau, 0, 0, z/\tau)$, where $\tau = (t^2 - z^2)^{1/2}$ is the proper time and $\eta = (1/2)\log((t+z)/(t-z))$ is spatial rapidity. In an expanding system, observe that the rapidity density of total momentum $G \equiv \int g\tau dx_\perp$, where the integral is over the transverse area of the two colliding nuclei. If one identifies spatial rapidity η with the momentum-space rapidity of particles, then G is observable and we look for correlations of the form

$$r_G = \langle G(\mathbf{x}_1)G(\mathbf{x}_2)\rangle - \langle G(\mathbf{x}_1)\rangle\langle G(\mathbf{x}_2)\rangle \tag{3}$$

Additionally we switch coordinates to relative $\eta_r = \eta_1 - \eta_2$ and average $\eta_a = (\eta_1+\eta_2)/2$ rapidity and finally we have our first order result

$$\left[\frac{\partial}{\partial \tau} - \frac{\nu}{\tau^2}\left(2\frac{\partial^2}{\partial \eta_r^2} + \frac{1}{2}\frac{\partial^2}{\partial \eta_a^2}\right)\right]\Delta r_G = 0. \tag{4}$$

Second order hydrodynamics is especially important for diffusive phenomena, where it renders the theory causal. In first order diffusion (2), a delta function perturbation instantaneously spreads into a Gaussian, with tails extending to infinity. New transport coefficients at second order include relaxation times for shear and bulk stresses, among other terms. Linearized forms of the second order equations are discussed in [4, 5]. To linear order the shear modes satisfy a Maxwell-Cattaneo equation

$$\left(\tau_\pi\frac{\partial^2}{\partial t^2} + \frac{\partial}{\partial t}\right)\mathbf{g} = \nu\nabla^2\mathbf{g} \tag{5}$$

where the transport coefficient τ_π is a relaxation time for the shear modes. As with the first order case, we look to solve (5) for the correlations (3). We find

$$\left[\frac{\tau_\pi}{2}\frac{\partial^2}{\partial \tau^2} + \frac{\partial}{\partial \tau} - \frac{\nu}{\tau^2}\left(2\frac{\partial^2}{\partial \eta_r^2} + \frac{1}{2}\frac{\partial^2}{\partial \eta_a^2}\right)\right]\Delta r_G = 0. \tag{6}$$

For a detailed derivation please refer to ref. [3]. Observe that (6) reduces to the first order result (4) when $\tau_\pi = 0$. Further observe that (6) has features of both wave and diffusion equations. As we will discuss in sec. 3, at early times, fluctuation signals propagate as waves while at later times diffusion fills in the space between the wave fronts. It is precisely this behavior that delays the growth of the rapidity correlation structure in comparison to the first order case.

The correlation function Δr_G is observable by measuring the covariance

$$\mathcal{C} = \langle N \rangle^{-2} \langle \sum_{a \neq b} p_{t,a} p_{t,b} \rangle - \langle p_t \rangle^2 = \langle N \rangle^{-2} \int \Delta r_G(\eta_r, \eta_a) d\eta_r d\eta_a, \tag{7}$$

where a and b label particles from each event, and the brackets represent the event average. The average transverse momentum is $\langle p_t \rangle \equiv \langle \sum_a p_{t,a} \rangle / \langle N \rangle$. In the absence of correlations $\mathcal{C} = 0$, as is the case for local equilibrium in an infinite system.

The STAR collaboration at RHIC reports a differential version of the quantity $\mathcal{C}$ as a function of relative pseudorapidity η_r and azimuthal angle ϕ_r of pairs:

$$\mathcal{C}(\eta_r, \phi_r) = (\langle N \rangle_1 \langle N \rangle_2)^{-1} \left\langle \sum_{a \neq b} p_{t,a} p_{t,b} \right\rangle_{1,2} - \langle p_t \rangle_1 \langle p_t \rangle_2, \tag{8}$$

where the numbers $\langle N \rangle_k$ and $\langle p_t \rangle_k$ refer to the particle number and transverse momentum in (η_k, ϕ_k) bins for particles $k = 1, 2$ [2]. We will discuss this observable in detail in the next section.

3. Results

We will now explore the rapidity dependence of transverse momentum fluctuations. In an earlier work we show that the relative rapidity width, σ, of (8) can be related to the shear viscosity of the system using first order diffusion [1]. Here we discuss how the diffusion equation derived from second order hydrodynamics, (6), modifies this result. In particular we demonstrate that the first order theory cannot simultaneously explain both the width and shape of (8) in relative rapidity; where the first order theory fails, the second order theory succeeds. Interestingly, while the kinematic viscosity $\nu = \eta/Ts$ remains the dominant factor in determining the width, the relaxation time of the second order theory, τ_π, is the primary influence on the shape.

In Au+Au collisions at the top RHIC energy, STAR has measured $\mathcal{C}$ in both its differential form (8) and its relative rapidity width [2]. The measured differential shape of (8) resembles that of similar measurements of the "ridge" without p_t weights. In Ref. [2] authors interpret the near side, $|\phi_r| < 1$, of (8) as a peak sitting on a flat pedestal and fit the ϕ_r integrated profile $\mathcal{C}(\eta_r)$ to extract the magnitude of the pedestal. The measured width of $\mathcal{C}(\eta_r)$ is shown in fig. 1. Solid circles represent the RMS width in η_r of (8) integrated over an azimuthal range of $|\phi_r| < 1$ radians. The gray error band in fig. 1 represents the uncertainty in their fit procedure. Figure 4 shows the $\mathcal{C}(\eta_r)$ profile from which these widths are calculated. Open stars are from Ref. [2] and solid circles are from Ref. [6]. In all cases we have subtracted the pedestal.

The flat pedestal might be interpreted as the long-range correlations usually associated with the ridge while shorter range correlations would influence the shape of the peak sitting on the pedestal. Our calculations here specifically address the short range correlations. As discussed in Ref. [1], the growth of the width of this peak from peripheral to central collisions can be used to extract the viscosity. To compute the width in the second order theory, we follow ref. [7] and multiply (6) by η_r^n. Next, we integrate over η_r and η_a and use $\int \eta_r^n \partial^2 \Delta r_G / \partial \eta_r^2 = n(n-1) \int \eta_r^{n-2} \Delta r_G$, which is nonzero only for $n \geq 2$. We find

$$\left(\frac{\tau_\pi}{2} \frac{d^2}{d\tau^2} + \frac{d}{d\tau} \right) A \langle \eta_r^n \rangle = \frac{2\nu}{\tau^2} n(n-1) A \langle \eta_r^{n-2} \rangle, \tag{9}$$

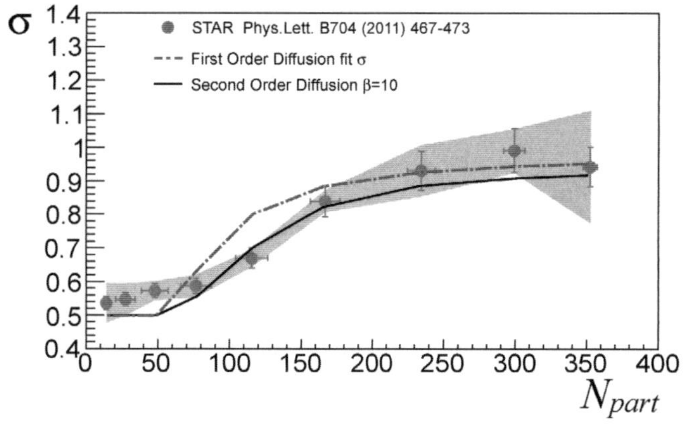

Figure 1. Rapidity width of $\mathcal{C}(\eta_r)$ as a function of the number of participants for second order momentum diffusion calculations (solid curve) compared to first order results (dot-dashed curve). Data (solid circles) from STAR include shaded area to denote the systematic uncertainty in the fit procedure [2].

where $\langle \eta_r^n \rangle = A^{-1} \int \eta_r^n \Delta r_G d\eta_r d\eta_a$ are the normalized moments of the rapidity correlation function. The amplitude A and the mean $\langle \eta_r \rangle$ both satisfy (9) with the right side equal to zero. The vanishing first moment $\langle \eta_r \rangle = 0$ makes physical sense if we assume a symmetric system. Therefore, the width can then be described as $\sigma^2 = \langle \eta_r^2 \rangle - \langle \eta_r \rangle^2 = \langle \eta_r^2 \rangle$ and following from (9) we obtain the differential equation for the width,

$$\left(\frac{\tau_\pi}{2} \frac{d^2}{d\tau^2} + \frac{d}{d\tau} \right) \sigma^2 = \frac{4\nu}{\tau^2}. \tag{10}$$

The solution of (10) yields the time dependence of the broadening of $\mathcal{C}(\eta_r)$. We can observe that we should expect that longer lived collisions have broader rapidity distribution by examining the first order case where $\tau_\pi = 0$. The solution of (10) is then

$$\sigma^2 = \sigma_0^2 + \frac{4\nu}{\tau_0} \left(1 - \frac{\tau_0}{\tau} \right), \tag{11}$$

as found in Ref. [1]. Here $\nu = \eta/Ts = const$ is the kinematic viscosity, τ_0 is the formation time, and σ_0 is the width in peripheral collisions. STAR used this method to estimate the average shear viscosity to entropy density ratio to be $\eta/s = 0.13 \pm 0.03$ [2].

To find the rapidity width for second order diffusion using (10) we must specify an initial condition for $d\sigma^2/d\tau \equiv \theta_0^2$ at $\tau = \tau_0$, the value of which is unknown. We examine two cases, one where $\theta_0^2 = 0$ which assumes that no modifications to the correlation function (3) occur prior to the formation time, and one that attempts to account for the effects of initial state expansion. In this proceedings we focus on the latter case and direct the reader to Ref. [3] for a detailed comparison of these two conditions. For (6), we take the initial correlation function to satisfy

$$\left. \frac{\partial \Delta r_G}{\partial \tau} \right|_{\tau=\tau_0} = \frac{\nu_0}{\tau_0^2} \left(2\frac{\partial^2}{\partial \eta_r^2} + \frac{1}{2}\frac{\partial^2}{\partial \eta_a^2} \right) \Delta r_G, \tag{12}$$

which corresponds to $\theta_0^2 = 4\nu/\tau_0^2$ in (10). Using (12) and solving (10) we find

$$\sigma^2 = \sigma_0^2 + \frac{\theta_0^2 \tau_\pi}{2} \left(1 - e^{-2(\tau-\tau_0)/\tau_\pi} \right) + \frac{8\nu}{\tau_\pi} \int_{\tau_0}^{\tau} du \int_{\tau_0}^{u} \frac{ds}{s^2} e^{2(s-u)/\tau_\pi}. \tag{13}$$

The solid black curve in fig. 1 shows the value of (13) at the freeze out time parameterized as

$$\tau_F - \tau_0 = K(R(N_{part}) - R_0)^2 \tag{14}$$

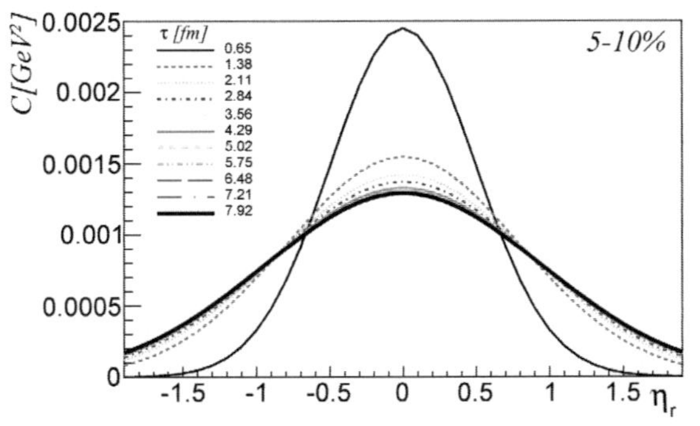

Figure 2. Time evolution of the rapidity structure of transverse momentum correlations from first order diffusion (4) for 5 − 10% central collisions.

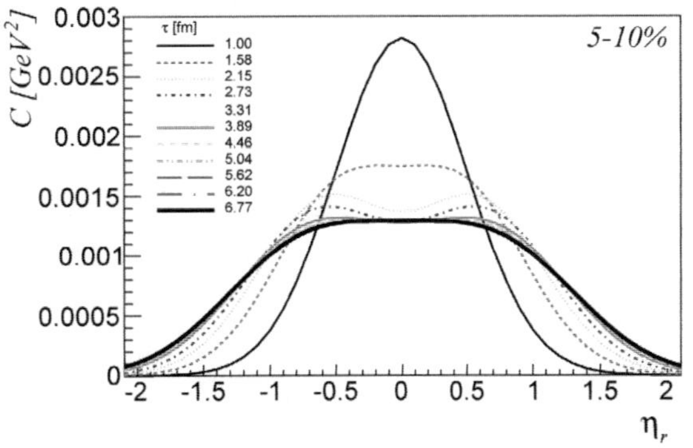

Figure 3. Time evolution of the rapidity structure of transverse momentum correlations from second order diffusion (6) for 5 − 10% central collisions.

where τ_0 is the formation time and R_0 is roughly the proton size. We compute N_{part} and R from a Glauber model and fix the constant K so that the freeze out time in the most central collisions has a specified value τ_{Fc}. Again we take $\nu = \eta/Ts = const$ for $\eta/s = 1/4\pi$, but now with $T = 150$ MeV for all centralities. We must now specify the second-order relaxation time $\tau_\pi = \beta\nu$, for which we take $\beta = 10$. The values $\tau_0 = 1.0$ fm and $\tau_{Fc} = 10$ fm then give superb agreement with data.

The dash-dot curve shows our best fit to this data using the first order result (11) evaluated at τ_F, eq. (14). Again $\eta/s = 1/4\pi$ but we take the freeze out temperature to be $T = 140$ MeV for all centralities. Values of the space time parameters $\tau_0 = 0.65$ fm and $\tau_{Fc} = 12$ fm then specify the lifetime (14). One might argue the rough agreement in fig. 1 for the first order result is compelling were it not for the fact that it consistently overestimates the data in the region where the data grows the fastest. It is a feature of first order diffusion that it occurs a-causally; correlations diffuse too rapidly. Hence, we see that collision systems that freeze out at earlier times disagree with the first order case while more central, longer lived collision systems match with the asymptotic limit of (11). For this reason the estimate of the shear viscosity in Ref. [2] is still on strong footing since authors used only the difference in central and most peripheral widths.

The asymptotic value of the first order solution (11) is

$$\sigma_\infty^2 = \sigma_0^2 + \frac{4\nu}{\tau_0}. \tag{15}$$

This saturation of the rapidity width to the value (15) is a straightforward consequence of Bjorken flow. In a stationary liquid, a spike in momentum diffuses over a range $\sim (2\nu t)^{1/2}$ that grows with time t. Bjorken expansion of the underlying fluid stretches the longitudinal scale $\propto t$, rapidly overtaking diffusion and "freezing in" the initial inhomogeneity. Correspondingly, one can observe this same behavior in (6). At long times the rightmost term vanishes, meaning the rapidity of particles ceases to change due to shear forces. Consequently, the rapidity width from second order diffusion, in the case of $\theta_0^2 = 0$, will reach the same asymptotic value. Importantly, the difference between the first and second order cases is that in the second order case the growth of the rapidity width, or equivalently the evolution of the correlation function (3), is delayed according to the relaxation time τ_π.

The freeze out time plays an important role in selecting the state of the correlation function measured by the experiment. For constant ν and τ_π all centralities follow the same trajectory

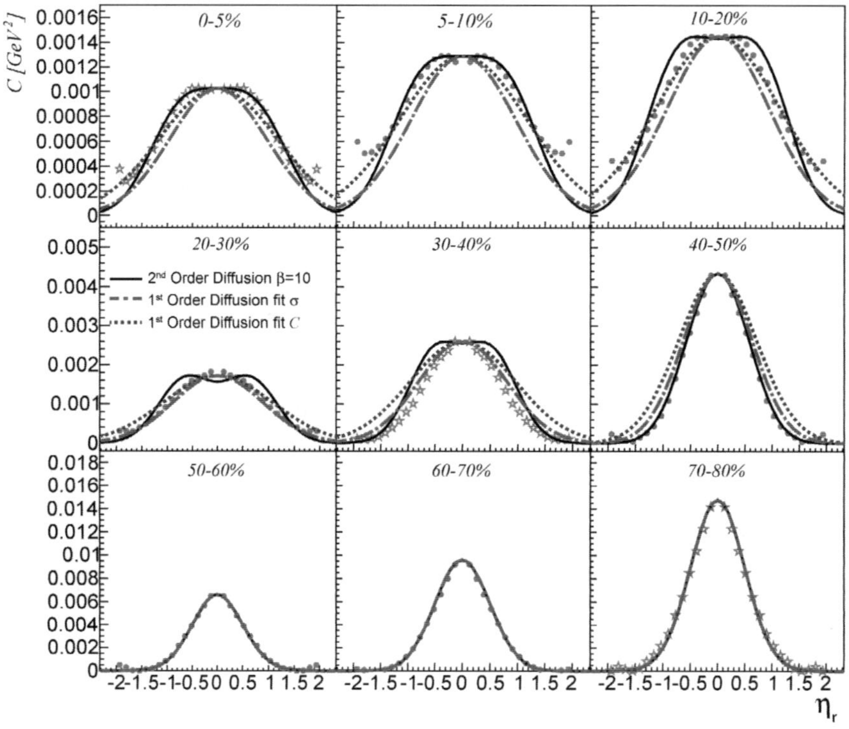

Figure 4. Second order momentum diffusion calculations (solid curves) compared to the rapidity dependence of the measured covariance (8). First order calculations are also compared for best fit to σ in fig. 1 (dot-dashed curves) and the best fit to $\mathcal{C}(\eta_r)$ (dashed curves). Data (open stars) are from [2] and (filled circles) from [6]. Percentages of the cross section indicate centrality, with each panel corresponding to a width measurement in fig. 1.

in the solution of (6) and consequently (10). If ν and τ_π change throughout the evolution, each collision would have a unique solution to (6), but we leave that to future work. Figures 2 and 3 show the time evolution of the correlation function for $5-10\%$ central collisions for the first and second order cases respectively. In both cases we assume the initial transverse momentum correlation function to be $\Delta r_G(\eta_r, \eta_a, \tau_0) = Ae^{-\eta_r^2/2\sigma_0^2}e^{-\eta_a^2/2\Sigma_0^2}$. This distribution is motivated by the rapidity dependence of measured correlation functions for multiplicity and net charge in pp collisions. We set the initial width in relative rapidity, σ_0 to fit the most peripheral distribution in fig. 4. Furthermore, we assume there is insufficient time for significant evolution in the three most peripheral cases in fig. 4. The data supports this claim and gives a consistent value of $\sigma_0 = 0.50$. The average pseudo-rapidity width $\Sigma_0 \approx 5-6$ units is assumed to be a "large" value relative to the size of experimental acceptance. We will take A to fit the peak value of the measured $\mathcal{C}$. This parameter has little impact on our current study, since we are only concerned with the shape of the function. We use (12) for the initial value of the first derivative.

The striking difference between figures 2 and 3 is our primary result. In the first order case, diffusion is the only process taking place; starting from an initial Gaussian shape the rapidity distribution of correlations will always remain Gaussian. In the second order case, fluctuation signals propagate in opposite directions with a wave speed of $v = \sqrt{\nu/\tau_\pi}$ in Cartesian coordinates. In rapidity coordinates, as longitudinal expansion overtakes the signals, fluctuations are frozen into the final distribution. Depending on what time this takes place a flattened or even

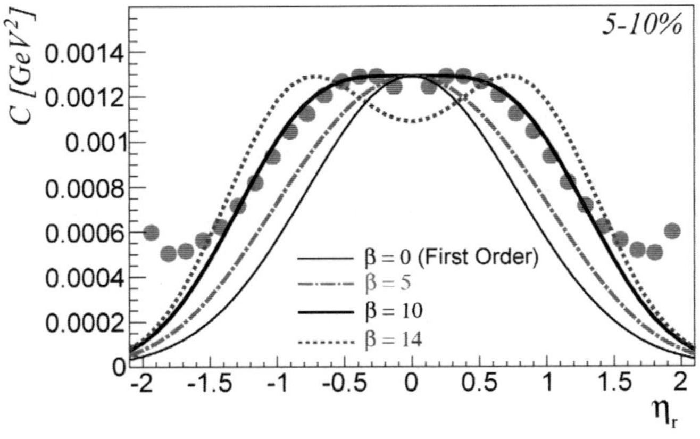

Figure 5. Rapidity structure of transverse momentum correlations at freeze out from second order diffusion (6). Different values of $\tau_\pi = \beta\nu$ are chosen by changing the parameter β. Results are selected for $5-10\%$ central collisions and compared to data [2, 6].

bimodal peak structure is visible. Indeed STAR's discovery of this type of rapidity structure in the measurement of (8), shown in fig. 4, is the motivation for this work.

In fig. 4 we show comparisons of first and second order cases to measured data (with the pedestal subtracted). The most noticeable feature is that in central collisions there is a flattening of the peak and even hints of a double bump structure. First order diffusion will never be able to achieve this shape regardless of choices of parameters. We illustrate this with two choices of parameter sets. For the dot-dashed curves we use the same parameters chosen to fit the width in fig. 1. Agreement is poor. Since the first order shape is always Gaussian, much of the value of the width will depend on how the tails are truncated to match the acceptance of the data. If we ignore the acceptance constraints and just try to match $C(\eta_r)$ in fig. 4 as best as possible we find the dashed curves for parameter values $\eta/s = 1/4\pi$, $T = 110$ MeV, $\tau_0 = 0.50$ fm, and $\tau_{Fc} = 10$ fm. Agreement with the measured shape is still quite poor. The solid black curves use the second order case with the same parameters as used in fig. 1. Agreement with data is much better than either of the first order cases and we see signs of the flattening of the peak in central collisions.

The flattening of the peak is definitively a second order effect that is modulated by the relaxation time $\tau_\pi = \beta\nu$. This suggests that the experiment has access to this relaxation time. In fig. 5 we examine the effect of τ_π on our $5 - 10\%$ centrality result. In all cases we use the same parameter choices as used in in figs. 1, 3, and 4 with the exception of β. Previously for all second order results we chose $\beta = 10$ since it yielded the most consistently strong agreement with all centralities. This solution is represented as the thick black line in fig. 5. Incidentally, kinetic theory of massless Boltzmann particles predicts a value of $\beta = 5$. Larger values of β correspond to slower propagation of fluctuation signals in the medium, a fact that could be used to characterize the nature of the medium. The slower signal propagation associated with increasing β means that the system takes longer to reach the asymptotic diffusion limit and longitudinal expansion "freezes" the correlation structure earlier in its development. As we can see in fig. 3, at earlier times the wave nature of the structure still dominates the correlation shape. So, for larger β values we should see an emergence of a more well defined double bump structure, as we see in the dashed line in fig. 5 corresponding to $\beta = 14$.

We remark that the resolution of the double peak structure also depends on the narrowness of the initial width σ_0. If the initial correlation structure is wide then we may never see a well defined double peak structure, but that structure will still be decidedly non-Gaussian. Conversely if the initial structure is very narrow we could still see a double peak structure with smaller β values.

Journal of Physics: Conference Series **736** (2016) 012020 doi:10.1088/1742-6596/736/1/012020

4. Conclusions

In this paper we study the rapidity correlation structure of transverse momentum correlations in nuclear collisions (7). In earlier work we suggested how these correlations could be used to study viscosity [1]. In an effort to obtain an estimate of the shear viscosity to entropy density ratio, STAR measured a differential version of these correlations (8) and, in doing so, discovered that the rapidity shape has a distinctive non-Gaussian flattening in the more central centrality classes. In ref. [3] we use second order hydrodynamics with stochastic noise to develop a second order diffusion equation (6) for the evolution of transverse momentum fluctuations that lead to the correlations (7).

Comparisons of first and second order diffusion results to data in figs. 1 and 4 suggest that first order diffusion cannot be used to simultaneously explain both the broadening of and the emergence of the non-Gaussian shape of the correlation structure from peripheral to central collisions. We attribute this failure to the a-causal nature of first order diffusion where fluctuation signals can propagate at infinite speeds throughout the medium. In second order diffusion fluctuation signals propagate as wave fronts with speed $\sqrt{\nu/\tau_\pi}$ where $\nu = \eta/Ts$ is the kinematic viscosity and $\tau_\pi = \beta\nu$ is the relaxation time related to the transport of the fluctuation signal through the medium. We solve the second order diffusion equation (6) for constant ν and τ_π to find excellent agreement with data using a value of $\beta = 10$. We further find that the kinematic viscosity is the primary factor in determining the broadening of the correlation structure and τ_π has significant influence over the flattening or even bimodal nature of the correlation peak.

The emergence of a flattened peak in the rapidity correlation structure and the natural way in which second order hydrodynamics accommodates this behavior suggests that experiments have the ability to measure the relaxation time τ_π. Since τ_π is the characteristic time for the dissipation of fluctuations in the medium, studying this quantity can yield information on the thermalization process.

Acknowledgments

We thank Rajendra Pokharel for collaboration in the early stages of this work. Special thanks to Monika Sharma and Claude Pruneau for discussing the STAR data. We thank Victoria Drolshagen, Mauricio Martinez, Jaki Noronha-Hostler, Jorge Noronha, Scott Pratt, and Clint Young for discussions. This work was supported in part by the U.S. NSF grant PHY-1207687.

References

[1] S. Gavin and M. Abdel-Aziz, "Measuring Shear Viscosity Using Transverse Momentum Correlations in Relativistic Nuclear Collisions," *Phys.Rev.Lett.* **97** (2006) 162302.

[2] **STAR** Collaboration, H. Agakishiev *et al.*, "Evolution of the differential transverse momentum correlation function with centrality in Au+Au collisions at $\sqrt{s_{NN}} = 200$ GeV," *Phys.Lett.* **B704** (2011) 467–473.

[3] S. Gavin, G. Moschelli, and C. Zin, "Rapidity Correlation Structure in Nuclear Collisions," arXiv:1606.02692 [nucl-th].

[4] P. Romatschke, "New Developments in Relativistic Viscous Hydrodynamics," *Int. J. Mod. Phys.* **E19** (2010) 1–53.

[5] C. Young, J. I. Kapusta, C. Gale, S. Jeon, and B. Schenke, "Thermally Fluctuating Second-Order Viscous Hydrodynamics and Heavy-Ion Collisions," *Phys. Rev.* **C91** no. 4, (2015) 044901.

[6] *M. Sharma and C. Pruneau (private communication, 2010).* .

[7] M. A. Aziz and S. Gavin, "Causal diffusion and the survival of charge fluctuations in nuclear collisions," *Phys. Rev.* **C70** (2004) 034905.

Journal of Physics: Conference Series **736** (2016) 012021 doi:10.1088/1742-6596/736/1/012021

Exploring Jet-Hadron correlations at the LHC with ALICE

Joel Mazer

Department of Physics, University of Tennessee, Knoxville, Tennessee 37996-1200, USA

E-mail: joel.mazer@cern.ch

Abstract. In relativistic heavy ion collisions at the Large Hadron Collider (LHC), the conditions are met to produce the hot and dense, strongly interacting medium known as the Quark Gluon Plasma (QGP). The QGP, a state of matter created shortly after the Big Bang, is a phase where the deconfinement of quarks and gluons is hypothesized. Jets, the collimated sprays of hadrons from fragmenting partons, are a key probe of the medium. The experimental methods used for jet measurements at ALICE to remove, reduce, and correct for the underlying background event will be presented. In pp collisions, jet production is well understood within the framework of perturbative QCD and acts as a rigorous baseline measurement for jet quenching measurements. By comparing to heavy ion collision systems, we can study the suppression of the number of jets seen and study the modification of the p_T or angular distributions of jet fragments. Azimuthal angular correlations of charged hadrons with respect to the axis of a full (charged + neutral) reconstructed (trigger) jet in Pb–Pb and pp collisions at $\sqrt{s_{NN}} = 2.76$ TeV in ALICE will be presented here. Newly developed combinatoric background subtraction methods and their improvement compared to prior techniques will be discussed.

1. Introduction

In heavy ion collisions, a hot and dense QCD medium is formed [1, 2]. Jets are an ideal probe because the hard-scattered partons which produce them are created very early in the collision, prior to formation of the medium, or Quark Gluon Plasma (QGP). Partons resulting from the hard scatterings in heavy ion collisions are modified in the presence of a medium through induced gluon radiation and collisional energy loss [3, 4]. This modification was observed at both RHIC and LHC energies via the suppression of high-momentum particles [5–8]. Jets can also be used as a probe by studying the suppression of the number of jets in Pb–Pb collisions compared to baseline measurements in pp or pPb collisions. A similar suppression can also be observed for high p_T di-hadrons correlations [9–12].

Measurements of jets in pp collisions offer a reference for comparing the medium induced jet modifications in heavy ion collisions. The production cross section of jets in pp collisions are calculable using perturbative Quantum Chromodynamics (pQCD). Jet measurements in pp collisions also allow for a quantitative understanding of the impact of hadronization on reconstructed jets [13].

2. Experimental setup

In ALICE, jets are reconstructed using charged particles measured with the central tracking system and clusters measured in the Electromagnetic Calorimeter (EMCal). Tracks from charged

Journal of Physics: Conference Series **736** (2016) 012021 doi:10.1088/1742-6596/736/1/012021

particles are reconstructed using information from the Time Projection Chamber (TPC) and Inner Tracking System (ITS). The TPC is the main tracking detector of the central barrel and provides particle identification (PID) and charged particle momentum measurements [14]. The ITS is a six-layer silicon detector with that provides a precise measurement of the primary vertex location. The two inner layers of the ITS, known as the Silicon Pixel Detector (SPD), are also used for selecting high quality tracks coming from the primary collision [15]. Tracks reconstructed for jet analyses come from mid-rapidity ($|\eta_{lab}| < 0.9$) and cover the full range in azimuth. The tracking detectors measure particles down to 0.15 GeV/c in transverse momentum (p_T). The EMCal is a Pb-scintillator sampling calorimeter which has an acceptance of $|\eta_{lab}| < 0.7$ and $|\Delta\phi| = 107^{\circ}$ [16]. For a complete description of the ALICE experiment detector system see [17].

3. Jet Reconstruction

There is no unique definition of a jet. Experimentally, we measure the jet through particles which are produced in the fragmentation of quarks and gluons and seen as tracks and clusters. These fragments are expected to reflect the kinematics and topology of the originating partons. We need a general definition to apply to both theory and experimental calculations which, consists of a jet-finding algorithm, its parameters, and a recombination scheme. ALICE uses the k_T and anti-k_T jet-finding algorithms from the FastJet package [18]. A jet-finding algorithm defines how tracks and clusters in the detector (experiment) or partons (theory) are grouped into jets. For the k_T and anti-k_T methods, the algorithms group nearby objects together pair-wise until all particles are clustered into jet candidates. In ALICE, we use the anti-k_T algorithm for our jet signal, which merges the highest p_T particles first. It has the advantage of creating a circular shape around the jet constituents, defined by the jet resolution parameter R, which is given by $R = \sqrt{(\Delta\eta)^2 + (\Delta\varphi)^2}$. R is used to set the maximum distance in pseudorapity η and azimuthal angle φ that constituent particles can be clustered together when reconstructing a jet [18]. For calculating the background energy from the underlying event uncorrelated to the hard scattering, we use the k_T algorithm which is not bound to a circular structure and begins by clustering the lowest p_T particles first. In ALICE, the boost-invariant sequential recombination scheme is used [18].

By studying fully reconstructed jets, several variables associated with jet reconstruction can be adjusted allowing for them to serve as very versatile triggers [19]. By selecting high p_T leading jets, it is expected that the jet production points are biased towards the surface of the interaction region. This is known as surface biasing and the initial energy of the parton on the opposing side can be approximated more accurately [19]. In principle, by adjusting the surface bias of the trigger jet through various cuts [19], we can adjust the pathlength traversed through the medium of the oppposing jet and perform true jet tomography [20]. Applying a constituent cut or imposing a leading track requirement on the reconstructed jet enhances the surface bias by forcing the jet to be harder and rejects combinatorial jets from the measurement. Combinatorial jets are reconstructed jets which do not originate from a hard process. These cuts don't affect the opposing jet being studied. [19]

4. Spectra results

4.1. Data sample

The data used for the spectra analyses was collected in 2011 by ALICE during the 2.76 TeV pp and 2.76 TeV Pb–Pb collision runs. The pp collisions include two different triggers to perform the spectra measurement. A combination of a minimum bias (MB) trigger and an EMCal trigger requiring a single shower with $E_T > 3$ GeV/c was used in order to allow for statistics to be collected over a wider p_T range of 20 to 125 GeV/c [13, 21, 22].

4.2. pp collisions

In order to study the properties of the QCD medium generated in Pb–Pb collisions at LHC energies, a reference collision system is needed. In pp collisions, a medium is not expected to exist. Jet measurements performed in pp collisions serve as a baseline reference when interpreting the Pb–Pb collision results. These measured jets are corrected bin-by-bin in p_T back to the particle level [21]. Additional analysis details can be found in [21]. Figure 1 shows the inclusive differential jet cross section with $R = 0.2$ at $\sqrt{s} = 2.76$ TeV. We compare our measurement to various Next-to-Leading-Order (NLO) pQCD calculations [23,24], with and without the effects of hadronization [25] and at parton level. A good agreement to NLO calcuations can be seen when including hadronization. This shows that we have a very good understanding of jet production in collisions where a medium is not produced.

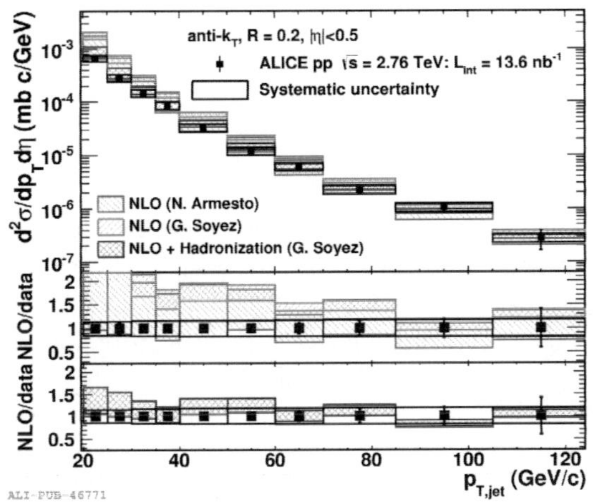

Figure 1. The inclusive differential $R = 0.2$ full jet cross section in pp collisions at $\sqrt{s} = 2.76$ TeV. A comparison of the experimental calculation is shown to NLO pQCD calculations both with and without the effects of hadronization. The systematic uncertainty is designated by the box around the data points. [21]

4.3. Pb–Pb collisions

The jets produced in heavy ion collisions sit on top of a large amount of soft ($p_T < 2.0$ GeV/c) background. The variable most often used to quantify this background is ρ, which refers to the underlying event background energy density. The procedure for obtaining ρ requires clustering all charged tracks in the event into groups using the k_T jet-finding algorithm provided by FastJet [18]. The two leading jets in the event are removed and the underlying event background density is expressed as the median of the summed p_T per area ($A_{k_T jet}$) of the remaining k_T clusters and is given by Eq. 1.

$$\rho_{ch} = median\left(p^{ch}_{T,k_T jet}/A_{k_T jet}\right) \tag{1}$$

In order to reduce the bias due on jet fragmentation, tracks with $p_T^{min} > 0.15$ GeV/c are used in this analysis. It is shown in [26] that the background density depends on event multiplicity and that the spread of ρ in a given centrality bin is considerable. This serves as motivation for an event-by-event background subtraction performed on a jet-by-jet basis.

When reconstructing full jets, the neutral energy component is compensated for by scaling the charged background density distribution by the EMCal scale factor, $\rho_{scaled} = \rho_{ch} \times s_{EMC}$, where s_{EMC} is defined by Eq. 2 and parametrized as a function of multiplicity.

$$s_{EMC} = \frac{\left(\sum p_T^{track} + \sum E_T^{cluster}\right)}{\sum p_T^{track}} \tag{2}$$

From the results of [8, 26], it can be seen that for the 10% most central events, the scaled background density amounts to roughly 200 GeV/c per unit area. The average background for a full jet in a central collision with $R = 0.2$ is 25 GeV/c. This is subtracted from of our reconstructed jet p_T via Eq. 3. For additional analysis details, see [8, 26].

$$p_{T,jet}^{unc} = p_{T,jet}^{rec} - \rho_{scaled} A_{jet} \tag{3}$$

Figure 2 shows the inclusive jet yields per event for $R = 0.2$ full jets at $\sqrt{s_{NN}} = 2.76$ TeV in both the 0-10% and 10-30% centrality classes compared to the measurement at the same $\sqrt{s}$ in pp collisions. A leading charged jet constituent with $p_T > 5GeV/$c is required to reduce the combinatorial background and improve unfolding stability. To compare the two datasets, the Pb–Pb results are divided by the number of binary collisions, N_{coll}, calculated in a Monte Carlo Glauber model [27] which assumes independent binary nucleon-nucleon collisions.

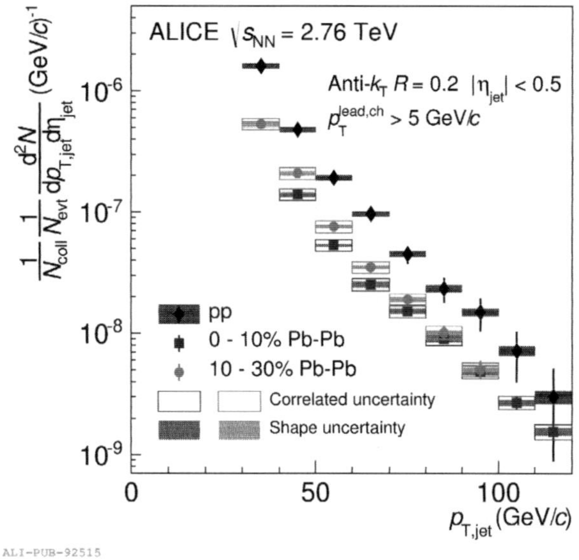

Figure 2. The inclusive spectra for $R = 0.2$ full jets in Pb–Pb collisions at $\sqrt{s_{NN}} = 2.76$ TeV reconstructed with a leading charged track bias of 5 GeV/c. For comparison, 0-10%, 0-30% are plotted with the results from pp collisions and normalized by N_{coll}. All three are normalized per event. [8]

The nuclear modification factor, R_{AA}, is defined by Eq. 4 and is the ratio of the Pb–Pb per-event yield divided by the pp cross section multiplied by $T_{AA} = N_{coll}/\sigma_{pp}^{inel}$ [27].

$$R_{AA} = \frac{\frac{1}{N_{events}}\frac{d^2N_{jets}}{dp_T d\eta}}{T_{AA}\frac{d^2\sigma}{dp_T d\eta}} \tag{4}$$

A value greater (less) than 1.0 indicates an enhancement (suppression) in Pb–Pb relative to pp collisional system. The R_{AA} for $R = 0.2$ full jets is shown in Fig. 3 for both 0-10% and 10-30% central events. Jet suppression is seen to be centrality dependent, and within both the statistical and systematic uncertainties the two measurements are independent of p_T^{jet} [8, 13]. The largest suppression occurs for the 0-10% most central events with an $R_{AA} = 0.28 \pm 0.04$ [8]. The suppression fairly agrees within error to both YaJEM [28] and JEWEL [29], which are different models for energy loss in the medium [8].

5. Correlation results

Correlation measurements are powerful tools which can be used to probe the jet-medium interactions in AA collisions, provide additional constraints on energy loss models and thereby

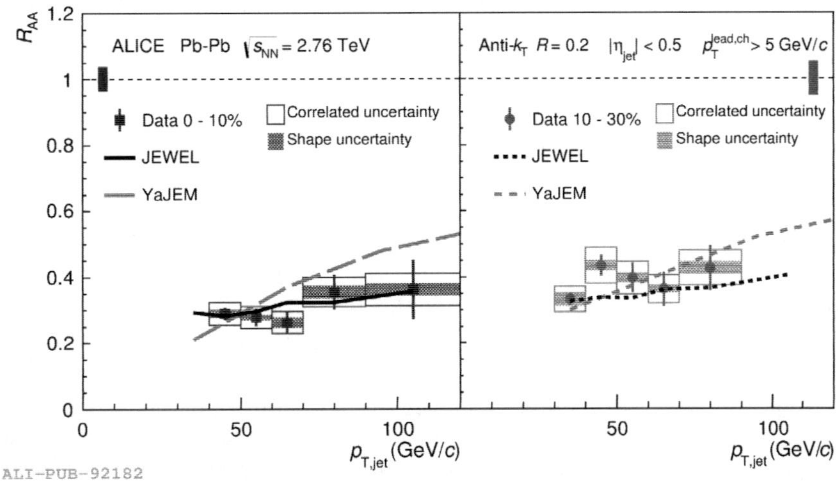

Figure 3. The nuclear modification factor, R_{AA}, for $R = 0.2$ full jets in 0-10% (left) and 10-30% (right) central events. Also shown are calculations from YaJEM [28] and JEWEL [29]. [8]

provide us with a better understanding of the properties of the QGP [19]. The jet-hadron correlation function is defined in Eq. 5 [19] as the number of same event pairs to the number of mixed event pairs.

$$\frac{1}{N_{trig}}\frac{dN}{d\Delta\varphi} = \frac{1}{\epsilon N_{trig}}\frac{\frac{dN^{same}_{pairs}}{d\Delta\varphi}}{\frac{dN^{mixed}_{pairs}}{d\Delta\varphi}} \tag{5}$$

Where N_{trig} corresponds to the number of trigger jets and η is the efficiency of the associated particles. The mixed event pair distribution, which corrects for the detector acceptance effects, is normalized such that the ($\Delta\varphi = 0$, $\Delta\eta = 0$) bin is 1.0. The resulting distribution is then projected to show the angular correlations between the jet and the associated particles given by $\Delta\varphi = \varphi_{jet} - \varphi_{assoc}$. When studying the correlations from baseline pp collisions, a peak is seen on the near side at $\Delta\varphi = 0$ and on away side at $\Delta\varphi = \pi$ [19]. The near-side is due to the associated particles of the trigger jet that defines the $\Delta\varphi = 0$ position. The away-side is then due to the particles associated with the opposite jet created from the initial hard scattering.

At RHIC, the STAR collaboration has studied jet-hadron correlations in Au+Au at 200 GeV [30]. They calculated the away-side energy balance function, D_{AA}, which measures the p_T difference of the associated particle yield between AuAu and pp collisions for a given p_T^{assoc} bin. The results show that the suppression of high-p_T hadron yield on the away side is in large part balanced by an enhancement at low-p_T. They also showed an away-side jet width measurement which suggests there is jet broadening to large angles at low-p_T when compared to pp collisions, but the significance is limited due to large uncertainties of the jet v_n used in the determination of the heavy ion background. These results imply that most modifications to jets occur at low hadron p_T. At LHC, we can study the effect at higher $\sqrt{s_{NN}}$ and access higher parton energies. In ALICE, our particle identification and reconstruction with the ALICE Time-Of-Flight (TOF) detector and the TPC works best at these lower transverse momenta and therefore allows us to measure identified particles in jets.

Journal of Physics: Conference Series **736** (2016) 012021 doi:10.1088/1742-6596/736/1/012021

5.1. *Data sample*

The data used for the correlations analyses was collected in 2011 by ALICE during the 2.76 TeV pp and 2.76 TeV Pb–Pb collision runs. The pp collisions includes events that fired the trigger used by the ALICE electromagnetic calorimeter (EMCal), which requires a shower containing more than 3 GeV of energy for acceptance [19]. This allows for enhanced statistics provided by the resulting data sample.

5.2. *pp correlations*

Figure 4 shows the jet-hadron measurement for pp collisions at $\sqrt{s} = 2.76$ TeV [19]. The $R = 0.4$ trigger jets are 20-60 GeV/c with constituent particles > 3 GeV/c and have a 6 GeV/c leading track bias. The correlations are shown for associated particles in three bins: [0.5, 1.0], [1.0, 2.0], and [2.0, 3.0] GeV/c. The background subtraction of the underlying event is estimated by using a flat pedestal since the jet-hadron correlations in the $\Delta\eta$ range shown in Fig. 4 ($|\Delta\eta| < 0.4$) were rather flat. The uncertainty from this subtraction is shown as the grey band about 0 [19]. The efficiency correction caused an overall uncertainty on the normalization of the correlation function of 7% [19]. We see a general decrease in the jet widths as we increase the associated track momentum and observe it to be less significant on the away-side [19]. These results agree with expectations [19] and serve as our baseline for heavy ion measurements.

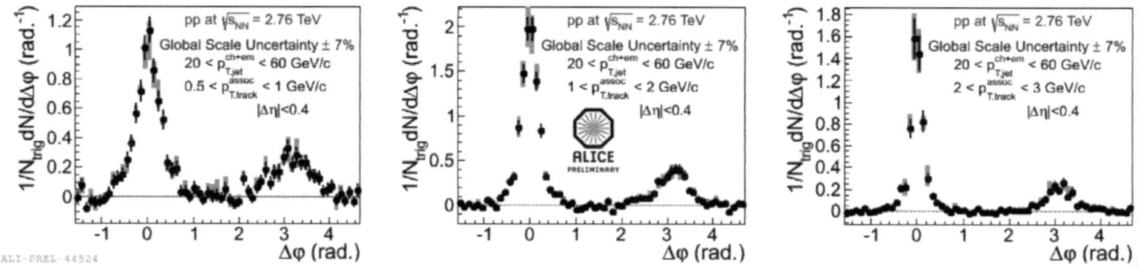

Figure 4. The pp baseline jet-hadron correlations measurement of ALICE for 3 associated momentum bins. The $\Delta\varphi$ distributions are shown for 20-60 GeV/c jets reconstructed using the anti-k_T algorithm. These reconstructed jets have a resolution parameter of $R = 0.4$, constituent particles > 3 GeV/c, and require a track bias of $p_T > 6$ GeV/c. The uncertainty from the pedestal background subtraction is shown as the grey band. [19]

5.3. *Pb–Pb correlations*

In heavy ion collisions our correlation function now takes on an additional term and is expressed by Eq. 6 [19].

$$\frac{1}{N_{trig}}\frac{dN}{d\Delta\varphi} = \frac{1}{\epsilon N_{trig}}\frac{\frac{dN^{same}_{pairs}}{d\Delta\varphi}}{\frac{dN^{mixed}_{pairs}}{d\Delta\varphi}} - b_0(1 + \sum v_n^{trig} v_n^{assoc}\cos(n\Delta\varphi)) \tag{6}$$

The second term accounts for the underlying event of heavy ion collisions which is now modulated by hydrodynamical flow. To subtract off this background, the knowledge or calculation of the Fourier coefficients (v_n terms) from the trigger jet and associated particles are needed. A first look towards jet-hadron correlations without the background subtraction or efficiency correction is seen in Fig. 5 for Pb–Pb collisions. It is for $R = 0.4$ jets of $20-60$ GeV/c, constituents with $p_T > 3$ GeV/c, and associated particles of $2 - 3$ GeV/c. There is a near-side peak observed, but not much can be concluded until the background is understood and subtracted.

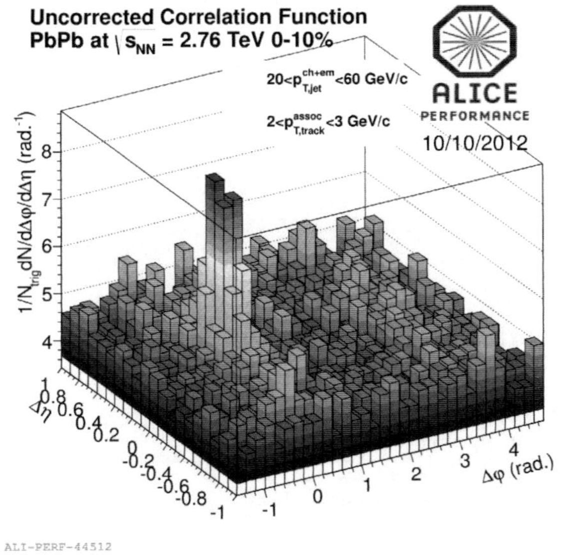

Figure 5. The uncorrected correlation function for full jets in Pb–Pb collisions at $\sqrt{s_{NN}} = 2.76$ TeV for the 0-10% most central events. These results have yet to have an efficiency correction applied or the flow background removed. [19]

6. Background subtraction

In order to analyze the jet-hadron correlation signal in Pb–Pb collisions, it becomes necessary to understand and subtract the large combinatoric background. In addition to the large background produced, collective flow effects contribute to the bulk particle production in a similar way to that of correlations strictly due to jet production and this dominates at low momenta. With the aim of improving the constraints of prior studies of the combinatoric background, various background methods were developed in [31]. Two key methods were the near-side fit (NSF) and the reaction plane fit (RPF). The NSF method works under the assumption that the signal is negligible in the large $\Delta\eta$ and small $\Delta\varphi$ region. This background dominated region is projected for $1.0 < |\Delta\varphi| < 1.4$ and Eq. 7 is fit up to n=4 for $|\Delta\varphi| < \pi/2$ and extrapolated over the full range of $\Delta\varphi$.

$$\frac{dN^{pairs}}{\pi\Delta\phi} = B(1 + \sum_{n=1} 2v_n^{trigger} v_n^{assoc} \cos(n\Delta\phi)) \tag{7}$$

The NSF method does not require independent measurements of v_n and is able to fit the background and therefore extract the signal with less bias and smaller errors than the ZYAM method [31]. When the trigger is restricted relative to the reaction plane, both the background level and effective v_n are modified [32]. The RPF method is an extension of the NSF method which uses the reaction plane dependence of the correlation function to determine the background to higher precision. It was shown by [31] that the RPF works very effectively while using fewer assumptions and has smaller errors than other background methods. For more method details, see [31].

7. Summary and Outlook

The jet spectra and jet-hadron correlations measurements in ALICE were shown for pp collisions at $\sqrt{s} = 2.76$ TeV. They establish baseline measurements needed for understanding modification to jets in the presence of a medium. The Pb–Pb spectra of the same collisional energy was shown and compared to the pp collision baseline in the R_{AA} measurement. It was seen that jets from Pb–Pb collisions are suppressed in a centrality dependent way, but nearly independent across p_T^{jet} bins to within both systematic and statistical uncertainties. This helps establish jets as a very critcal probe toward understanding the QGP. We then gave an early look at the jet-hadron correlations in Pb–Pb collisions. Higher statistics and improved background subtraction techniques allow higher precision measurements of jet widths and yields. Jet-hadron correlations relative to the event plane are in progress.

References

[1] Adcox K *et al.* (PHENIX) 2005 *Nucl. Phys.* **A757** 184–283 (*Preprint* `nucl-ex/0410003`)
[2] Adams J *et al.* (STAR) 2005 *Nucl. Phys.* **A757** 102–183 (*Preprint* `nucl-ex/0501009`)
[3] Qin G Y, Ruppert J, Gale C, Jeon S, Moore G D and Mustafa M G 2008 *Phys. Rev. Lett.* **100** 072301 (*Preprint* `0710.0605`)
[4] Zakharov B G 2007 *JETP Lett.* **86** 444–450 (*Preprint* `0708.0816`)
[5] Adare A *et al.* (PHENIX) 2008 *Phys.Rev.* **C77** 011901 (*Preprint* `0705.3238`)
[6] Adams J *et al.* (STAR) 2003 *Phys. Rev. Lett.* **91** 172302 (*Preprint* `nucl-ex/0305015`)
[7] Adare A *et al.* (PHENIX) 2011 *Phys.Rev.* **C84** 024904 (*Preprint* `1010.1521`)
[8] Adam J *et al.* (ALICE) 2015 *Phys. Lett.* **B746** 1–14 (*Preprint* `1502.01689`)
[9] Adams J *et al.* (STAR Collaboration) 2005 *Phys. Rev. Lett.* **95**(15) 152301
[10] Abelev B I *et al.* (STAR Collaboration) 2009 *Phys. Rev. C* **80**(6) 064912
[11] Agakishiev G *et al.* (STAR Collaboration) 2012 *Phys. Rev. C* **85**(1) 014903
[12] Aggarwal M M *et al.* (STAR Collaboration) 2010 *Phys. Rev. C* **82**(2) 024912
[13] Reed R and Collaboration A 2015 *Journal of Physics: Conference Series* **636** 012010
[14] Alme J *et al.* 2010 *Nuclear Instruments and Methods in Physics Research Section A: Accelerators, Spectrometers, Detectors and Associated Equipment* **622** 316 – 367 ISSN 0168-9002
[15] Abelev B *et al.* (ALICE) 2014 *J. Phys.* **G41** 087002
[16] Abeysekara U *et al.* (ALICE EMCal) 2010 (*Preprint* `1008.0413`)
[17] Abelev B B *et al.* (ALICE) 2014 *Int. J. Mod. Phys.* **A29** 1430044 (*Preprint* `1402.4476`)
[18] Cacciari M, Salam G P and Soyez G 2012 *Eur.Phys.J.* **C72** 1896 (*Preprint* `1111.6097`)
[19] Connors M 2013 *Journal of Physics: Conference Series* **446** 012009
[20] Renk T 2013 *Phys. Rev.* **C87** 024905 (*Preprint* `1210.1330`)
[21] Abelev B *et al.* (ALICE) 2013 *Phys. Lett.* **B722** 262–272 (*Preprint* `1301.3475`)
[22] Ma R (ALICE) 2013 *Nucl. Phys.* **A910-911** 319–322 (*Preprint* `1208.4537`)
[23] Frixione S, Kunszt Z and Signer A 1996 *Nucl. Phys.* **B467** 399–442 (*Preprint* `hep-ph/9512328`)
[24] Frixione S 1997 *Nucl. Phys.* **B507** 295–314 (*Preprint* `hep-ph/9706545`)
[25] Cacciari M, Rojo J, Salam G P and Soyez G 2011 *Eur. Phys. J.* **C71** 1539 (*Preprint* `1010.1759`)
[26] Abelev B *et al.* (ALICE) 2012 *JHEP* **1203** 053 (*Preprint* `1201.2423`)
[27] Abelev B *et al.* (ALICE) 2013 *Phys. Rev.* **C88** 044909 (*Preprint* `1301.4361`)
[28] Renk T 2013 *Phys. Rev.* **C88** 014905 (*Preprint* `1302.3710`)
[29] Zapp K C 2014 *Phys. Lett.* **B735** 157–163 (*Preprint* `1312.5536`)
[30] Adamczyk L *et al.* (STAR) 2014 *Phys.Rev.Lett.* **112** 122301 (*Preprint* `1302.6184`)
[31] Sharma N, Mazer J, Stuart M and Nattrass C 2016 *Phys. Rev.* **C93** 044915 (*Preprint* `1509.04732`)
[32] Bielcikova J, Esumi S, Filimonov K, Voloshin S and Wurm J 2004 *Phys.Rev.* **C69** 021901 (*Preprint* `nucl-ex/0311007`)

Journal of Physics: Conference Series **736** (2016) 012022 doi:10.1088/1742-6596/736/1/012022

Study of hard core repulsive interactions in an hadronic gas from a comparison with lattice QCD

P Alba[1], **V.Vovchenko**[1,2,3], **H.Stoecker**[1,2,4]

[1] Frankfurt Institute for Advanced Studies, Goethe Universität Frankfurt, D-60438 Frankfurt am Main, Germany

[2] Institut für Theoretische Physik, Goethe Universität Frankfurt, D-60438 Frankfurt am Main, Germany

[3] Department of Physics, Taras Shevchenko National University of Kiev, 03022 Kiev, Ukraine

[4] GSI Helmholtzzentrum für Schwerionenforschung GmbH, D-64291 Darmstadt, Germany

E-mail: `alba@fias.uni-frankfurt.de`

Abstract. We study the influence of hard-core repulsive interactions within the Hadron-Resonace Gas model in comparison to first principle calculation performed on a lattice. We check the effect of a bag-like parametrization for particle eigenvolume on flavor correlators, looking for an extension of the agreement with lattice simulations up to higher temperatures, as was yet pointed out in an analysis of hadron yields measured by the ALICE experiment. Hints for a flavor depending eigenvolume are present.

1. Introduction

One of the main topics of the Heavy-Ion Collision field has been the study of particle production in order to understand the inner mechanism of strong interactions which happen in the early stages of particle collision. The Hadron-Resonance Gas (HRG) model has proven to be an essential tool in giving preliminar predictions both for experimental hadron production [1, 2, 3, 4, 5, 6, 7, 8, 9], and first principle QCD calculations on a lattice [10], and the community now agrees on the idea that at vanishing net-baryon density the confinement transition develops at a pseudo-critical temperature of about 155 MeV, with a smooth crossover from partonic-to-hadronic degrees of freedom [11].
However the exact mechanism is still not understood, and there is an open debate on different issues, e.g for possible sources of such a fast equilibration [12, 13], or on the existence of a Critical End Point at high densities, where the crossover should turn into a first order transition [14, 15]. We show that the inclusion of hard core repulsive interactions among hadrons leads to a better description of available experimental data and lattice results, leading to the question on the effective nature of confinement transition on a broad range of temperatures which formerly seemed unaccessible [16].

The paper is organized as follow, we first give an introduction on general aspects of the HRG model together with the explanation of the hard core repulsions, and then we briefly summarize the observables which could be analyzed in this framework; following the conclusions.

Journal of Physics: Conference Series **736** (2016) 012022 doi:10.1088/1742-6596/736/1/012022

2. The Hadron-Resonance Gas model

The nowadays well established HRG model consists basically on the idea that a system of interacting hadrons can be described by a non-interacting gas of hadrons and resonances, where the latter mediate the strong interactions among the particle constituents [1].
The thermodynamics of such equilibrated system is described by the sum of independent single particle contributions, e.g. for the pressure:

$$p(T,\{\mu_h\}) = \sum_i p^{id}(T,\mu_i) = \sum_i (-1)^{B_i+1} \frac{d_i\, T}{(2\pi)^3} \int d^3p \,\ln\left[1 + (-1)^{B_i+1} e^{-(\sqrt{\vec{p}^2+m_i^2}-\mu_i)/T}\right] , \tag{1}$$

where the sum runs over all the available hadronic spectrum, and p^{id} is the pressure of a gas of i-species particles in the grand-canonical ensemble, which depends on the system temperature and on the chemical potentials corresponding to the conserved net-charges (baryon number, electric charge and strangeness in the case of strong interactions).
We use a particle list from the Particle Data Group (PDG) [17] with particles with masses up to 2 GeV, without the inclusion of charm degrees of freedom which however could play a role in such a high temperature regime.
Then we also consider as an additional effect the finite resonance widths, with the usual convolution with the Breit-Wigner distribution, and the feed-down to hadron yields from resonance decay. For simplicity all the results are obtained with Boltzmann statistics.

2.1. The repulsive interaction

In order to provide a more realistic description of the produced system, has been proposed to account for the short-range repulsive Van deer Waals interactions directly in the HRG model [18, 19, 20]; the most common version is the one presented in [21, 22], where each particle species i occupies an eigenvolume given by:

$$v_i = 4 * \frac{4}{3}\pi\, r_i^3 \tag{2}$$

which must be subtracted from the total available system volume.
The Excluded Volume (EV) procedure can be accounted for by solving the following trascendental equation for the pressure:

$$p(T,\{\mu_h\}) = \sum_i p^{id}(T,\widetilde{\mu}_i) \tag{3}$$

which is essentially the same as (1), with a shift in the particle chemical potential given by:

$$\widetilde{\mu}_i = \mu_i - v_i\, p(T,\{\mu_h\})\,. \tag{4}$$

The other quantities are modified accordingly to the thermodynamical consistency formulas, e.g. the single particle density reads:

$$n_i(T,\{\mu_h\}) = \frac{n_i^{id}(T,\mu_i)}{1+\sum_j v_j n_j^{id}(T,\mu_i)} \tag{5}$$

It is worth to notice here that if the same EV volume is considered for all particle species, there is no big change for volume-independent observables, since the suppression factors due to the hadronic finite size is canceled at leading order, with the general effect of increasing the total system volume with respect to the point-like case.

However has been shown that considering different eigenradii leads to a sizeable effect [23] in extracting the freeze-out parameters.
The multi-component EV leaves room for different choices in the particle eigenvolume; in the following we employed a bag like parameterization where the hadron eigenvolume is proportional to the mass $v_i = m_i/\epsilon_0$, the bag-constant being fixed in order to reproduce a fixed value for the proton radius, as well as a special case with point-like pions which is relevant to reproduce the limit for the speed of sound.
It has been recently pointed out that the inclusion of a bag-like parametrization leads to a substantially improvement in describing the hadron yields measured by the ALICE collaboration [24], and together with a Hagedorn spectrum may lead to an improvement in the agreement with lattice data.

2.2. Crossterms from the virial expansion theorem

To properly account the multi-body repulsive forces one should account for next-to leading order coefficients in the virial expansion of the thermodynamics of gas of hard spheres. Up to the second order the pressure reads:

$$p(T,n) = T\sum_i n_i + T\sum_{i,j} b_{ij}n_in_j + \mathcal{O}(n^3) \tag{6}$$

where

$$b_{ij} = \frac{2\pi}{3}(r_i + r_j)^3 \tag{7}$$

are the crossterms of the interactions among different hadron species. Note that for i=j we obtain the same eigenvolume as in the model presented in eq. (9), which from now will be referred as diagonal.
Although the diagonal formulation catches the essential nature of a system of hadron with differente sizes, the fact that this is not consistent with the virial expansion may lead to a difficult interpretation of the eigenradius r_i.
For this reason we here employ also the crossterms formulation, which leads to the following formula in the Boltzmann approximation in the GCE:

$$p(T,\mu) = \sum_i p_i^{id}(T,\widetilde{\mu}_i) = T\sum_R \frac{n_i^{id}(T,\widetilde{\mu}_i)}{1+\sum_j \widetilde{b}_{ij}n_j^{id}(T,\widetilde{\mu}_i)} \tag{8}$$

where now the shifted effective chemical potential reads as:

$$\widetilde{\mu}_i = \mu_i - \sum_j \widetilde{b}_{ij}\, p_j(T,\{\mu_h\}) \tag{9}$$

which essentially leads to a system of coupled trascendental equations, one for each considered particle species [25, 26].

3. Comparison to the lattice

Lattice data have reached an unprecedent accuracy, and now are performed with physical values of quark masses allowing for a meaningful comparison with HRG predictions for full QCD.
In Fig. 3 we show the pressure and energy densities obtained from lattice simulations [11], compared with HRG calculations for different choices of proton eigenradius. Here it is clearly shown that the usual point-particle description fails in describing data above about 160 MeV,

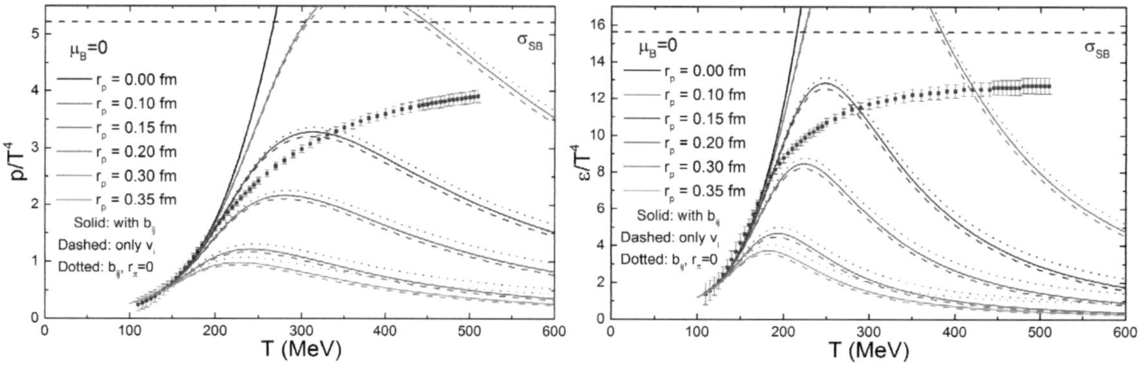

Figure 1. (Color online) Lattice simulation results [11] for pressure (left panel) and energy density (right panel) compared to HRG model calculations with and without EV.

while the EV can extend the description up to about 240 MeV, with reasonable values for the proton eigenradius in the range of 0.15÷0.20 fm.
It is worth to notice that in the low temperature regime the available uncertainities do not allow to distinguish among the different curves, and that an important suppression is seen for radii bigger than 0.3 fm, mainly due to density suppression.
In general we note that for thermodynamical variables there is no qualitative difference between the diagonal and crossterm models, as well as considering point-like pions, and the main role is played by the running eigenvolume with the particle mass.

Susceptibilities are expected, to be more sensitive to fine details of the hadronic spectrum [27, 28, 29]; they are defined as:

$$\chi_{lmn}^{BSQ} = \frac{\partial^{\,l+m+n}\left(p/T^4\right)}{\partial\left(\mu_B/T\right)^l \partial\left(\mu_S/T\right)^m \partial\left(\mu_Q/T\right)^n}. \tag{10}$$

they have recently triggered a lot of attention from the community, leading to the first experimental measurements [30, 31], which seems to be in agreement within the HRG framework once the experimental setup is properly accounted [32].

In Fig. 3 we show the second order fluctuations for net-baryon, net-electric charge and net-strangeness with diagonal and crossterm EV HRG and point-like mesons, in comparison with the corresponding lattice data [33].
We find that considering baryon number and electric charge, the region of agreement between the statistical model and lattice could be easily improved up to about 240 MeV, but this is not the same for strangeness which seems to suffer too much the EV suppression.
Besides the fact that this behavior can be simply connected with a lack in the strange sector of the hadronic spectrum [29], this may hint to a flavor dependent eigenvolume. Investigations in this direction are in progress [34].

Anyway the inclusion of Hagedorn (strange-)states in the analysis of lattice is mandatory, whose importance in the interplay with the EV was pointed out yet [24].

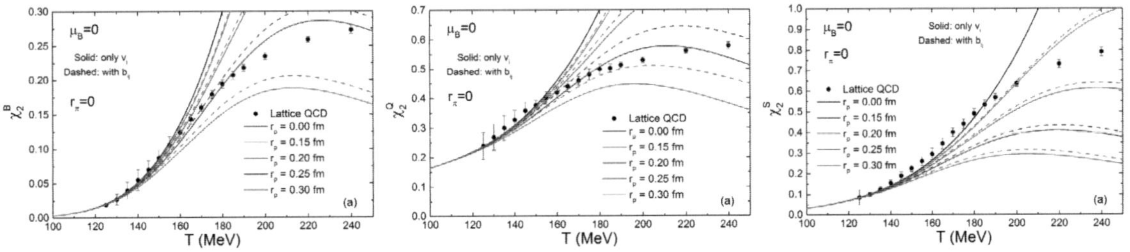

Figure 2. (Color online) Lattice simulation results [33] for the second moment of baryon number (left panel), electric charge (central panel) and strangeness (right panel) net distributions, compared to HRG model calculations with finite eigenvolumes.

4. Check at LHC energies

The consistency with experimental measurements is an important check for the HRG model. We perform a thermal fit to the midrapidity yields of $\pi^{+,-}$, $K^{+,-}$, (anti-)protons, Λ, (anti-)Ξ^-, Ω+anti-Ω, K^0_S and ϕ, measured by the ALICE collaboration in the 0-5% most central Pb+Pb collisions at $\sqrt{s_{NN}}$=2.76 TeV [35, 36, 37, 38, 39].
In Fig. 4 we show the χ^2 profile with temperature at vanishing μ_B, with χ^2 being:

$$\frac{\chi^2}{N_{\rm dof}} = \frac{1}{N_{\rm dof}} \sum_{h=1}^{N} \frac{\left(\langle N_h^{\rm exp}\rangle - \langle N_h\rangle\right)^2}{\sigma_h^2}, \tag{11}$$

where $\langle N_h^{\rm exp}\rangle$ and $\langle N_h\rangle$ are the experimental and theoretical hadron yields, respectively, $N_{\rm dof}$ is the number of degrees of freedom, that is the number of the data points minus the number of fitting parameters, and σ_h is the experimental uncertainty on $\langle N_h^{\rm exp}\rangle$.
We show that the experimental analysis is strongly dependent on the EV assumption, basically related to a difference in the particle eigenvolumes. This can be seen simply considering different parameters between mesons and baryons, but the effect is much stronger when considering a running eigenvolume for which a flat χ^2 region appears for temperatures up to 300 MeV, concurrently improving the description of experimental data.
It is worth to note that there is a strong reduction of the thermal "proton anomaly" [40], and the EV could be a candidate to solve this issue as well as missing [41] and final state interactions [39]. A systematic analysis of such a flavor dependence for strange particles is in preparation [34].

5. Conclusions

We investigated the influence of repulsive interactions within the HRG model. Both a comparison with lattice and experiment give clear indication that for a proper choice of the suppression parameters it is possible to extend the description to higher temperatures, together with a noticeably improvement in the description for all measured hadron yields, giving special attention to the proton anomaly.
This opens interesting options, like the role of glueballs and the corresponding first order phase transition, as it was proposed in [42].
A systematic study of the influence for a mass dependent eigenvolume, as well as a flavor dependence is in progress [34].

Acknowledgments

Talk presented by HSt, who acknowledges support through the Judah M. Eisenberg professur laureatus for theoretical physics. This work was supported by HIC for FAIR within the LOEWE

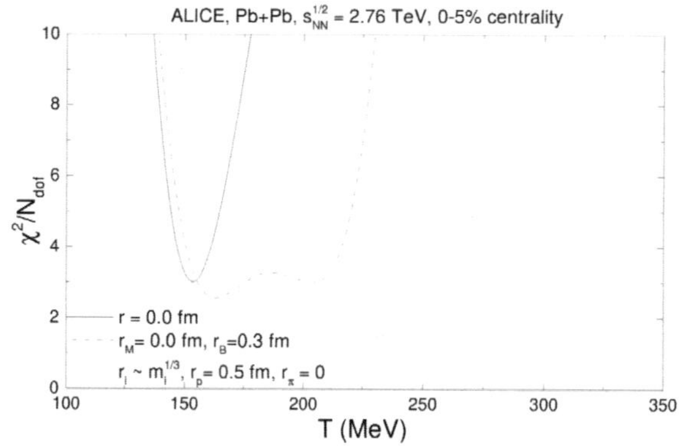

Figure 3. (Color online) χ^2 profile with temperature at vanishing μ_B, obtained from an analysis of hadron yields measured by the ALICE collaboration at $\sqrt{s_{NN}}$ of 2.76 TeV, for point-like particles (black solid line) in comparison with EV assumptions for meson-to-baryon different eigenvolumes (green dashed line) and running eigenvolume with particle mass (red dotted line).

program of the State of Hesse. V.V. acknowledges the support from HGS-HIRe.

References

[1] Braun-Munzinger P, Redlich K and Stachel J, *In *Hwa, R.C. (ed.) et al.: Quark gluon plasma** 491-599
[2] Cleymans J and Satz H (1993) *Z. Phys. C* **57** 135.
[3] Yen G D, Gorenstein M I, Greiner W and Yang S N 1997 *Phys.Rev. C* **56** 2210.
[4] Becattini F, Cleymans J, Keranen A, Suhonen E and Redlich K 2001 *Phys. Rev. C* **64** 024901.
[5] Braun-Munzinger P, Magestro D, Redlich K and Stachel J 2001 *Phys. Lett. B* **518** 41.
[6] Rafelski J and Letessier J 2003 *Nucl. Phys. A* **715** 98c.
[7] Andronic A, Braun-Munzinger P and Stachel J 2006 *Nucl. Phys. A* **772**, 167.
[8] Becattini F, Manninen J and Gazdzicki M 2006 *Phys. Rev. C* **73**, 044905.
[9] Andronic A, Braun-Munzinger P and Stachel J 2009 *Phys. Lett. B* **673** 142.
[10] Karsch F, Redlich K and Tawfik A 2003 *Phys. Lett. B* **571** 67
[11] Borsanyi S, Endrodi G, Fodor Z, Jakovac A, Katz S D, Krieg S, Ratti C and Szabo K K 2010 *JHEP* **1011** 077
[12] Ruggieri M, Puglisi A, Oliva L, Plumari S, Scardina F and Greco V 2015 *Phys. Rev. C* **92** 064904
[13] Epelbaum T and Gelis T 2013 *Phys. Rev. Lett.* **111** 232301
[14] Stephanov M A, Rajagopal K and Shuryak E V 1998 *Phys. Rev. Lett.* **81** 4816
[15] Hatta Y and Ikeda T 2003 *Phys. Rev. D* **67** 014028
[16] Vovchenko V and Stoecker H *Preprint* hep-ph/1512.08046
[17] Olive K A *et al.* [Particle Data Group Collaboration] 2014 *Chin. Phys. C* **38** 090001.
[18] Hagedorn R and Rafelski J 1980 *Phys. Lett. B* **97** 136.
[19] Gorenstein M I, Petrov V K and Zinovev G M 1981 *Phys. Lett. B* **106** 327.
[20] Kapusta J I and Olive K A 1983 *Nucl. Phys. A* **408** 478.
[21] Rischke D H, Gorenstein M I, Stoecker H and Greiner W 1991 *Z. Phys. C* **51** 485.
[22] Yen G D, Gorenstein M I, Greiner W and Yang S N 1997 *Phys. Rev. C* **56** 2210
[23] Yen G D and Gorenstein M I 1999 *Phys. Rev. C* **59** 2788
[24] Vovchenko V, Anchishkin D V and Gorenstein M I 2015 *Phys. Rev. C* **91** no.2, 024905
[25] Gorenstein M I, Kostyuk A P and Krivenko Y D 1999 *J. Phys. G* **25** L75
[26] Bugaev K A, Oliinychenko D R, Sorin A S and Zinovjev G M 2013 *Eur. Phys. J. A* **49** 30
[27] Athanasiou C, Rajagopal K and Stephanov M 2010 *Phys. Rev. D* **82** 074008
[28] Alba P, Bellwied R, Bluhm M, Mantovani Sarti V, Nahrgang M and Ratti C 2015 *Phys. Rev. C* **92** no.6, 064910
[29] Bazavov A *et al.* 2014 *Phys. Rev. Lett.* **113** no.7, 072001
[30] Adamczyk L *et al.* [STAR Collaboration] 2014 *Phys. Rev. Lett.* **112** 032302
[31] Adamczyk L *et al.* [STAR Collaboration] 2014 *Phys. Rev. Lett.* **113** 092301

[32] Alba P, Alberico W, Bellwied R, Bluhm M, Mantovani Sarti V, Nahrgang M and Ratti C 2014 *Phys. Lett. B* **738** 305
[33] Borsanyi S, Fodor Z, Katz S D, Krieg S, Ratti C and K. Szabo 2012 *JHEP* **1201** 138
[34] Alba P, Vovchenko V, Gorenstein M I and H Stoecker, *Preprint* hep-ph/1606.06542
[35] Abelev B *et al.* [ALICE Collaboration] 2013 *Phys. Rev. C* **88** 044910
[36] Abelev B *et al.* [ALICE Collaboration] 2013 *Phys. Rev. Lett.* **111** 222301
[37] Abelev B *et al.* [ALICE Collaboration] 2014 *Phys. Lett. B* **728** 216
[38] Abelev B *et al.* [ALICE Collaboration] 2015 *Phys. Rev. C* **91** 024609
[39] Becattini F, Grossi E, Bleicher M, Steinheimer J and Stock R 2014 *Phys. Rev. C* **90** no.5, 054907
[40] Preghenella R [ALICE Collaboration] 2012 *Acta Phys. Polon.* B **43** 555
[41] Noronha-Hostler J and Greiner C 2014 *Nucl. Phys. A* **931** 1108
[42] Stoecker H *et al.* 2016 *J. Phys. G* **43** no.1, 015105

Journal of Physics: Conference Series **736** (2016) 012023 doi:10.1088/1742-6596/736/1/012023

Event-Shape Engineering and Jet Quenching

Peter Christiansen
Lund University, Division of Particle Physics, Sweden
E-mail: peter.christiansen@hep.lu.se

Abstract. Event-Shape Engineering (ESE) is a tool that enables some control of the initial geometry in heavy-ion collisions in a similar way as the centrality enables some control of the number of participants. Utilizing ESE, the path length in and out-of plane can be varied while keeping the medium properties (centrality) fixed. In this proceeding it is argued that this provides additional experimental information about jet quenching. Finally, it is suggested that if ESE studies are done in parallel for light and heavy quarks one can determine, in a model independent way, if the path-length dependence of their quenching differs.

1. Introduction

Jet quenching offers a possibility to determine properties of the QGP but it requires that the energy loss mechanism and the effects of the expanding medium are under control. Results from the LHC have demonstrated that by going to high p_{T}, $p_{\mathrm{T}} > 10\,\mathrm{GeV}/c$, the p_{T} spectra are dominated by (quenched) jets and so one avoids complicated overlap effects with soft and intermediate p_{T} physics processes. In approximately this p_{T} region, PHENIX has shown [1] that energy-loss models in general fails at describing both the nuclear modification factor, R_{AA}, and the elliptic-flow coefficient, v_2 (at high p_{T}, v_2 is expected to be entirely due to jet quenching reflecting the azimuthal asymmetry of the path length). Recently, some models have been able to describe both [2, 3], especially for $p_{\mathrm{T}} > 20\,\mathrm{GeV}/c$. However, in this author's opinion, it is a complication that a realistic model must necessarily involve also the initial energy density and the expansion of the medium, and so one might ask if it is possible to construct better experimental discriminators. The goal of this proceeding is to point out that Event-Shape Engineering (ESE) might be such a tool.

This proceeding is outlined as follows. First, the idea is outlined. Second, a concrete prediction is given for a simple energy-loss-scaling model previously developed [4]. Finally, some general ideas are given on how these results can be extended.

2. ESE as a jet quenching tool

The ESE technique [5] relies on the observation that the QGP flows like a nearly ideal (reversible) fluid. This means that, as the initial state ellipticity, ε_2, varies event-by-event, the hydrodynamic flow at low p_{T}, $p_{\mathrm{T}} < 2\,\mathrm{GeV}/c$, will be directly proportional to the ε_2. For a narrow centrality range we therefore have in each event for $p_{\mathrm{T}} < 2\,\mathrm{GeV}/c$ that

$$v_2(p_{\mathrm{T}}) = k_{\mathrm{flow}}(p_{\mathrm{T}})\varepsilon_2, \tag{1}$$

where $k_{\mathrm{flow}}(p_{\mathrm{T}})$ is independent of ε_2 and derivable from viscous hydrodynamic modeling of the QGP.

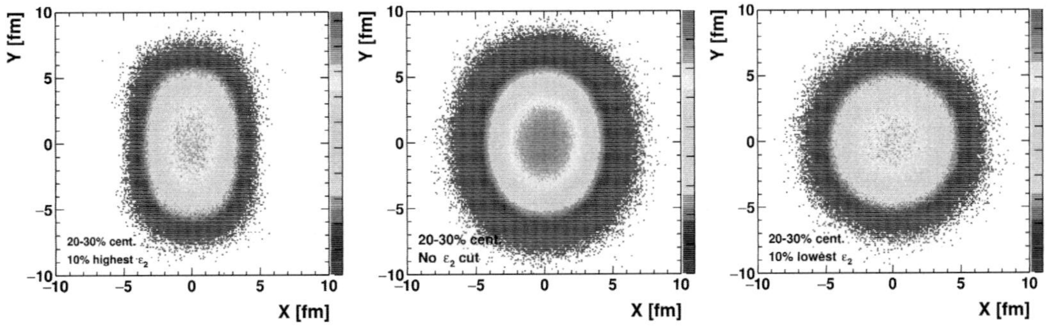

Figure 1. The initial state energy density as given by the N_{part} distribution for a Glauber calculation of 20-30% central Pb-Pb collisions at $\sqrt{s_{\text{NN}}} = 2.76\,\text{TeV}$. The central plot shows the unbiased distribution while the left (right) shows the distribution for the 10% of the events with the highest (lowest) ε_2 that can be studied using ESE.

Figure 1 shows Glauber model calculations for 20-30% central Pb-Pb collisions at $\sqrt{s_{\text{NN}}} = 2.76\,\text{TeV}$ where in two cases a selection has been done on ε_2. As can be seen, this changes the geometry dramatically while keeping the "area" of the medium approximately fixed, which suggests that one does not change the average energy density of the QGP [1] but only the path length azimuthal asymmetry. By cutting on a variable related to the final measured integrated v_2 (the length of the 2nd-order flow vector) one can select (ESE) events with these extreme geometries. Importantly, these ideas have been tested and verified by experiments [7, 6].

By being able to vary the path length, L, while keeping the medium properties fixed, e.g. $\hat{q}$, one constrains the model significantly more than by R_{AA} and v_2 alone. I hope to convince the reader of this in the following.

A caveat of many measurements is that detectors have finite resolution and so selecting the top 10% ε_2 in data will not correspond to the top 10% in models. However, by comparing two ESE selections, a and b, we can construct the ratio, R_{flow}:

$$R_{\text{flow}} \equiv \frac{\langle v_{2,a}(p_{\text{T}})\rangle}{\langle v_{2,b}(p_{\text{T}})\rangle}(p_{\text{T}} < 2\,\text{GeV}/c) = \frac{k_{\text{flow}}(p_{\text{T}})\langle \varepsilon_{2,a}\rangle}{k_{\text{flow}}(p_{\text{T}})\langle \varepsilon_{2,b}\rangle} = \frac{\langle \varepsilon_{2,a}\rangle}{\langle \varepsilon_{2,b}\rangle}, \tag{2}$$

which is, to first order, *independent* of p_{T} for hydrodynamic flow (i.e., for $p_{\text{T}} < 2\,\text{GeV}/c$). We note that this ratio gives us an experimental measure of the observed geometry variation that can be directly compared to geometry variations in model calculations, i.e., finite-resolution effects will make R_{flow} closer to unity, but that can be mimicked in models by selecting broader ε_2 intervals. In Fig. 1, $R_{\text{flow}} = 1.73$ (0.31) for the highest (lowest) ε_2 samples relative to the unbiased sample.

Now we turn to the high p_{T} region and the question of what we expect for the jet ratio

$$R_{\text{quench}} \equiv \frac{\langle v_{2,a}(p_{\text{T}})\rangle}{\langle v_{2,b}(p_{\text{T}})\rangle}(p_{\text{T}} > 10\,\text{GeV}/c). \tag{3}$$

If the jet quenching v_2 grows faster than the flow v_2 when increasing ε_2, then $R_{\text{quench}} > R_{\text{flow}}$ when we compare more asymmetric collisions to unbiased collisions, but $R_{\text{quench}} < R_{\text{flow}}$ when we compare more symmetric collisions to unbiased collisions.

[1] For the concrete model in Sec. 3 the decrease is 10% going from $\varepsilon_2 \approx 0$ to $\varepsilon_2 \approx 0.6$. It seems possible to reduce this bias further by doing the ESE in narrower centrality bins, see e.g. [6].

Journal of Physics: Conference Series **736** (2016) 012023 doi:10.1088/1742-6596/736/1/012023

The basic understanding of energy loss is that in different limits the path-length dependence can vary. If interference between adjacent interactions are important, as is expected for radiation energy loss, then the so-called Landau-Pomeranchuk-Migdal (LPM) effect will give a quadratic (L^2) pathlength dependence. However, if collisional energy loss plays a large role, or we are in the Bethe-Heitler (BH) limit where there is no interference, one expects a linear (L) dependence. Finally, in some strongly interacting QCD-like theories that can be solved via the Maldacena conjecture (the so called AdS-CFT correspondence) one finds a cubic (L^3) path-length dependence. In the following I will try to explain/justify that the jet-quenching v_2 can be approximately factorized like this:

$$v_2(p_\mathrm{T}) \approx k_\mathrm{medium}(p_\mathrm{T}) k_\mathrm{pathlength}(\varepsilon_2), \tag{4}$$

where $k_\mathrm{medium}(p_\mathrm{T})$ encodes the strength of the coupling to the medium and $k_\mathrm{pathlength}(\varepsilon_2)$ encodes the path-length dependence. If we have a light quark and a gluon propagating through the medium then the gluon will in most models couple stronger to the QGP than the quark but the path-length dependence of the energy loss will be the same. Even the energy loss of gluons is therefore larger, the idea here is that, the relative variation of the energy loss (and therefore to some approximation R_AA and v_2) when doing the ESE is the same. For light and heavy quarks the color factor is the same but there can be kinematic differences at a fixed p_T due to the deadcone effect on the radiative energy loss of heavy quarks and the quark-mass difference for collisional and radiative energy loss. An example of the resulting energy-loss differences for light and heavy quarks can be found in Fig. 2 in [8]. But as long as this does not change the path-length dependence then one would again expect the relative changes to be the same. However, the factorization will break down if the path-length dependence varies with p_T (or in a more positive formulation, one can test this factorization with experimental data).

If Eq. 4 is a good approximation then:

$$R_\mathrm{quench} \approx \frac{k_\mathrm{pathlength}(\langle \varepsilon_{2,a} \rangle)}{k_\mathrm{pathlength}(\langle \varepsilon_{2,b} \rangle)}, \tag{5}$$

where we expect that $k_\mathrm{pathlength}(\varepsilon_2)$ is a monotonically increasing function of ε_2.

The claim is therefore that by measuring the ratio of $v_2(p_\mathrm{T})$ at low and high p_T for two ESE classes and taking their ratio one can get rid of the complicated functions $k_\mathrm{flow}(p_\mathrm{T})$ and $k_\mathrm{medium}(p_\mathrm{T})$ to narrow down the path-length dependence, $k_\mathrm{pathlength}(\varepsilon_2)$.

3. Estimating the ratio R_quench-to-R_flow in a simple model

In [4], a simple model that could describe R_AA and v_2 at high p_T was developed. The model describes high-p_T charged hadrons with $p_\mathrm{T} \approx 10\,\mathrm{GeV}/c$ based on the following assumptions/simplifications:

- The initial geometry is obtained from a Glauber calculation. For each event the participating nucleons are centered at (0, 0) and rotated so that the 2nd-order symmetry plane $\Psi_2 = 0$, see e.g. Fig 1.
- Jets are assumed to propagate from (0, 0) and the path length, L, is calculated as the standard deviation of the source in the direction of propagation: σ_X in plane and σ_Y out-of-plane.
- The source is assumed to be static and to have a uniform density of

$$\rho = k \frac{\mathrm{d}N/\mathrm{d}\eta}{4\pi\sigma_X\sigma_Y}, \tag{6}$$

where k is an unspecified constant, and $4\pi\sigma_X\sigma_Y$ is an estimate for the area of the medium.

- Only jets propagating in-plane and out-of-plane are considered.

The underlying idea is, just as in the previous section, that jet quenching depends mainly on two effects: medium density, ρ, and path length, L.

The R_{AA} and v_2 can be combined to estimate the R_{AA} in and out-of plane:

$$R_{AA,in}(p_T) \approx (1 + 2v_2(p_T))R_{AA}(p_T) \quad (7)$$

$$R_{AA,out}(p_T) \approx (1 - 2v_2(p_T))R_{AA}(p_T), \quad (8)$$

using results from ALICE [9] and ATLAS [10]. The first step was to attempt to find a scaling variable $\rho^a L^b$ so that all the $R_{AA,in}$ and $R_{AA,out}$ would follow the same curve. At first, we were unsure if this was possible because one could expect that L is not a good variable for the energy loss in plane where the source is expected to expand. We found that this was possible and took this as an indication that there is little or no effect on the effective path length of the transverse expansion, i.e., the path lengths evaluated for the initial state are meaningful variables for describing jet quenching.

A second issue turned out to be that, once a scaling variable is found the same scaling variable to any power also works. To find a unique solution, we demanded that the energy loss is roughly linear in this scaling variable. The p_T loss has been estimated in several PHENIX publications, e.g. [11], assuming that the difference between the expected and observed p_T spectra is mainly due to a p_T shift (loss). If one can parameterize the pp spectrum dN/dp_T by a power law, $k_1 p_T^{k_2}$, in the relevant p_T range, then one obtains a simple expression for the p_T loss:

$$\Delta p_T = p_{T,0} - p_{T,final} = \left(1 - R_{AA}^{-1/(k_2+1)}\right) p_{T,0}, \quad (9)$$

where $p_{T,0}$ is the initial p_T. This non-linear relation between energy loss and R_{AA} complicates the intuitive understanding of how the function $k_{pathlength}(\varepsilon_2)$ should behave because increasing the energy loss dramatically will eventually lead to much smaller changes in R_{AA}.

In the estimate used here we take into account that quenching not only shifts but also compresses the spectrum.

Figure 2 left shows the main result of [4]. The found scaling relation raises numerous questions that are beyond the scope of this proceeding. Here, I just want to point out that for symmetric systems, where $L_{in} \approx L_{out}$, the scaling variable is *independent* of L, $\Delta p_T \propto \sqrt{dN/d\eta}$. A similar scaling was recently found by PHENIX [13]. Interestingly, if this result can be trusted, this would suggest that small systems produced in pp and p–Pb collisions should also exhibit jet quenching.

The main point of the scaling relation for this proceding is that, once you have such a model then you can further test it by fixing the medium properties (ρ) and varying the path length according to Fig. 1. Figure 2 right shows the predicted p_T loss from this model. To obtain R_{quench} one simply estimates $R_{AA,in}$ and $R_{AA,out}$ using Eq. 9 and then determines v_2 from Eq. 7 and 8. For the concrete model, taking into account that ρ changes slightly when we vary ε_2, we find:

$$10\% \text{ highest } \varepsilon_2: \quad R_{quench} \approx 1.05 R_{flow} \quad (10)$$

$$10\% \text{ lowest } \varepsilon_2: \quad R_{quench} \approx 0.97 R_{flow} \quad (11)$$

This means that jet quenching and flow have similar sensitivity to the azimuthal asymmetry of the initial overlap region.

ATLAS has measured v_2 as a function of p_T up to 15 GeV/c for different ESE classes [10]. Within the large statistical uncertainties the ratio is flat (Fig. 3 top left in the paper), but for a definitive answer one will need the higher statistics of Run 2.

Note that recently similar calculations, as have been outlined in this section, have been done in a realistic model [14].

Journal of Physics: Conference Series **736** (2016) 012023 doi:10.1088/1742-6596/736/1/012023

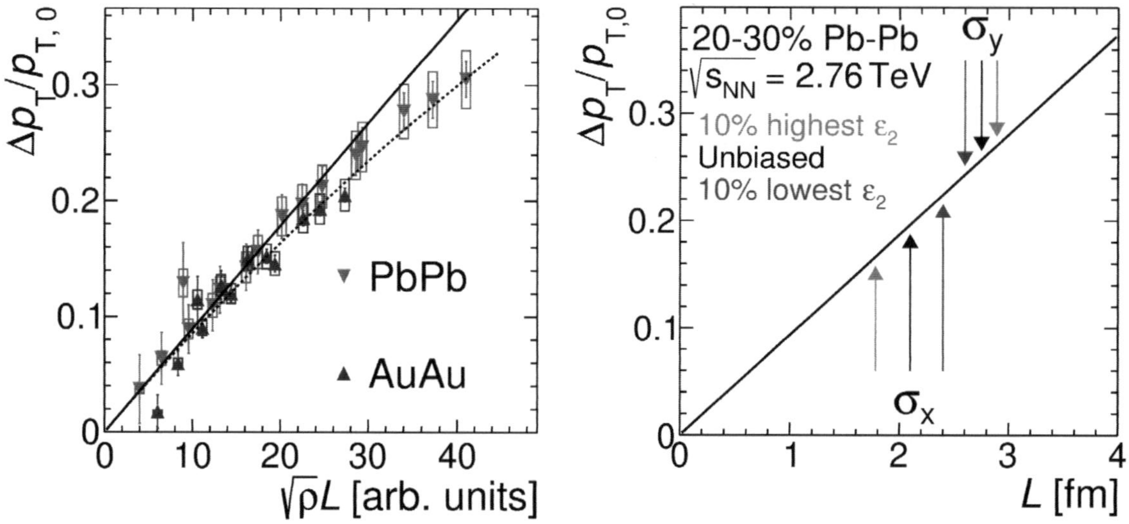

Figure 2. Left: the p_T loss is found to scale with $\sqrt{\rho}L$ in exactly the same way at $\sqrt{s_{NN}} = 2.76$ TeV (PbPb), where the scaling relation was derived, and at $\sqrt{s_{NN}} = 200$ GeV (AuAu, data taken from [12]). The solid line is a linear fit. The dashed line is a fit that takes into account the decreasing p_T of the propagating particle leading to a small non-linear correction for large p_T losses. Right: the p_T losses corresponding to the initial geometries in Fig. 1 (when the small medium density variation is neglected).

4. Conclusions and outlook

The $R_{AA,in}$ and $R_{AA,out}$ have proven to be difficult observables to describe for jet-quenching models. By doing these measurements as a function of the eccentricity one constrains the underlying geometry in a way that simplifies the direct interpretation of the results and that allows further tests of models.

Still, one does not avoid the need of a model to interpret the data. To be able to avoid models one would need two different probes. If one would measure R_{quench} for light *and* heavy quarks then one could in principle compare these directly to understand if the effective path-length dependence of the energy loss is different, as expected from theoretical models. There is no reason to expect that the jet quenching v_2 of light and heavy quarks are the same because the coupling to the medium, $k_{medium}(p_T)$, can be different, but if the path-length dependence, $k_{pathlength}(p_T)$, is the same then one would to first order expect that R_{quench} is the same (due to the non-linearity of energy loss and R_{AA}, small differences can be expected) and so one can in a model independent way test if the path-length dependence, which is directly related to the physics mechanism of the energy loss, is the same or not.

Acknowledgments

The author would like to thank the organizers for a wonderful conference, Vytautas Vislavicius for making the additional Glauber-model calculations for these studies, and Jamie Nagle for comments to the proceeding.

References

[1] **PHENIX** Collaboration, A. Adare *et al.*, "Azimuthal anisotropy of neutral pion production in Au+Au collisions at $\sqrt{s_{NN}} = 200$ GeV: Path-length dependence of jet quenching and the role of initial

geometry," *Phys. Rev. Lett.* **105** (2010) 142301, arXiv:1006.3740 [nucl-ex].

[2] B. Betz and M. Gyulassy, "Examining a reduced jet-medium coupling in Pb+Pb collisions at the Large Hadron Collider," *Phys. Rev. C* **86** (2012) 024903, arXiv:1201.0281 [nucl-th].

[3] B. Betz and M. Gyulassy, "Constraints on the Path-Length Dependence of Jet Quenching in Nuclear Collisions at RHIC and LHC," *JHEP* **08** (2014) 090, arXiv:1404.6378 [hep-ph]. [Erratum: JHEP10,043(2014)].

[4] P. Christiansen, K. Tywoniuk, and V. Vislavicius, "Universal scaling dependence of QCD energy loss from data driven studies," *Phys. Rev. C* **89** (2014) 034912, arXiv:1311.1173 [hep-ph].

[5] J. Schukraft, A. Timmins, and S. A. Voloshin, "Ultra-relativistic nuclear collisions: event shape engineering," *Phys. Lett. B* **719** (2013) 394–398, arXiv:1208.4563 [nucl-ex].

[6] **ALICE** Collaboration, J. Adam *et al.*, "Event shape engineering for inclusive spectra and elliptic flow in Pb-Pb collisions at $\sqrt{s_{\rm NN}} = 2.76$ TeV," *Phys. Rev. C* **93** (2016) 034916, arXiv:1507.06194 [nucl-ex].

[7] **ATLAS** Collaboration, G. Aad *et al.*, "Measurement of the correlation between flow harmonics of different order in lead-lead collisions at $\sqrt{s_{\rm NN}}$=2.76 TeV with the ATLAS detector," *Phys. Rev. C* **92** (2015) 034903, arXiv:1504.01289 [hep-ex].

[8] S. Wicks, W. Horowitz, M. Djordjevic, and M. Gyulassy, "Elastic, inelastic, and path length fluctuations in jet tomography," *Nucl. Phys.* **A784** (2007) 426–442, arXiv:nucl-th/0512076 [nucl-th].

[9] **ALICE** Collaboration, B. Abelev *et al.*, "Centrality Dependence of Charged Particle Production at Large Transverse Momentum in Pb–Pb Collisions at $\sqrt{s_{\rm NN}} = 2.76$ TeV," *Phys. Lett. B* **720** (2013) 52–62, arXiv:1208.2711 [hep-ex].

[10] **ATLAS** Collaboration, G. Aad *et al.*, "Measurement of the pseudorapidity and transverse momentum dependence of the elliptic flow of charged particles in lead-lead collisions at $\sqrt{s_{NN}} = 2.76$ TeV with the ATLAS detector," *Phys. Lett. B* **707** (2012) 330–348, arXiv:1108.6018 [hep-ex].

[11] **PHENIX** Collaboration, S. S. Adler *et al.*, "A Detailed Study of High-p(T) Neutral Pion Suppression and Azimuthal Anisotropy in Au+Au Collisions at s(NN)**(1/2) = 200-GeV," *Phys. Rev. C* **76** (2007) 034904, arXiv:nucl-ex/0611007 [nucl-ex].

[12] **PHENIX** Collaboration, A. Adare *et al.*, "Neutral pion production with respect to centrality and reaction plane in Au+Au collisions at $\sqrt{s_{NN}}$=200 GeV," *Phys. Rev. C* **87** (2013) 034911, arXiv:1208.2254 [nucl-ex].

[13] **PHENIX** Collaboration, A. Adare *et al.*, "Scaling properties of fractional momentum loss of high-p_T hadrons in nucleus-nucleus collisions at $\sqrt{s_{NN}}$ from 62.4 GeV to 2.76 TeV," *Phys. Rev.* **C93** (2016) 024911, arXiv:1509.06735 [nucl-ex].

[14] J. Noronha-Hostler, B. Betz, J. Noronha, and M. Gyulassy, "Soft-Hard Event Engineering in Ultrarelativistic Heavy Ion Collisions," arXiv:1602.03788 [nucl-th].

32nd Winter Workshop on Nuclear Dynamics IOP Publishing
Journal of Physics: Conference Series **736** (2016) 012024 doi:10.1088/1742-6596/736/1/012024

HYDRO + JETS (HYDJET++) event generator for Pb+Pb collisions at LHC

L Bravina[1,4,5], **B H Brusheim Johansson**[1], **J Crkovská**[3], **G Eyyubova**[2], **V Korotkikh**[2], **I Lokhtin**[2], **L Malinina**[2], **E Nazarova**[2], **S Petrushanko**[2], **A Snigirev**[2], **E Zabrodin**[1,2,4,5]

[1] Department of Physics, University of Oslo, Norway
[2] Skobeltsyn Institute of Nuclear Physics, Lomonosov Moscow State University, Moscow, Russia
[3] Institut de Physique Nucléaire, CNRS-IN2P3, Univ. Paris-Sud, Université Paris-Saclay, France
[4] Frankfurt Institute for Advanced Studies, D-60438 Frankfurt a.M., Germany
[5] National Research Nuclear University "MEPhI" (Moscow Engineering Physics Institute), Moscow, Russia

E-mail: larissa.bravina@fys.uio.no

Abstract. The Monte Carlo event generator HYDJET++ is one of the few generators, designed for the calculations of heavy-ion collisions at ultrarelativistic energies, which combine treatment of soft hydro-like processes with the description of jets traversing the hot and dense partonic medium. The model is employed to study the azimuthal anisotropy phenomena, dihadron angular correlations and event-by-event (EbyE) fluctuations of the anisotropic flow in Pb+Pb collisions at $\sqrt{s_{\rm NN}} = 2.76$ TeV. The interplay of soft and hard processes describes the violation of the mass hierarchy of meson and baryon elliptic and triangular flows at $p_{\rm T} \geq 2$ GeV/c, the fall-off of the flow harmonics at intermediate transverse momenta, and the worsening of the number-of-constituent-quark (NCQ) scaling of elliptic/triangular flow at LHC compared to RHIC energies. The cross-talk of v_2 and v_3 leads to emergence of higher order harmonics in the model and to appearance of the *ridge* structure in dihadron angular correlations in a broad pseudorapidity range. HYDJET++ possesses also the dynamical EbyE fluctuations of the anisotropic flow. The model results agree well with the experimental data.

1. Introduction. HYDJET++ model

The Monte Carlo event generator HYDJET++ (HYDrodynamics with JETs) [1] consists of two parts describing the soft and the hard processes, respectively. The generator of soft processes FASTMC [2] was originally designed to help the experimentalists in simulation of significant data samples of heavy-ion collisions at energies of RHIC and LHC. The demands to the model were formulated as follows: (i) it should be able to simulate tens of thousands of central gold-gold or lead-lead events at energies from $\sqrt{s_{\rm NN}} = 200$ GeV to several TeV within relatively modest CPU time; (ii) the yields and transverse momentum spectra of most abundant particles should be very close to real data. To meet these requirements the authors opted for parameterised ideal hydrodynamics. The model was extended soon to non-central nuclear interactions [3]. The first step is the calculation of the effective volume of the fireball $V_{\rm eff}$ which is a subject of the mean number of participating nucleons at given collision centrality. Particle composition is frozen at chemical freeze-out, but the fireball expands further and breaks down at thermal freeze-out temperature, where the contact between hadrons is lost. The final-state interactions (FSI) assume two- and three-body decays of resonances. The table of particles in the model contains more than 360 meson and baryon states including also the charmed ones. This approach is close to the THERMINATOR model [4]. Obviously, the ideal hydrodynamic description of particle spectra is justified for transverse momenta below 2 GeV/c. At higher $p_{\rm T}$ one has to take into account hard processes. These processes are governed by the PYQUEN routine [5], which propagates the hard partons through the hot and dense medium, most presumably quark-gluon plasma. The partons experience collisional and radiative energy losses. At the end of the rescattering stage all emitted quarks and gluons are hadronized according to the Lund model. The number of produced jets in HYDJET++ is proportional to the number of binary nucleon-nucleon collisions at a given impact parameter and the integral cross section of the hard processes in NN collision with minimal transverse momentum transfer.

The synergy between the soft processes and quenched jets became obvious soon after the merging of two independent generators, FASTMC and PYQUEN, into the model called HYDJET++ [1]. Several examples concerning the interplay of hard and soft processes will be discussed in Sec. 2. Then, the model was further upgraded. Namely, after the publication of first data on proton-proton collisions at $\sqrt{s_{\rm NN}} = 7$ TeV it became clear that the standard version of PYTHIA_6.4 [6] should be adjusted. Several tunes have been proposed and the HYDJET++ group has opted for Pro-Q20 tune in the new release of the model [7]. Next was the extension of the model to triangular flow [8].

Recall that the azimuthal distribution of particles can be cast [9, 10] in the form of Fourier series

$$E\frac{d^3N}{d^3p} = \frac{1}{2\pi}\frac{d^2N}{p_{\rm T}dp_{\rm T}dy}\left\{1 + \sum_{n=1}^{\infty} 2v_n \cos\left[n(\phi - \Psi_n)\right]\right\} . \tag{1}$$

Here ϕ, $p_{\rm T}$ and y are the azimuthal angle, the transverse momentum and the rapidity of a particle, respectively. Ψ_n is the azimuth of the corresponding event plane, and the sum of harmonics in the rhs of Eq. (1) represents anisotropic flow. The coefficients v_n are dubbed directed flow v_1, elliptic flow v_2, triangular flow v_3 and so forth. In HYDJET++ elliptic and triangular flows arise because of the corresponding spatial eccentricities of the fireball. The radii of the elliptic and triangular spatial eccentricities are defined as function of the impact parameter b and azimuthal angle ϕ as follows [3, 8] :

$$R_{\rm ell}(b,\phi) = R_{\rm fr}\left[\frac{1-\varepsilon^2(b)}{1+\varepsilon^2(b)\cos 2(\phi-\Psi_2)}\right]^{1/2} , \tag{2}$$

$$R_{\rm trian}(b,\phi) = R_{\rm ell}(b,\phi)\{1+\varepsilon_3(b)\cos 3(\phi-\Psi_3)\} , \tag{3}$$

where $R_{\rm fr}(b) = R_0\sqrt{1-\varepsilon(b)}$, and R_0 is the freeze-out radius of the fireball in a central collision. Two free parameters, $\varepsilon(b)$ and $\varepsilon_3(b)$, control the ellipticity and triangularity of the fireball. Ψ_2 and Ψ_3 are the azimuths of the corresponding event planes randomly distributed with respect to each other. The pressure gradients are stronger in the direction of short axis of the ellipsoid, however, the momentum anisotropy angle $\phi_{\rm fl}$ is related to the spatial anisotropy ϕ via

$$\frac{\tan\phi_{\rm fl}}{\tan\phi} = \left[\frac{1-\delta(b)}{1+\delta(b)}\right]^{1/2} , \tag{4}$$

containing the third (and the last) free parameter, $\delta(b)$, responsible for the formation of anisotropic flow in HYDJET++. The cross-talk of elliptic and triangular harmonics leads to appearance of both even and odd higher harmonics of the anisotropic flow in the model because of the nonlinear contributions to v_n from v_2, v_3 or their product v_2v_3. This interplay explains also the formation of ridge in long-range dihadron correlations, including the characteristic double-bump profile of the ridge at the away-side [11]. The last developments of the model deal with its extension to open and hidden charm production [12] and to EbyE fluctuations of the anisotropic flow [13]. The aspects of the calculations are presented in Sec. 2. Conclusions are drawn in Sec. 3.

2. Interplay of soft processes, jets and final-state interactions

Consequences for elliptic and triangular flows. Hadrons produced in soft processes at the freeze-out hypersurface should carry collective flow, whereas the flow assigned to jet particles with intermediate transverse momenta is essentially zero. At $p_T \geq 4$ GeV/c hadrons decoupled from jets can develop a weak anisotropic flow because of the well-known effect of jet quenching. Decays of resonances also modify the p_T distributions of the flow excitation functions. In HYDJET++ these effects were studied in non-central heavy-ion collisions at RHIC and LHC energies in [14, 15] for the elliptic flow and in [16, 17] for the triangular flow, respectively.

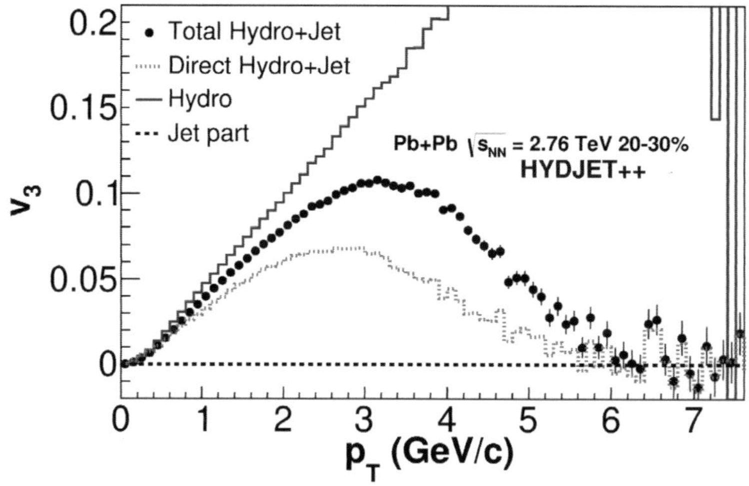

Figure 1. The p_T dependence of the triangular flow of charged particles produced in soft+hard (circles) and soft processes (solid line), from the jets (dashed line) and direct particles (dotted line) in HYDJET++ in Pb + Pb collisions at $\sqrt{s_{NN}} = 2.76$ TeV at centrality 20-30%.

To illustrate the typical features of the development of the flow components we plot in Fig. 1 triangular flow of charged hadrons produced in Pb+Pb collisions at $\sqrt{s_{NN}} = 2.76$ TeV at centrality $\sigma/\sigma_{geo} = 20 - 30\%$. The partial flow of jet hadrons is absent, while the flow of particles governed by hydrodynamics increases with rising p_T. Hadrons from the soft processes dominate the particle spectrum at $p_T \leq 2$ GeV/c, and the total v_3 increases in this p_T range. At higher transverse momenta the fraction of jet hadrons prevails over the "soft" ones. Thus, the triangular flow in HYDJET++ decreases after a certain p_T. The lighter the hadron, the smaller is the value of transverse momentum where the spectra of hadrons originated from the soft and the hard processes cross each other. This circumstance explains the violation of the mass ordering of hadron elliptic and triangular flows. One can also see in Fig. 1 that the decays of resonances increase the maximum of the $v_3(p_T)$ distribution by about 20% and shift its position to higher p_T. However, 90% of hadrons have the transverse momenta less than 1 GeV/c; therefore, the p_T-integrated values of v_3 at selected centralities are changed insignificantly [17].

Another important result of the interplay of soft and hard processes is the worsening of the number-of-constituent-quark (NCQ) scaling for both elliptic and triangular flow at LHC energies compared to the RHIC ones. The $v_2(KE_T/n_q)/n_q$ distributions of most abundant hadron species, where $KE_T = m_T - m_0$ is the transverse kinetic energy and n_q is the number of constituent quarks, are shown in Fig. 2. Experimental results of ALICE collaboration [18] are plotted onto the calculations as well. We see that the model provides a fair description of the data.

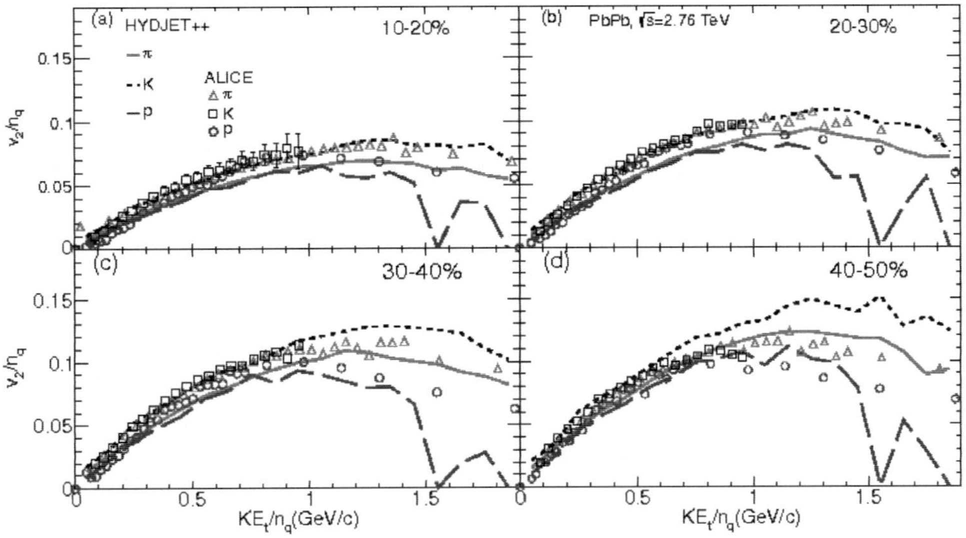

Figure 2. The KE_T/n_q dependence of elliptic flow of pions, kaons and protons in Pb+Pb collisions at $\sqrt{s_\mathrm{NN}} = 2.76$ TeV at different centralities. Curves denote the model results, ALICE data from [18] are shown by symbols.

The scaling is approximately fulfilled within the 20% accuracy limit in the interval $0.1 \leq KE_\mathrm{T} \leq 0.7$ GeV. The worsening of the NCQ scaling conditions for v_2 at LHC was predicted in HYDJET++ in [14, 15]. Our study shows that the effect takes place for both v_2 [14, 19] and v_3 [16, 17] distributions. Scaling definitely holds in the hydrodynamic sector of the model. Moreover, FSI work towards its fulfilment, because many light hadrons, especially pions, get the feeddown from the decays of heavy resonances. But jets, which are more influential at LHC energies compared to that of RHIC, is the main reason causing the NCQ-scaling violation with increasing collision energy.

Cross-talk of elliptic and triangular flows. The present version of HYDJET++ contains no genuine flow harmonics related to the eccentricities of order higher than three. Higher flow harmonics v_n, $n \geq 4$ arise due to nonlinear contributions of v_2 and v_3 [8, 20]. The detailed comparison of model predictions with the data shows [8] that HYDJET++ underpredicts the magnitude of $v_4(p_\mathrm{T})$ - $v_6(p_\mathrm{T})$ signals in central events 0-5%, but for more peripheral collisions the agreement between the model results and the data is much better. It means that nonlinear contributions of v_2 and v_3 to higher harmonics dominate over the intrinsic momentum anisotropy v_n caused by spatial eccentricity ε_n. Of particular interest is the hexagonal flow v_6, because in pure hydrodynamic approximation this harmonic depends on independent contributions coming from v_2 and v_3 [20]:

$$v_6 \approx \frac{1}{6}v_2^3 + \frac{1}{2}v_3^2 \tag{5}$$

The ratio $v_n^{1/n}/v_2^{1/2}$ was proposed in [21] to check the possible scaling trends. We employed this ratio for the hexagonal flow, and used the v_6 defined either in Ψ_2 or Ψ_3 plane, but not in the own Ψ_6 plane. Results [20] are displayed in Fig. 3 for $v_6^{1/6}/v_2^{1/2}$ and in Fig. 4 for $v_6^{1/6}/v_3^{1/3}$ distributions.

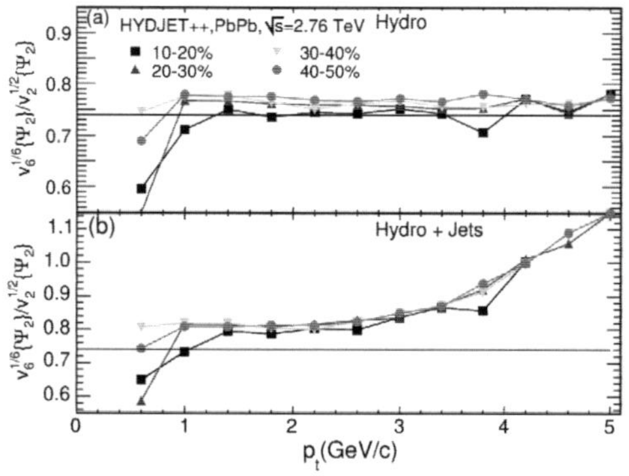

Figure 3. (a) Ratio $v_6^{1/6}/v_2^{1/2}$ in the Ψ_2 event plane for charged hadrons from soft processes calculated in HYDJET++ for Pb+Pb collisions at 2.76 TeV at several centralities. (b) The same as (a) but for both soft and hard processes.

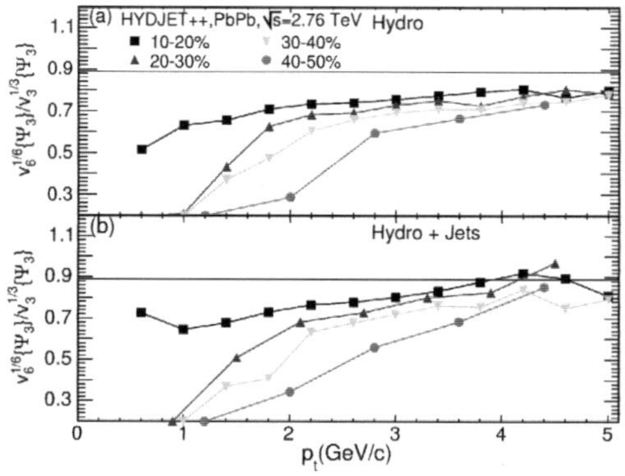

Figure 4. (a) Ratio $v_6^{1/6}/v_3^{1/3}$ in the Ψ_3 event plane for charged hadrons from soft processes calculated in HYDJET++ for Pb+Pb collisions at 2.76 TeV at several centralities. (b) The same as (a) but for both soft and hard processes.

For the $v_6^{1/6}/v_2^{1/2}$ distributions in Fig. 3 the independence of the ratio on centrality of the collision is observed at $p_T \geq 1$ GeV/c. All curves are on the top of each other. For the particles produced solely in soft processes this ratio is indeed close to $(1/6)^{1/6}$, whereas jets increase it by about 10% and lead to the rise of high-p_T tails of the distributions at $p_T \geq 3$ GeV/c. For the $v_6^{1/6}/v_3^{1/3}$ ratios the centrality hierarchy is revealed instead of the scaling. Here the ratio drops as the reactions become more peripheral. This may be explained by the significant rise of the elliptic flow with increasing impact parameter b, while the rise of the triangular flow in more peripheral collisions is not so dramatic. As a result, the event plane Ψ_6 becomes more correlated with the plane Ψ_2 rather than with Ψ_3. This is in line with the experimental observations [22].

Next interesting issue is the study of two-particle angular correlations. The two-particle correlation function is typically defined as the ratio of pair distribution in the event to the combinatorial background of uncorrelated particles. In the flow dominated regime the pair angular distribution reads [cf. Eq.(1)]

$$\frac{dN^{pairs}}{d\Delta\varphi} \propto 1 + 2\sum_{n=1}^{\infty} V_n(p_T^{tr}, p_T^{a}) \cos n(\Delta\varphi) \ , \tag{6}$$

where $\Delta\varphi = \varphi^{tr} - \varphi^{a}$, and indices "tr" and "a" indicate the so-called "trigger" and "associated" particle, respectively. The study of angular dihadron correlations in relativistic heavy-ion collisions revealed the long-range correlations dubbed "ridge" [23, 24]. Many interesting options have been proposed for the description of the ridge phenomenon, for instance Cerenkov gluon radiation or Mach-cone of shock waves. The authors of [25] suggested that the triangular flow should be important for understanding of this signal. HYDJET++ is ideally suited for such a check, because the long-range correlations in the model appear merely due to the collective flow.

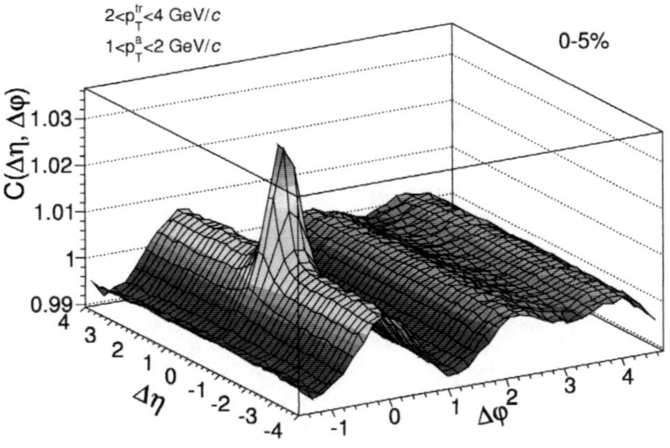

Figure 5. 2D correlation function in HYDJET++ in Pb+Pb collisions at $\sqrt{s} = 2.76$ TeV for $2 < p_{\mathrm{T}}^{\mathrm{tr}} < 4$ GeV/c and $1 < p_{\mathrm{T}}^{\mathrm{a}} < 2$ GeV/c for centrality 0-5% with both v_2 and v_3 present.

As was shown in [11], no long range azimuthal correlations at the near-side or away-side ($\Delta\varphi \approx \pi$) appear in the case of perfect central collision with $b = 0$, when both elliptic and triangular flows are absent. If only the elliptic flow is present, whereas the triangular flow is switched off, the long range correlations start to appear at the both sides. However, only the presence of the triangular flow in addition to the elliptic one leads to development of ridge at near-side and, simultaneously, to formation of characteristic double-hump structure at the away-side, as seen in Fig. 5, in full agreement with the experimental observations.

Event-by-Event (EbyE) fluctuations. The EbyE distributions of harmonics of anisotropic flow in lead-lead collisions at LHC were studied, e.g., by ATLAS Collaboration in [26]. The results were obtained after the application of the so-called *unfolding* procedure [27] in order to extract the "true" value of the flow vector and get rid of the nonflow effects caused by the finite event multiplicities, jet fragmentation and decays of resonances. The procedure is cumbersome, so often people simply rescale their predictions to make a comparison with the data. Our analysis shows [13] that such a simplistic approach is not always justified.

As an input, one selects the spectra of charged particles with $p_{\mathrm{T}} \geq 0.5$ GeV/c and $|\eta| < 2.5$, corresponding to ATLAS kinematic cuts. Angular distribution of particles is modified as

$$\frac{dN}{d\varphi} \propto 1 + 2\sum_{n=1}^{\infty} V_n^{\mathrm{obs}} \cos\left[n(\varphi - \Psi_n^{\mathrm{obs}})\right] = 1 + 2\sum_{n=1}^{\infty} \left(V_{n,x}^{\mathrm{obs}} \cos n\varphi + V_{n,y}^{\mathrm{obs}} \sin n\varphi\right) , \quad (7)$$

with V_n^{obs} being the magnitude of the observed per-particle flow vector, whereas Ψ_n^{obs} represents the azimuth of the observed event plane. Then, the single-particle event-by-event distributions are constructed

$$\begin{aligned} V_n^{\mathrm{obs}} &= \sqrt{(V_{n,x}^{\mathrm{obs}})^2 + (V_{n,y}^{\mathrm{obs}})^2} , \\ V_{n,x}^{\mathrm{obs}} &= V_n^{\mathrm{obs}} \cos n\Psi_n^{\mathrm{obs}} = \langle \cos n\varphi \rangle , \\ V_{n,y}^{\mathrm{obs}} &= V_n^{\mathrm{obs}} \sin n\Psi_n^{\mathrm{obs}} = \langle \sin n\varphi \rangle . \end{aligned} \quad (8)$$

The two sub-events (2SE) method subdivides the event sample further into two sub-groups containing charged particles emitted in forward and backward hemispheres in the c.m. system.

The difference between the EbyE flow vectors (to exclude the collective flow) of the two sub-events is fitted to the Gaussian with the width $\delta_{2\mathrm{SE}} = 2\delta$, which enters the response function [26]

$$P(V_n^{\mathrm{obs}}|V_n) \propto V_n^{\mathrm{obs}} \exp\left[-\frac{(V_n^{\mathrm{obs}})^2 + V_n^2}{2\delta^2}\right] I_0\left(\frac{V_n^{\mathrm{obs}} V_n}{\delta^2}\right) . \quad (9)$$

The obtained response function is then used as an input to the iteration procedure [27] allowing us to find the Bayesian unfolding. The effects of finite multiplicity and nonflow processes give rise to the nonzero value of $\delta_{2\mathrm{SE}}$. The EbyE unfolding procedure significantly subtracts these contributions and leave the dynamical flow fluctuations only.

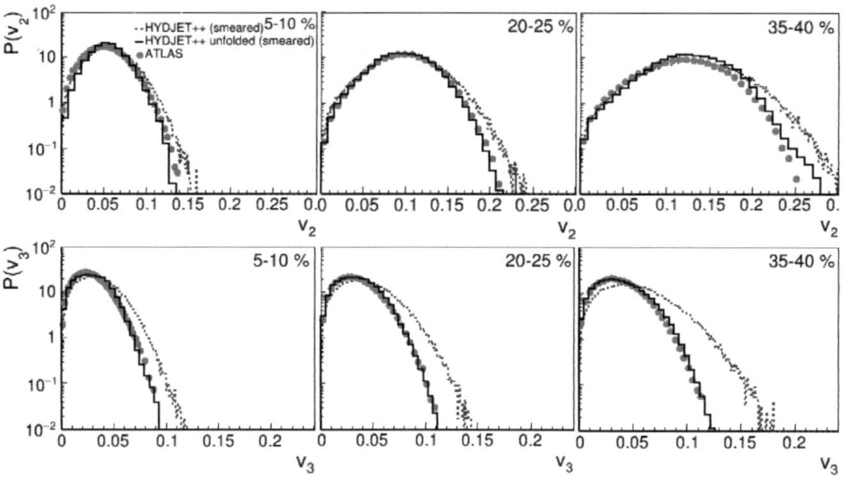

Figure 6. The probability density distributions of elliptic flow V_2 (upper row) and triangular flow V_3 (bottom row) in three centrality intervals: $5-10\%$ (left), $20-25\%$ (middle) and $35-40\%$ (right). Dashed and solid histograms present the results for simulated HYDJET++ events before and after the unfolding procedure, respectively. The full circles are the ATLAS data from [26].

Figure 6 displays the probability density distributions of elliptic and triangular EbyE flows obtained in three centrality intervals: $\sigma/\sigma_{geo} = 5-10\%$, $20-25\%$, and $35-40\%$. To describe the data we allow for variations of $\varepsilon(b)$ and $\varepsilon_3(b)$ [13]. Now the values of both parameters are smeared normally around their previously fixed values. The width proportionality coefficients are tuned to fit the data at a *single arbitrary centrality*, say $10-15\%$ or $20-25\%$, and the obtained values are used then for all other centralities. The initial $P(V_n)$ distributions are broader than the unfolded ones only in the areas of relatively high flow values. The agreement of unfolded spectra with the data is very good. Since the unfolding suppresses strongly the non-flow fluctuations, Fig. 6 confirms the dynamical origin of the flow fluctuations in HYDJET++.

3. Conclusions

Anisotropic flow, azimuthal dihadron correlations and event-by-event fluctuations of the flow harmonics in Pb+Pb collisions at $\sqrt{s_{\mathrm{NN}}} = 2.76$ TeV are studied within the hybrid hydro+jets model HYDJET++. Several features are observed. Hadrons, produced in jet fragmentation, display very weak flow because of the jet quenching effect. These hadrons dominate particle spectrum at certain p_{T}; the heavier the particle, the larger the transverse momentum. Such an interplay of soft and hard processes explains (i) the breaking of mass ordering of p_{T} distributions of anisotropic flow; (ii) the falloff of the flow harmonics at intermediate transverse momenta; (iii) violation of the NCQ scaling at LHC energies compared to RHIC ones, because hard processes become more abundant with rising collision energy. The cross-talk of two harmonics, v_2 and v_3, leads to (i) long-range azimuthal dihadron correlations (ridge); (ii) formation of the characteristic double-hump structure at the away side; (iii) nonlinear contributions to higher order harmonics, e.g., to hexagonal flow, which is correlated with the triangular flow in central collisions and becomes more correlated with the elliptic flow in more peripheral ones. Analysis of EbyE flow fluctuations by means of the unfolding procedure reveals their dynamical nature in HYDJET++. Its origin is traced to the correlations between the coordinates and momenta of the hadrons and the velocities of the fluid cells.

Acknowledgments
LB acknowledges financial support of the Alexander von Humboldt Foundation.

References
[1] Lokhtin I P, Malinina L V, Petrushanko S V, Snigirev A M, Arsene I and Tywoniuk K 2009 *Comput. Phys. Commun.* **180** 779
[2] Amelin N S, Lednicky R, Pocheptsov T A, Lokhtin I P, Malinina L V, Snigirev A M, Karpenko Iu A and Sinyukov Yu M 2006 *Phys. Rev.* C **74** 064901
[3] Amelin N S, Lednicky R, Lokhtin I P, Malinina L V, Snigirev A M, Karpenko Iu A, Sinyukov Yu M, Arsene I and Bravina L V 2008 *Phys. Rev.* C **77** 014903
[4] Kisiel A, Taluc T, Broniowski W and Florkowski W 2006 *Comput. Phys. Commun.* **174** 669
[5] Lokhtin I P and Snigirev A M 2006 *Eur. Phys. J.* C **46** 211
[6] Sjostrand T, Mrenna S and Skands P Z 2006 *J. High Energy Phys.* JHEP05(2006)026
[7] Lokhtin I P, Belyaev A V, Malinina L V, Petrushanko S V, Rogochaya E R and Snigirev A M 2012 *Eur. Phys. J.* C **72** 2045
[8] Bravina L V, Brusheim Johansson B H, Eyyubova G Kh, Korotkikh V L, Lokhtin I P, Malinina L V, Petrushanko S V, Snigirev A M and Zabrodin E E 2014 *Eur. Phys. J.* C **74** (2014) 2807
[9] Voloshin S and Zhang Y 1996 *Z. Phys.* C **70** 665
[10] Poskanzer A M and Voloshin S A 1998 *Phys. Rev.* C **58** 1671
[11] Eyyubova G Kh, Korotkikh V L, Lokhtin I P, Petrushanko S V, Snigirev A M, Bravina L V and Zabrodin E E 2015 *Phys. Rev.* C **91** 064907
[12] Lokhtin I P, Belyaev A V, Eyyubova G Kh, Ponimatkin G and Pronina E Yu 2016 Charmed meson production pattern in PbPb collisions at the LHC *Preprint* hep-ph/1601.00799
[13] Bravina L V, Fotina E S, Korotkikh V L, Lokhtin I P, Malinina L V, Nazarova E N, Petrushanko S V, Snigirev A M and Zabrodin E E 2015 *Eur. Phys. J.* C **75** 588
[14] G. Eyyubova G, Bravina, Korotkih V L, Lokhtin I P, Malinina L V, Petrushanko S V, Snigirev A M and Zabrodin E 2009 *Phys. Rev.* C **80** 064907
[15] Zabrodin E E, Bravina L V, Eyyubova G Kh, Lokhtin I P, Malinina L V, Petrushanko S V and Snigirev A M 2010 *J. Phys.* G **37** 094060
[16] Zabrodin E E, Bravina L V, Brusheim Johansson B H, Crkovská J, Eyyubova G Kh, Korotkikh V L, Lokhtin I P, Malinina L V, Petrushanko S V and Snigirev A M 2016 *J. Phys.: Conf. Ser.* **668** 012099
[17] Crkovská J *et al* Influence of jets and final-state interactions on the triangular flow in ultrarelativistic heavy-ion collisions 2016 *Preprint* hep-ph/1603.09621
[18] Noferini F *et al* (ALICE Collaboration) 2013 *Nucl. Phys.* A **904-905** 438c
[19] Bravina L, Brusheim Johansson B H, Eyyubova G and Zabrodin E 2013 *Phys. Rev.* C **87** 034901
[20] Bravina L V, Brusheim Johansson B H, Eyyubova G Kh, Korotkikh V L, Lokhtin I P, Malinina L V, Petrushanko S V, Snigirev A M and Zabrodin E E 2014 *Phys. Rev.* C **89** 024909
[21] Aad G *et al* (ATLAS Collaboration) 2012 *Phys. Rev.* C **86** 014907
[22] Jia J *et al* (ATLAS Collaboration) 2013 *Nucl. Phys.* A **910-911** 276
[23] Adare A *et al* (PHENIX Collaboration) 2008 *Phys. Rev.* C **78** 014901
[24] Aamodt K *et al* (ALICE Collaboration) 2011 *Phys. Rev. Lett.* **107** 032301
[25] Alver B and Roland G 2010 *Phys. Rev.* C **81** 054905
[26] Aad G *et al* (ATLAS Collaboration) 2013 *J. High Energy Phys.* JHEP11(2013)183
[27] Adye T 2011 Unfolding algorithms and tests using RooUnfold *Preprint* physics.data-an/1105.1160

Journal of Physics: Conference Series **736** (2016) 012025 doi:10.1088/1742-6596/736/1/012025

The color glass condensate and the average transverse momentum in proton-nucleus collisions

F.O. Durães[1], A.V. Giannini[2], V.P. Gonçalves[3] and F.S. Navarra[2]

[1] Curso de Física, Escola de Engenharia, Universidade Presbiteriana Mackenzie, CEP 01302-907, São Paulo, Brazil

[2] Instituto de Física, Universidade de São Paulo, CEP 05315-970 São Paulo, SP, Brazil

[3] Instituto de Física e Matemática, Universidade Federal de Pelotas, CEP 96010-900, Pelotas, RS, Brazil

E-mail: duraes@mackenzie.br

E-mail: avgiannini@usp.br

E-mail: victorpbg@gmail.com

E-mail: navarra@if.usp.br

Abstract.
We compute the energy and rapidity dependence of the average transverse momentum $\langle p_T \rangle$ in pp and pA collisions at RHIC and LHC energies in the framework of the Color Glass Condensate (CGC) formalism. We update previous predictions for the p_T - spectra using the hybrid formalism of the CGC approach and two phenomenological models for the dipole - target scattering amplitude. We demonstrate that these models are able to describe the RHIC and LHC data for the hadron production in pp, dAu and pPb collisions at $p_T \leq 20$ GeV. Moreover, we present our predictions for $\langle p_T \rangle$ and demonstrate that the ratio $\langle p_T(y) \rangle / \langle p_T(y=0) \rangle$ decreases with the rapidity and has a behavior similar to that predicted by hydrodynamical calculations.

1. Introduction

In the LHC experiments we can study several aspects of *forward physics* as, for instance, soft and hard diffraction, exclusive production of new mass states, low-x dynamics and other important topics [1]. Forward physics is characterized by the production of particles with relatively small transverse momentum, being traditionally associated with soft particle production, which is intrinsically non perturbative and not amenable to first-principles analysis. However, in particle production at large energies and forward rapidities, the wave function of one of the projectiles is probed at large Bjorken x and that of the other at very small x. The latter is characterized by a large number of gluons, which is expected to form a new state of matter - the Color Glass Condensate (CGC) - where the gluon distribution saturates and non linear phenomena dominate. Such a system is endowed with a new dynamical momentum scale, the saturation scale Q_s, which controls the main features of particle production. At large energies and rapidities, Q_s is expected to become much larger than the QCD confinement scale Λ_{QCD}. Furthermore, the saturation scale is expected to determine the typical transverse momentum of the produced partons in the interaction.

Journal of Physics: Conference Series **736** (2016) 012025 doi:10.1088/1742-6596/736/1/012025

In a recent paper [2] it was suggested that the analysis of $\langle p_T \rangle$ in pp and pA collisions can be used to disentangle the hydrodynamical and the CGC descriptions of the "ridge" effect observed in high multiplicity events in small colliding systems such as pp and $p(d)A$. While the ridge-type structure previously observed in heavy-ion collisions at RHIC and the LHC was considered as an evidence of the hydrodynamical nature of the quark-gluon-plasma, there is no compelling reason why small systems should also exhibit a hydrodynamical behavior even though a hydro approach is able to describe the experimental data. On the other hand, the CGC approach also provides a qualitatively good description of the same data. Therefore, the origin of the ridge in pp and pA collisions is still an open question. Like the ridge effect, the azimuthal asymmetries observed in pPb collisions at the LHC energies can also be understood with different theoretical explanations. While in the hydro approaches those anisotropies emerge as a final state feature due to the hydrodynamic flow, in the CGC approach they are described as a initial state anisotropies which are present at the earliest stages of the collision.

In Ref. [2], the authors have studied the rapidity (y) dependence of the average transverse momentum of charged particles using very general arguments that lead to simple analytical expressions. In particular, the Golec - Biernat – Wusthoff (GBW) model [3] was used to describe the unintegrated gluon distribution and the fragmentation of the partons into final state particles was neglected. The authors of [2] have found that the average transverse momentum $\langle p_T \rangle$ in the CGC approach grows with rapidity, in contrast to what is expected from a collective expansion. Indeed, the hydrodynamical model predicts a decrease of the average transverse momentum when going from midrapidity, $y = 0$, to the proton side, owing to a decreasing number of produced particles. These different predictions of the hydrodynamical and CGC approaches motivate us to perform a more careful analysis of the energy and rapidity dependencies of $\langle p_T \rangle$. As the GBW model does not describe the p_T - spectra measured in pp/pA collisions, in our study we will consider two more realistic phenomenological saturation models that are able to reproduce the experimental data in the region of small transverse momenta. This is the region that determines the behavior of the average transverse momentum. Moreover, we will analyse the impact of the inclusion of parton fragmentation in the rapidity dependence of $\langle p_T \rangle$. With these improvements, we are able to present realistic predictions for $\langle p_T \rangle$ based on the CGC results that are able to describe the current experimental data on hadron production in hadronic collisions.

2. Particle production in the CGC hybrid formalism

Forward hadron production in hadron-hadron collisions is a typical example of a dilute-dense process [4] which is an ideal system to study the small-x components of the target wave function. In this case the cross section is expressed as a convolution of the standard parton distributions for the dilute projectile, the dipole-hadron scattering amplitude (which includes the high-density effects) and the parton fragmentation functions. Basically, assuming this generalized dense-dilute factorization, the minimum bias invariant yield for single-inclusive hadron production in hadron-hadron processes is described in the CGC formalism by [5]

$$\frac{dN_h}{dyd^2p_T} = \frac{K(y)}{(2\pi)^2}\int_{x_F}^{1} dx_1 \, \frac{x_1}{x_F}\left[f_{q/p}(x_1,\mu^2)\,\widetilde{\mathcal{N}}_F\left(\frac{x_1}{x_F}p_T, x_2\right) D_{h/q}\left(\frac{x_F}{x_1},\mu^2\right) \right. \\ \left. + \, f_{g/p}(x_1,\mu^2)\,\widetilde{\mathcal{N}}_A\left(\frac{x_1}{x_F}p_T, x_2\right) D_{h/g}\left(\frac{x_F}{x_1},\mu^2\right)\right] , \tag{1}$$

where p_T, y and x_F are the transverse momentum, rapidity and the Feynman-x of the produced hadron, respectively. The $K(y)$-factor mimics the effect of higher order corrections and, effectively, that of other dynamical effects not included in the CGC formulation. The variable x_1 denotes the momentum fraction of a projectile parton, $f(x_1, \mu^2)$ the projectile parton distribution

Journal of Physics: Conference Series **736** (2016) 012025 doi:10.1088/1742-6596/736/1/012025

functions and $D(z, \mu^2)$ the parton fragmentation functions into hadrons. These quantities evolve according to the DGLAP evolution equations and obey the momentum sum-rule. It is useful to assume $\mu^2 = p_T^2$. Moreover, $x_F = \frac{p_T}{\sqrt{s}} e^y$ and the momentum fraction of the target partons is given by $x_2 = x_1 e^{-2y}$ (For details see e.g. [5]). In Eq. (1), $\widetilde{\mathcal{N}}_F(x,k)$ and $\widetilde{\mathcal{N}}_A(x,k)$ are the fundamental and adjoint representations of the forward dipole amplitude in momentum space and are given by

$$\widetilde{\mathcal{N}}_{A,F}(x, p_T) = \int d^2r \, e^{i\vec{p_T}\cdot\vec{r}} \left[1 - \mathcal{N}_{A,\mathcal{F}}(x,r)\right] \,, \tag{2}$$

where $\mathcal{N}_{A,\mathcal{F}}(x,r)$ encodes all the information about the hadronic scattering, and thus about the non-linear and quantum effects in the hadron wave function. Following [6], we will assume in what follows that $\mathcal{N}_F(x,r)$ can be obtained from $\mathcal{N}_A(x,r)$ after rescaling the saturation scale by $Q^2_{s,F} = (C_F/C_A)Q^2_{s,A}$ where $C_F/C_A = 4/9$. We will consider two different phenomenological models based on the analytical solutions of this equation. This allows us to investigate the possibility of getting a first insight on whether or not the LHC data are sensitive to geometric scaling violations at high values of p_T. Moreover, as these phenomenological models differ from the GBW model in the dependence of the anomalous dimension with the momentum scale (see below), it becomes possible to clarify the origin of the differences between the predictions.

Several groups have constructed phenomenological models for the dipole scattering amplitude using the RHIC and/or HERA data to fix the free parameters [5, 6, 7, 8]. In general, it is assumed that $\mathcal{N}$ can be modelled through a simple Glauber-like formula,

$$\mathcal{N}(x, r_T) = 1 - \exp\left[-\frac{1}{4}(r_T^2 Q_s^2)^\gamma\right] \,, \tag{3}$$

where γ is the anomalous dimension of the target gluon distribution. The speed with which we move from the non linear regime to the extended geometric scaling regime and then from the latter to the linear regime is what differs one phenomenological model from another. This transition speed is dictated by the behavior of the anomalous dimension $\gamma(x, r_T^2)$. In the GBW model, γ is assumed to be constant and equal to one. In this paper we will consider the dipole models proposed in Refs. [5, 6] to describe the p_T spectra of particles produced at RHIC. In the DHJ model [5], the anomalous dimension is given by

$$\gamma(x, r_T)_{DHJ} = \gamma_s + (1-\gamma_s)\frac{|\log(1/r_T^2 Q_s^2)|}{\lambda y + d\sqrt{y} + |\log(1/r_T^2 Q_s^2)|}. \tag{4}$$

with $Q_s^2 = A^{1/3} Q_0^2 (x_0/x_2)^\lambda$, $\gamma_s = 0.628$, $Q_0^2 = 1.0$ GeV^2, $x_0 = 3.0 \cdot 10^{-4}$, $\lambda = 0.288$ and $d = 1.2$. This model was designed to describe the forward dAu data at the highest RHIC energy taking into account geometric scaling violations characterized by terms depending on the target rapidity, $y = \log(1/x_2)$, in its parametrization of the anomalous dimension, with the parameter d controlling the strength of the subleading term in y. In contrast, in the BUW model [6] the anomalous dimension is given by

$$\gamma(\omega = q_T/Q_s)_{BUW} = \gamma_s + (1-\gamma_s)\frac{(\omega^a - 1)}{(\omega^a - 1) + b} \,, \tag{5}$$

where $q_T = p_T/z$ is the parton momentum. The parameters of the model ($\gamma_s = 0.628$, $a = 2.82$ and $b = 168$) have been fixed by fitting the p_T-spectra of the produced hadrons measured in pp and dAu collisions at RHIC energies. With these parameters the model was also able to describe the ep HERA data for the proton structure function if the light quark masses are neglected. An important feature of this model is the fact that it explicitly satisfies the property of geometric scaling. Since the forward RHIC data on p_T-spectra are reproduced by both models [6, 5], it was not possible to say whether experimental data show violations of the geometric scaling or not. In principle, it is expected that by considering the transverse momentum distribution of produced hadrons measured at the LHC energies it should be possible to address this question since the new data are taken at a wider range of p_T when compared to the RHIC data.

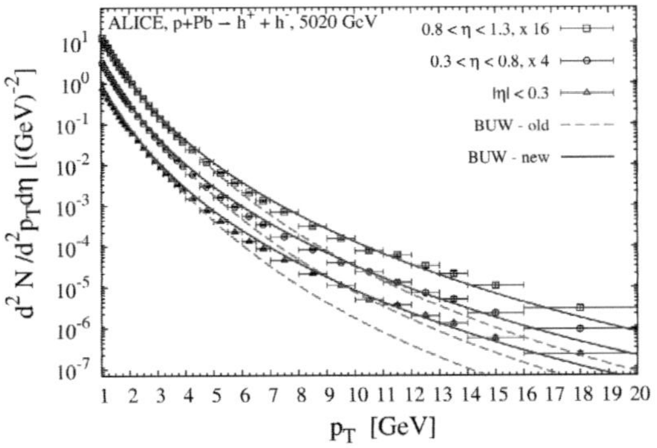

Figure 1. Comparison between the BUW predictions for the transverse momentum p_T-spectra of charged particles produced in pPb collisions and the ALICE data [9].

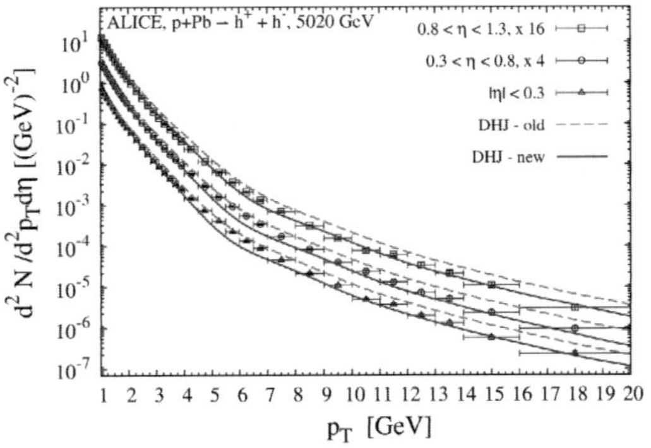

Figure 2. Comparison between the DHJ predictions for the transverse momentum p_T-spectra of charged particles produced in pPb collisions and the ALICE data [9].

3. Results and discussion

In what follows we will present our results for the average transverse momentum $\langle p_T \rangle$ defined by

$$\langle p_T \rangle = \frac{\int d^2p_T \, p_T \, \frac{dN_h}{dyd^2p_T}}{\int d^2p_T \, \frac{dN_h}{dyd^2p_T}} \qquad (6)$$

which is rapidity and energy dependent, i.e. $\langle p_T \rangle = \langle p_T(y, \sqrt{s}) \rangle$. In order to obtain realistic predictions at LHC energies it is fundamental to use a model, which can describe the experimental data on the p_T – spectra of produced particles, as input in the calculations. Consequently, as the first step we will initially compare the DHJ and BUW predictions with the recent LHC data. In Figs. 1 and 2 we present a comparison of these predictions using the original parameters, denoted "DHJ old" and "BUW old" in the figures, with the LHC data on the p_T – spectra of charged particles in pPb collisions at $\sqrt{s} = 5.02$ TeV and different rapidities [9]. We use the CTEQ5L parton distribution functions [10] and the KKP fragmentation functions [11]. Moreover, we compute Eq. (1) using the central values of η in the pseudorapidity ranges used in the experiment and choose $A \equiv A_{\text{min. bias}} = 20\,(18.5)$ for pPb (dAu) collisions. We find that these models are not able to describe the ALICE data [9] at large transverse momentum with their original parameters and a reanalysis of data is required. We have then determined the free parameters of the BUW and the DHJ dipole scattering amplitudes by fitting the p_T spectra of charged particles measured in pPb collisions at $\sqrt{s} = 5020$ GeV and then compared the updated models with the experimental data on pp collisions at other energies and rapidities. Moreover, differently from the authors of Refs. [5, 6], who have assumed that $\gamma_s \approx 0.63$, which is the value obtained from the leading order BFKL kernel, we will consider γ_s as a free parameter. The resulting fits are shown in Figs. 1 and 2 for the following parameters: $a = 2.0$, $b = 125$ and $\gamma_s = 0.74$ for the BUW model and $d = 1.0$ and $\gamma_s = 0.7$ for the DHJ model. The data are better described if we assume larger values of $\gamma_s \geq 0.7$, which is consistent with the results obtained using the renormalization group improved BFKL kernels at next-to-leading order and fixed running coupling. As it can be seen, with these parameter sets our curves agree well with the experimental data. In the range $4 < p_T < 7$ GeV the DHJ curves show an "edgy" behaviour which is a reminiscence of the numerical Fourier transform. This is not a big effect and can be considered as part of the theoretical error in our calculations. It is important to emphasize that $\langle p_T \rangle$ is only marginally affected by these small oscillations (see below) and the fits presented here for both models considered are sufficient to get a realistic prediction for this observable since it is dominated by the low p_T region.

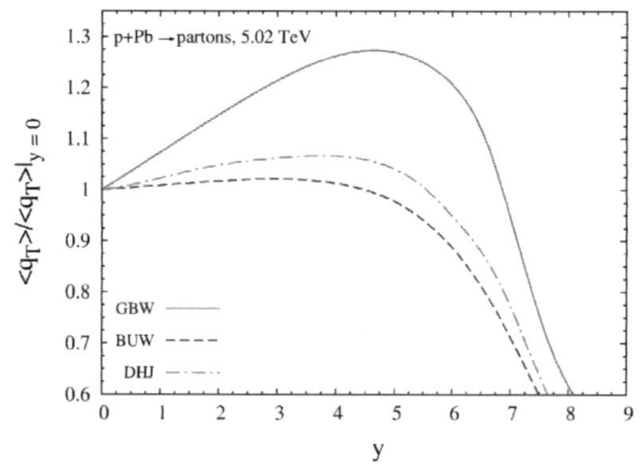

Figure 3. Ratio $\langle p_T(y, \sqrt{s}) \rangle / \langle p_T(0, \sqrt{s}) \rangle$ with different models of forward scattering amplitude in pPb collisions. Parton fragmentation is not included.

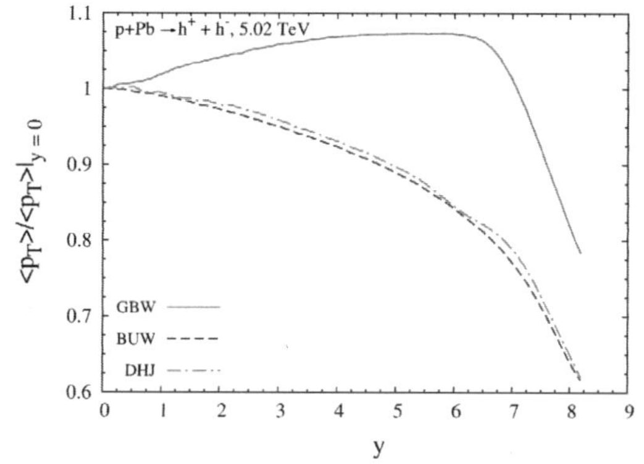

Figure 4. Ratio $\langle p_T(y, \sqrt{s}) \rangle / \langle p_T(0, \sqrt{s}) \rangle$ with different models of forward scattering amplitude in pPb collisions. Parton fragmentation is included.

The results presented in Figs. 1 and 2 make us confident to obtain realistic predictions for the average transverse momentum. In what follows we will study the energy and rapidity dependencies of the ratio

$$R = \frac{\langle p_T(y, \sqrt{s}) \rangle}{\langle p_T(0, \sqrt{s}) \rangle} \tag{7}$$

where the denominator represents the average transverse momentum at zero rapidity. Initially, let us analyse the dependence of our predictions on the model used to describe the forward scattering amplitudes $\mathcal{N}_{\mathcal{A},\mathcal{F}}(x, r)$ and the impact of the inclusion of parton fragmentation. In Figs. 3 and 4 we compare the predictions of the BUW and DHJ models with those from the GBW model [3], obtained assuming $p_{T,min} = 1$ GeV. The GBW model is not able to describe the experimental data on hadron production in hadronic collisions. However, as this model is usually considered to obtain analytical results for several observables, we would like to verify if its predictions for $\langle p_T \rangle$ are realistic. In Fig. 3 we present our predictions disregarding parton fragmentation, while in Fig. 4 fragmentation is included. It is important to emphasize that our results for the GBW model without fragmentation, obtained using the hybrid formalism are similar to those obtained in Ref. [2] with the k_T - factorization approach. We can see that the DHJ and BUW predictions are similar (to each other) and differ significantly from the GBW one. While the GBW model predicts a growth of the ratio for $y \leq 6$, the BUW and DHJ models predict that this ratio is almost constant or decreases with rapidity. The inclusion of parton fragmentation modifies the rapidity dependence, implying a smaller growth of the GBW prediction. In the case of the DHJ and BUW predictions, the inclusion of fragmentation implies that the fall of the ratio begins at smaller rapidities. Our results demonstrate that the inclusion of fragmentation has an important impact on the behavior of $\langle p_T \rangle$. However, the main difference between our predictions and those presented in Ref. [2] comes from the model used to describe the QCD dynamics at high energies. This distinct behavior is present for pp and pPb collisions, with the behavior of the ratio at very large rapidities being determined by kinematical constraints associated to the limited phase space.

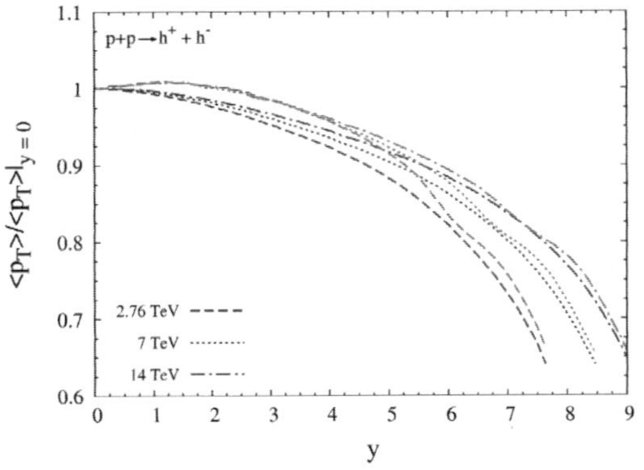

Figure 5. Rapidity dependence of the ratio $R = \langle p_T(y, \sqrt{s}) \rangle / \langle p_T(0, \sqrt{s}) \rangle$ in pp collisions for different energies. The BUW (DHJ) predictions are represented by blue (red) lines.

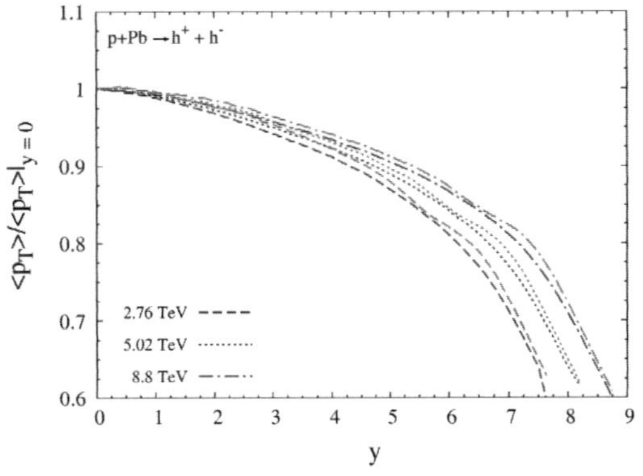

Figure 6. Rapidity dependence of the ratio $R = \langle p_T(y, \sqrt{s}) \rangle / \langle p_T(0, \sqrt{s}) \rangle$ in pPb collisions for different energies. The BUW (DHJ) predictions are represented by blue (red) lines.

In Figs. 5 and 6 we present the behavior of the ratio $\langle p_T(y, \sqrt{s}) \rangle / \langle p_T(0, \sqrt{s}) \rangle$ for pp and pPb collisions considering different center of mass energies. We find that the predictions of the DHJ (red lines) and BUW (blue lines) are similar, with the DHJ being slightly larger than the BUW, and that the ratio increases with energy. The results from Figs. 5 and 6 indicate that the ratio R decreases with the rapidity in pPb collisions for the energies probed by LHC, presenting a behavior similar to that obtained using a hydrodynamical approach, which implies that this observable cannot be used to discriminate the CGC and hydrodynamical approaches for the description of the high multiplicity events. This conclusion is opposite to that obtained in Ref. [2]. This difference comes from several facts. First, the CGC results in Ref. [2] were obtained using an analytical approximation for a particular unintegrated gluon distribution that does not describe (even at a qualitative level) the experimentally measured p_T-spectra. Second, the calculation presented in [2] does not include the important contribution of the fragmentation processes to the average transverse momentum. Finally, kinematical constraints associated with phase space restrictions at large rapidities were not included in [2] and, even in a partonic scenario, they play an important role at very large rapidities. In contrast, in our analysis we have calculated the ratio R using two different models for the forward scattering amplitude that are able to describe the current experimental data on charged hadron and pion p_T spectra measured in pp and pPb collisions at LHC. We have included the effects of parton fragmentation and phase space restrictions. It is important to emphasize that although we have used the hybrid formalism instead of the k_T - factorization approach, we have verified that both approaches imply a similar behavior for the ratio R when the GBW model is used as input and the parton fragmentation is not taken into account. Our results demonstrate that the main difference comes from the treatment of the QCD dynamics at high energies.

Journal of Physics: Conference Series **736** (2016) 012025 doi:10.1088/1742-6596/736/1/012025

4. Conclusions

We have considered the hybrid formalism to study the behavior of the average p_T with respect to the rapidity in pp and pPb collisions at several energies in the CGC picture of high energy collisions. In order to obtain realistic predictions we have updated previous phenomenological models for the forward scattering amplitude, one with and other without geometric scaling violations. After constraining their parameters with the most recent data on the p_T spectra of charged particles measured in pPb collisions at the LHC, we have demonstrated (see details in Ref.[12]) that they are able to describe the recent pp data on the charged hadron and pion p_T spectra measured at the LHC in the kinematical range of $p_T \leq 20$ GeV. Using these models as input, we have calculated the average transverse momentum $\langle p_T(y, \sqrt{s})\rangle$ in pp and pPb collisions, and estimated the energy and rapidity dependencies of the ratio $R = \frac{\langle p_T(y,\sqrt{s})\rangle}{\langle p_T(0,\sqrt{s})\rangle}$. We have demonstrated that this ratio increases with the energy for a fixed rapidity and decreases with the rapidity for a fixed energy, which is similar to the predication of hydrodynamical approaches for high multiplicity events. Our results indicated that this decreasing comes from the treatment of the QCD dynamics at high energies and the inclusion of the fragmentation process and kinematical constraints associated to the phase space restrictions at very large rapidities.

Acknowledgments

This work was partially financed by the Brazilian funding agencies CAPES, CNPq, FAPESP and FAPERGS.

References

[1] F. Gelis, E. Iancu, J. Jalilian-Marian and R. Venugopalan, Ann. Rev. Nucl. Part. Sci. **60**, 463 (2010); K. J. Golec-Biernat and M. Wusthoff, Phys. Rev. D **59**, 014017 (1998); M. S. Kugeratski, V. P. Goncalves and F. S. Navarra, Eur. Phys. J. C **46**, 413 (2006).
[2] P. Bozek, A. Bzdak and V. Skokov, Phys. Lett. B **728**, 662 (2014).
[3] K. J. Golec-Biernat, M. Wusthoff, Phys. Rev. D **59**, 014017 (1998).
[4] E. Iancu, C. Marquet and G. Soyez, Nucl. Phys. A **780**, 52 (2006).
[5] A. Dumitru, A. Hayashigaki and J. Jalilian-Marian, Nucl. Phys. A **765**, 464 (2006); Nucl. Phys. A **770**, 57 (2006).
[6] D. Boer, A. Utermann, E. Wessels, Phys. Rev. D **77**, 054014 (2008).
[7] D. Kharzeev, Y.V. Kovchegov and K. Tuchin, Phys. Lett. B **599**, 23 (2004).
[8] J. Bartels, K. Golec-Biernat, H. Kowalski, Phys. Rev. D **66**, 014001 (2002); H. Kowalski and D. Teaney, Phys. Rev. D **68**, 114005 (2003); E. Iancu, K. Itakura, S. Munier, Phys. Lett. B **590**, 199 (2004); H. Kowalski, L. Motyka and G. Watt, Phys. Rev. D **74**, 074016 (2006); V. P. Goncalves, M. S. Kugeratski, M. V. T. Machado and F. S. Navarra, Phys. Lett. B **643**, 273 (2006); C. Marquet, R. Peschanski and G. Soyez, Phys. Rev. D **76**, 034011 (2007); G. Soyez, Phys. Lett. B **655**, 32 (2007); G. Watt and H. Kowalski, Phys. Rev. D **78**, 014016 (2008.)
[9] B. Abelev *et al.* [ALICE Collaboration], Phys. Rev. Lett. **110**, 082302 (2013).
[10] H. L. Lai *et al.* [CTEQ Collaboration], Eur. Phys. J. C **12**, 375 (2000).
[11] B.A. Kniehl, G. Kramer and B. Pötter, Nucl. Phys. B **582**, 514 (2000).
[12] F. O. Durães, A. V. Giannini, V. P. Goncalves and F. S. Navarra, arXiv:1510.04737 [hep-ph].

Journal of Physics: Conference Series **736** (2016) 012026 doi:10.1088/1742-6596/736/1/012026

Determining the structure of $X(3872)$ in heavy ion collisions

L. M. Abreu[1], K. P. Khemchandani[2], A. Martínez Torres[3], F. S. Navarra[3], M. Nielsen[3]

[1] Instituto de Física, Universidade Federal da Bahia, 40210-340, Salvador, BA, Brazil
[2] Faculdade de Tecnologia, Universidade do Estado do Rio de Janeiro,
Rod. Presidente Dutra Km 298, Pólo Industrial, 27537-000 , Resende, RJ, Brazil
[3] Instituto de Física, Universidade de São Paulo
Rua do Matão Travessa R, 187, 055508-090, São Paulo, SP, Brazil

Abstract.
We study the time evolution of the $X(3872)$ abundance in the hot hadron gas produced in the late stage of heavy ion collisions. We use effective field Lagrangians to obtain the production and dissociation cross sections of $X(3872)$. In this evaluation we include diagrams involving the anomalous couplings $\pi D^*\bar{D}^*$ and $X\bar{D}^*D^*$ and also the couplings of the $X(3872)$ with charged D and D^* mesons. With these new terms the $X(3872)$ interaction cross sections are much larger than those found in previous works. Using these cross sections as input in rate equations, we conclude that during the expansion and cooling of the hadronic gas, the number of $X(3872)$, originally produced at the end of the mixed QGP/hadron gas phase, is reduced by a factor of 4.

1. Introduction

Several new charmonium states have been observed [1, 2, 3, 4] during the last decade. Among these new states, the most well known is the $X(3872)$. The structure of this new state has been subject of an intense debate. Calculations using constituent quark models give masses for possible charmonium states with $J^{PC} = 1^{++}$ quantum numbers, which are too large: $2\,^3P_1(3990)$ and $3\,^3P_1(4290)$ [5]. These results motivated the conjecture that these objects are multiquark states. The vicinity of the X mass to the $\bar{D}D^*$ threshold inspired the proposal that the $X(3872)$ could be a molecular $\bar{D}D^*$ bound state with a small binding energy [6, 7]. Another interpretation of the $X(3872)$ is that it can be a tetraquark state resulting from the binding of a diquark and an antidiquark [8, 9]. There are other proposals as well [2, 3, 4, 10, 11].

Hadronic collisions seem to be a new testing ground for ideas about the structure of the new states. It has been shown [3, 12] that it is extremely difficult to produce meson molecules in p p collisions. In the molecular approach the estimated cross section for $X(3872)$ production is two orders of magnitude smaller than the measured one. On the other hand, in Ref. [13] a simple model was proposed to compute the X production cross section in p p collisions in the tetraquark approach.

As pointed out in Refs. [14, 15], high energy heavy ion collisions offer an interesting scenario to study the production of multiquark states. In these processes, a quite large number of heavy

Journal of Physics: Conference Series **736** (2016) 012026 doi:10.1088/1742-6596/736/1/012026

quarks is expected to be produced, reaching as much as 20 $c\bar{c}$ pairs per unit rapidity in Pb + Pb collisions at the LHC. If the production mechanism is coalescence, then the production yield of these exotic hadrons at the moment of their formation strongly reflects their internal structure. In particular, it was shown that in the coalescence model the production yield of the $X(3872)$, at RHIC or LHC energies, is almost 20 times bigger for a molecular structure than for a tetraquark configuration.

In the last phase of heavy ion collisions, the $X(3872)$ interacts with other hadrons during the expansion of the hadronic matter. It can be destroyed in collisions with the comoving light mesons and it can also be produced through the inverse reactions.

The interactions of the X in a hadronic medium have been studied in the framework of $SU(4)$ effective Lagrangians in Ref. [16], where it was found that the absorption cross section is two orders of magnitude larger than the production one. In particular, it was found that $< \sigma v >_{\pi X \to D^* \bar{D}^*} \simeq 30 < \sigma v >_{D^* \bar{D}^* \to \pi X}$. In spite of this difference, the authors of Ref. [16] arrived at the intriguing conclusion that the number of X's stays approximately constant during the hadronic phase. In Ref. [16] the coupling of the $X(3872)$ with charged charm mesons (such as $D^- D^{*+}$) was neglected and only neutral mesons were considered (such as $D^0 \bar{D}^{0*}$). Moreover, the terms with anomalous couplings were not included in the calculations. In Ref. [17] we showed that the inclusion of the couplings of the $X(3872)$ to charged D's and D^*'s and those of the anomalous vertices, $\pi D^* \bar{D}^*$ and $X D^* \bar{D}^*$, increases the cross sections by more than one order of magnitude. Similar results were also observed in the case of $J/\psi\pi$ cross section [18]. These anomalous vertices also give rise to new reaction channels, namely, $\bar{D} + D^* \to \pi + X$ and $\pi + X \to \bar{D} + D^*$. Thus it is important to evaluate the changes that the above mentioned contributions can produce in the X abundance (and in its time evolution) in reactions as those considered in Ref. [16]. This was done in [19].

The formalism used in Refs. [16] and [17] was originally developed to study the interaction of the J/ψ with light mesons in a hot hadron gas many years ago [18, 20].

2. Thermal cross sections

In this section we calculate the cross sections averaged over the thermal distributions for the processes $\bar{D}D \to \pi X$, $\bar{D}^* D \to \pi X$ and $\bar{D}^* D^* \to \pi X$, and for the inverse reactions. In Fig. 1 we show the different diagrams contributing to each process. In Ref. [16] it was shown that the contribution from the reactions involving the ρ meson is very small compared to the reactions with pions and thus we neglect the former in what follows. The cross sections for the processes shown in Fig. 1 were obtained in Ref. [17]. For more details about the calculations, we refer the reader to Ref. [17]. As explained in that paper, we use monopole form factors to obtain the cross sections of the processes in Fig. 1. The thermally averaged cross section for a process $ab \to cd$ can be calculated using the expression

$$\langle \sigma_{ab\to cd}\, v_{ab} \rangle = \frac{\int d^3\mathbf{p}_a d^3\mathbf{p}_b\, f_a(\mathbf{p}_a)\, f_b(\mathbf{p}_b)\, \sigma_{ab\to cd}\, v_{ab}}{\int d^3\mathbf{p}_a d^3\mathbf{p}_b\, f_a(\mathbf{p}_a)\, f_b(\mathbf{p}_b)} \tag{1}$$

where f_a and f_b are Bose-Einstein distributions, $\sigma_{ab\to cd}$ are the cross sections computed in [17], v_{ab} represents the relative velocity of the two interacting particles (a and b).

In Fig. 2 we show the total thermally averaged cross sections for the processes involving the production of the $X(3872)$ state, i.e., $\bar{D}D \to \pi X$, $\bar{D}^* D \to \pi X$ and $\bar{D}^* D^* \to \pi X$ reactions, while in Fig. 3 we show the inverse processes, i.e., the dissociation of $X(3872)$ through the reactions $\pi X \to \bar{D}D$, $\pi X \to \bar{D}^* D$, $\pi X \to \bar{D}^* D^*$, respectively. For the latter cases, we use the principle of detailed balance to determine the corresponding cross sections. Figs. 2 and 3 should be compared with the Fig. 3 of Ref. [16]. Our cross sections are a factor 100 larger in

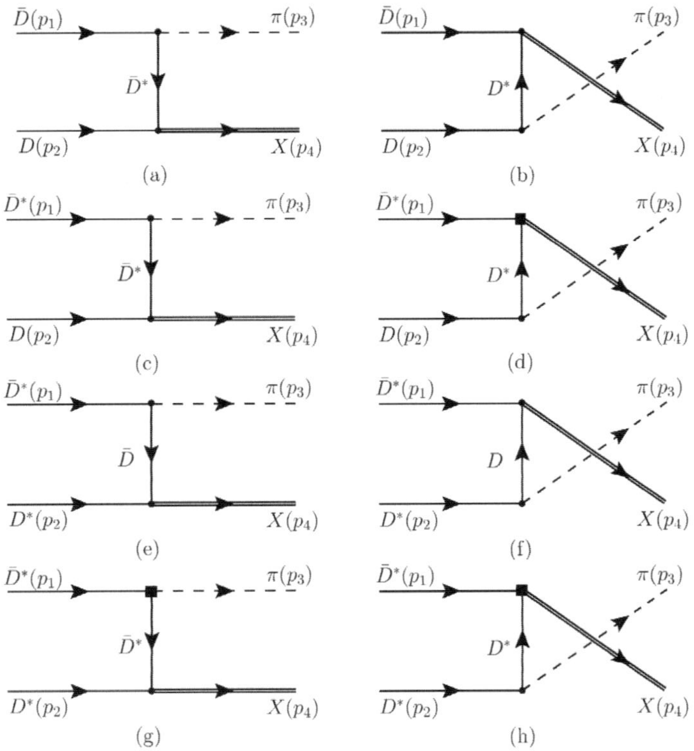

Figure 1. Diagrams contributing to the processes $\bar{D}D \to \pi X$ [(a) and (b)], $\bar{D}^*D \to \pi X$ [(c) and (d)] and $\bar{D}^*D^* \to \pi X$ [(e), (f), (g) and (h)]. The filled box in the diagrams (d), (g) and (h) represents the anomalous vertex $XD^*\bar{D}^*$, which was evaluated in Ref. [17].

reactions with $D^*\bar{D}^*$ in the initial or final state This can be attributed mostly to the inclusion of the anomalous terms. Moreover, our cross sections are a factor 10 larger in the case of DD mesons in the initial or final state. The difference comes from the inclusion of the coupling of the X to charged charged D's and D^*'s, which was not considered in Ref. [16].

3. Time evolution of the $X(3872)$ abundance

In order to compute the time evolution of the X abundance, we need to know how the temperature changes with time and this is highly model dependent. Fortunately, as one can see in Figs. 2 and 3 the dependence of $< \sigma v >$ on the temperature is relatively weak.

We study the yield of $X(3872)$ in central Au-Au collisions at $\sqrt{s_{NN}} = 200$ GeV at RHIC. By using the thermally averaged cross sections obtained in the previous section, we can now analyze the time evolution of the $X(3872)$ abundance in hadronic matter, which depends on the densities and abundances of the particles involved in the processes of Fig. 1, as well as the cross sections associated with these reactions (and the corresponding inverse reactions), Figs. 2 and 3. The momentum-integrated evolution equation has the form [16, 21, 22, 23]

$$\frac{dN_X(\tau)}{d\tau} = R_{QGP}(\tau) + \sum_{c,c'} \left[\langle \sigma_{cc' \to \pi X} v_{cc'} \rangle n_c(\tau) N_{c'}(\tau) - \langle \sigma_{\pi X \to cc'} v_{\pi X} \rangle n_\pi(\tau) N_X(\tau) \right], \quad (2)$$

where $N_X(\tau)$, $N_{c'}(\tau)$, $n_c(\tau)$ and $n_\pi(\tau)$ are the abundances of $X(3872)$, of charmed mesons of type c', of charmed mesons of type c and of pions at proper time τ, respectively. The term

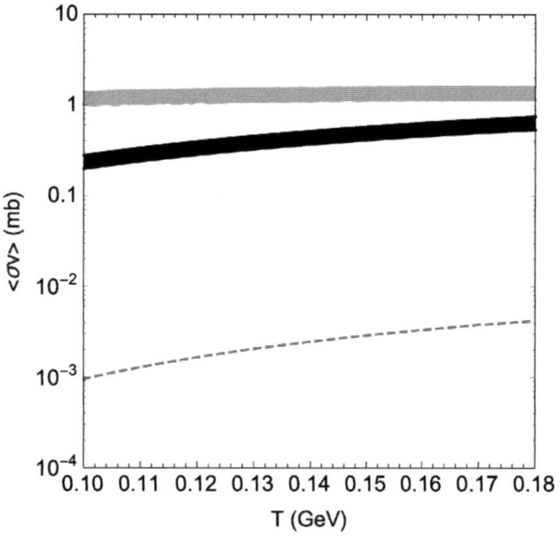

Figure 2. Thermally averaged cross sections. a) $\bar{D}D \to \pi X(3872)$ (dashed line), $\bar{D}^*D \to \pi X(3872)$ (dark-shaded region) and $\bar{D}^*D^* \to \pi X(3872)$ (light-shaded region)

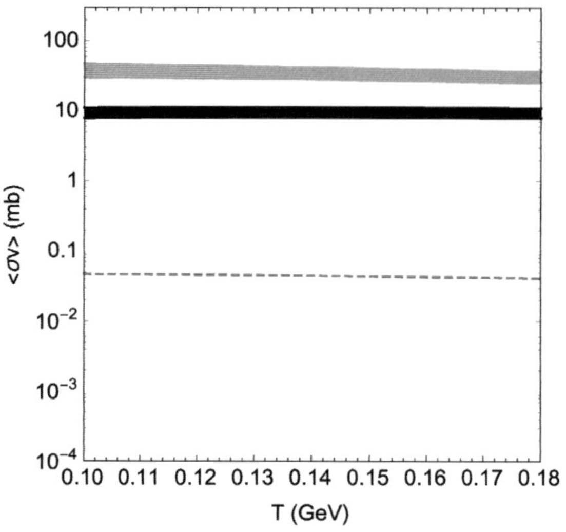

Figure 3. Thermally averaged cross sections. $\pi X(3872) \to \bar{D}D$ (dashed line), $\pi X(3872) \to \bar{D}^*D$ (dark-shaded region) and $\pi X(3872) \to \bar{D}^*D^*$ (light-shaded region).

$R_{QGP}(\tau)$ in Eq. (2) represents the X yield from the quark-gluon plasma in the mixed phase, since the hadronization of the QGP takes a finite time, and it is given by [16, 23]:

$$\begin{aligned} R_{QGP}(\tau) &= \frac{N_X^0}{\tau_H - \tau_C} \quad \text{if} \ \tau_C < \tau < \tau_H \\ R_{QGP}(\tau) &= 0 \ \text{otherwise} \end{aligned} \tag{3}$$

where the times $\tau_C = 5.0$ fm/c and $\tau_H = 7.5$ fm/c determine the beginning and the end of the mixed phase respectively. The constant N_X^0 corresponds to the total number of $X(3872)$ produced from quark-gluon plasma. To solve Eq. (2) we assume that the pions and charmed mesons in the reactions contributing to the abundance of X are in equilibrium. Therefore $N_{c'}(\tau)$, $n_c(\tau)$ and $n_\pi(\tau)$ can be written as [16, 21, 22, 23]

$$\begin{aligned} N_{c'}(\tau) &\approx \frac{1}{2\pi^2}\gamma_C\, g_D\, m_{D^{(*)}}^2\, T(\tau)\, V(\tau)\, K_2\left(\frac{m_{D^{(*)}}}{T(\tau)}\right), \\ n_c(\tau) &\approx \frac{1}{2\pi^2}\gamma_C\, g_D\, m_{D^{(*)}}^2\, T(\tau)\, K_2\left(\frac{m_{D^{(*)}}}{T(\tau)}\right), \\ n_\pi(\tau) &\approx \frac{1}{2\pi^2}\gamma_\pi\, g_\pi\, m_\pi^2\, T(\tau)\, K_2\left(\frac{m_\pi}{T(\tau)}\right), \end{aligned} \tag{4}$$

where γ_i and g_i are the fugacity factor and the spin degeneracy of particle i respectively. As can be seen in Eq. (4), the time dependence in Eq. (2) enters through the parametrization of the temperature $T(\tau)$ and volume $V(\tau)$ profiles suitable to describe the dynamics of the hot hadron gas after the end of the quark-gluon plasma phase.

Following Refs. [16, 22, 23], we assume the τ dependence of T and V to be given by [16, 22, 23]

$$\begin{aligned} T(\tau) &= T_C - (T_H - T_F)\left(\frac{\tau - \tau_H}{\tau_F - \tau_H}\right)^{\frac{4}{5}}, \\ V(\tau) &= \pi\left[R_C + v_C\,(\tau - \tau_C) + \frac{a_C}{2}\,(\tau - \tau_C)^2\right]^2 \tau_C. \end{aligned} \tag{5}$$

Journal of Physics: Conference Series **736** (2016) 012026 doi:10.1088/1742-6596/736/1/012026

In the above equation $R_C = 8.0$ fm denotes the final size of the quark-gluon plasma, while $v_C = 0.4\,c$ and $a_C = 0.02\,c^2/\text{fm}$ are its transverse flow velocity and transverse acceleration at τ_C respectively. The critical temperature of the (quark gluon plasma to hadronic matter) transition is $T_C = 175$ MeV; $T_H = T_C = 175$ MeV is the temperature of the hadronic matter at the end of the mixed phase. The freeze-out takes place at the freeze-out time $\tau_F = 17.3$ fm/c, when the temperature drops to $T_F = 125$ MeV.

To solve Eq. (2), we assume that the total number of charm quarks in charm hadrons is conserved during the production and dissociation reactions, and that the total number of charm quark pairs produced at the initial stage of the collisions at RHIC is 3, yielding the charm quark fugacity factor $\gamma_C \approx 6.4$ in Eq. (4) [16, 15]. In the case of pions, we follow Ref. [23] and work with the assumption that their total number at freeze-out is 926, which fixes the value of γ_π appearing in Eq. (4) to be ~ 1.725 .

Assuming that the state in question is a tetraquark with $J^{PC} = 1^{++}$, the number of $X(3872)$ produced at the end of the mixed phase is [15]:

$$N^0_{X(4q)} = N_{X(4q)}(\tau_H) \approx 4.0 \times 10^{-5}. \tag{6}$$

In order to determine the time evolution of the $X(3872)$ abundance we solve Eq. (2) starting at the end of the mixed phase, i.e. at $\tau_H = 7.5$ fm/c, and assuming that the $X(3872)$ is a tetraquark. The initial condition is given by Eq. (6). We use this initial abundance to integrate Eq. (2) and we show the result in Fig. 4. In the figure the solid line represents the result obtained using the same approximations as those made in Ref. [16]. Our curve is slightly different from that of Ref. [16] because we did not include the contribution of the ρ mesons, as discussed earlier. The dashed line shows the result when we include the couplings of the $X(3872)$ to charged D's and D^*'s. The light-shaded band shows the results obtained with the further inclusion of the diagrams containing the anomalous vertices. The band reflects the uncertainty in the $X\bar{D}^*D^*$ coupling constant, which is $g_{X\bar{D}^*D^*} = 12.5 \pm 1.0$ [17].

As can be seen, without the inclusion of the anomalous coupling terms, the abundance of X remains basically constant. This is because the magnitude of the thermally averaged cross sections for the X production and absorption reactions obtained within this approximation is so small that the second term in the right hand side of Eq. (2) is completely negligible compared to the first term. When including the coupling of the X to charged D's and D^*'s we basically do not find any important change for the time evolution of the X abundance, since, as it was shown in [19], the thermally averaged cross sections do not change drastically in this case. On the other hand, the inclusion of the anomalous coupling terms, depicted in Figs. 1c, 1d, 1f, 1g and 1h, modifies the behavior of the $X(3872)$ yield, producing a fast decrease of the X abundance with time. We emphasize that the $X(3872)$ abundance is the one which comes from the QGP and is what could be observed if the $X(3872)$ is a tetraquark state. However, if the $X(3872)$ is a molecular state, it will be formed by hadron coalescence at the end of the hadronic phase. According to Ref. [16], at this time the average number of X's, considering it as a $D\bar{D}^*$ molecule, is

$$N_{X(mol)} \approx 7.8 \times 10^{-4} \ , \tag{7}$$

which is about 80 times larger than the contribution for a tetraquark state at the end of the hadronic phase (see Fig. 4). We can then conclude that the QGP contribution for the $X(3872)$ production (as a tetraquark state and from quark coalescence) after being suppressed by hadronic interactions, becomes insignificant at the end of the hadronic phase.

4. Conclusions

We have presented a calculation of the time evolution of the $X(3872)$ abundance in heavy ion collisions. If the $X(3872)$ is a tetraquark state it will be produced at the mixed phase by

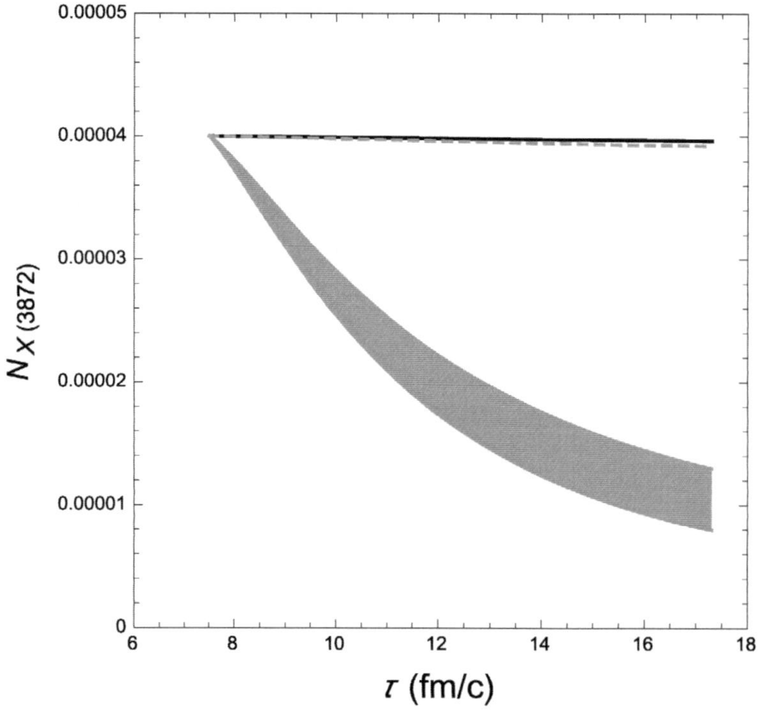

Figure 4. Time evolution of the $X(3872)$ abundance as a function of the proper time τ in central Au-Au collisions at $\sqrt{s_{NN}} = 200$ GeV. The solid line, the dashed line and the light-shaded region represent the results obtained considering only the neutral D's and D^*'s, adding the contribution from charged D's and D^*'s and including contributions from the anomalous vertices respectively. The initial condition is the abundance of the $X(3872)$ considered as a tetraquark Eq. (6).

quark coalescence. It can be destroyed in collisions with the comoving light mesons, such as $X + \pi \to \bar{D} + D$, $X + \pi \to \bar{D}^* + D^*$ but it can also be produced through the inverse reactions, such as $D + \bar{D} \to X + \pi$, $\bar{D}^* + D^* \to \pi + X$. We have considered the contributions of anomalous vertices, $\pi D^* \bar{D}^*$ and $X\bar{D}^* D^*$, and the contributions from charged D and D^* mesons to the $X(3872)$ production and dissociation cross sections. These vertices, apart from enhancing the cross sections associated with the $\bar{D}^* D^*$ channel, give rise to additional production/absorption mechanisms of X, which are found to be relevant.

The cross sections, averaged over the thermal distribution, have been used to analyze the time evolution of the $X(3872)$ abundance in hadronic matter. We have found that the abundance of a tetraquark X drops from $N_{X(4q)} \approx 4.0 \times 10^{-5}$ at the begining of the hadronic phase [16] to $N_{X(4q)} \sim 1.0 \times 10^{-5}$ at the end of the hadronic phase.

On the other hand, if the $X(3872)$ is a molecular state it will be produced by hadron coalescence at the end of the hadronic phase. According to Ref. [16], at this time the average number of X's, considering it as a $D\bar{D}^*$ molecule, is $N_{X(mol)} \approx 7.8 \times 10^{-4}$, which is about 80 times larger than $N_{X(4q)}$.

As expected, the results show that the X multiplicity in relativistic ion collisions depends on the structure of $X(3872)$. Our main conclusion is that the contribution from the anomalous vertices play an important role in determining the time evolution of the $X(3872)$ abundance and they lead to strong suppression of this state during the hadronic phase. Therefore, within the uncertainties of our calculation we can say that if the $X(3872)$ is observed in a heavy ion collision it must have been produced at the end of the hadronic phase and, hence, it must be a molecular state.

Acknowledgments

The authors would like to thank the Brazilian funding agencies CNPq and FAPESP for financial support. We also thank C. Greiner and J. Noronha-Hostler for fruitful discussions.

References

[1] K.A. Olive *et al.* (Particle Data Group), Chin. Phys. C **38**, 090001 (2014).
[2] A. Hosaka, T. Iijima, K. Miyabayashi, Y. Sakai and S. Yasui, arXiv:1603.09229 [hep-ph].
[3] A. Esposito, A. L. Guerrieri, F. Piccinini, A. Pilloni and A. D. Polosa, Int. J. Mod. Phys. A **30**, 1530002 (2014).
[4] M. Nielsen and F. S. Navarra, Mod. Phys. Lett. A **29**, 1430005 (2014); M. Nielsen, F. S. Navarra and S. H. Lee, Phys. Rept. **497**, 41 (2010); F. S. Navarra, M. Nielsen and S. H. Lee, Phys. Lett. B **649**, 166 (2007).
[5] T. Barnes and S. Godfrey, Phys. Rev. D **69**, 054008 (2004); T. Barnes, S. Godfrey and E.S. Swanson, Phys. Rev. D **72**, 054026 (2005).
[6] E.S. Swanson, Phys. Rept. **429**, 243 (2006).
[7] D. Gamermann and E. Oset, Phys. Rev. D **80**, 014003 (2009).
[8] L. Maiani, V. Riquer, F. Piccinini and A.D. Polosa, Phy. Rev. D **71**, 014028 (2005).
[9] R. D. Matheus, S. Narison, M. Nielsen and J. M. Richard, Phys. Rev. D **75**, 014005 (2007).
[10] R. D. Matheus, F. S. Navarra, M. Nielsen and C. M. Zanetti, Phys. Rev. D **80**, 056002 (2009).
[11] D. Gamermann, J. Nieves, E. Oset and E. Ruiz Arriola, Phys. Rev. D **81**, 014029 (2010).
[12] C. Bignamini, B. Grinstein, F. Piccinini, A. D. Polosa, and C. Sabelli, Phys. Rev. Lett. **103**, 162001 (2009).
[13] F. Carvalho, E. R. Cazaroto, V. P. Gonçalves and F. S. Navarra, Phys. Rev. D **93**, 034004 (2016).
[14] S. Cho *et al.* (ExHIC Collaboration), Phys. Rev. Lett. **106**, 212001 (2011).
[15] S. Cho *et al.* (ExHIC Collaboration), Phys. Rev. C **84**, 064910 (2011).
[16] S. Cho, S. H. Lee, Phys. Rev. C **88**, 054901 (2013).
[17] A. M. Torres, K. P. Khemchandani, F. S. Navarra, M. Nielsen and L. M. Abreu, Phys. Rev. D **90**, 114023 (2014); Erratum: [Phys. Rev. D **93**, 059902 (2016)].
[18] F. O. Duraes, S. H. Lee, F. S. Navarra and M. Nielsen, Phys. Lett. B **564**, 97 (2003); F. O. Duraes, H. c. Kim, S. H. Lee, F. S. Navarra and M. Nielsen, Phys. Rev. C **68**, 035208 (2003); Y. s. Oh, T. Song and S. H. Lee, Phys. Rev. C **63**, 034901 (2001); C. M. Ko, B. Zhang, X. N. Wang and X. F. Zhang, Phys. Lett. B **444**, 237 (1998).
[19] L. M. Abreu, K. P. Khemchandani, A. M. Torres, F. S. Navarra and M. Nielsen, arXiv:1604.07716 [hep-ph].
[20] For a comprehensive review see R. Rapp, D. Blaschke and P. Crochet, Prog. Part. Nucl. Phys. **65**, 209 (2010).
[21] P. Koch, B. Muller and J. Rafelski, Phys. Rep. **142**, 167 (1986).
[22] L. W. Chen, V. Greco, C. M. Ko, S. H. Lee and W. Liu, Phys. Lett B **601**, 34 (2004).
[23] L. W. Chen, C. M. Ko, W. Liu and M. Nielsen, Phys. Rev. C **76**, 014906 (2007).

Fulfilling the RHIC mission with sPHENIX

Megan Connors for the sPHENIX Collaboration

Georgia State University, Atlanta GA USA and RIKEN BNL Research Center, Brookhaven National Laboratory, Upton, NY, USA

E-mail: mconnors@gsu.edu

Abstract. RHIC has made groundbreaking contributions to the understanding of QCD under extreme conditions with the discovery of the quark gluon plasma (QGP) as a perfect fluid and first observations of energy loss. It continues to play a crucial role in understanding and quantifying the properties of the QGP as well as mapping out the QCD phase diagram. However, detailed questions concerning partonic energy loss in the QGP remain. There is a need to build a new detector at RHIC to measure important rare probes of the QGP. A new detector will benefit from advances in reconstructing jets in heavy ion collisions and the increased luminosity achievable with RHIC. Constraining models at RHIC and LHC energies are crucial for extracting the temperature dependence of transport properties of the QGP. To measure newly developed observables made at the LHC with high precision at RHIC, a detector with full azimuthal coverage and spanning a pseudorapidity range between -1.1 and 1.1, known as sPHENIX, has been proposed. The capabilities of the new detector will allow for a full understanding of jet energy loss and upsilon suppression. The goals for sPHENIX and route to achieving these goals along with the current status of the detector will be presented on behalf of the new collaboration.

1. The Motivation for sPHENIX

The 2015 Nuclear Science Long Range Plan explicitly states the need for "a state-of-the-art jet detector at RHIC, called sPHENIX" in its recommendations [1]. As the Relativistic Heavy Ion Collider (RHIC) completes its scientific mission, sPHENIX will play a crucial role in understanding the microscopic properties of the Quark Gluon Plasma (QGP). Fully reconstructed jets at RHIC probe the medium near the critical temperature, where the coupling is the strongest. A detector capable of detailed measurements of the modifications to the jet structure at these collision energies will greatly enhance our understanding of energy loss in the QGP. Capabilities for heavy flavor jet measurements will provide the ability to study the flavor dependence of energy loss, which will clarify the roles of radiative and collisional energy loss in the medium. In addition, the sPHENIX detector should be capable of separating the three states of the upsilon (Υ), which can be used to study the density dependence of the color screening effect.

In two years of running, sPHENIX will collect a large statistics data set as a result of its high data acquisition rate (15 kHz) combined with the high luminosity RHIC can deliver after more than 15 years of accelerator developments. This will increase the kinematic reach of previous measurements and the accessibility of rare probes in 200 GeV Au+Au collisions.

The Long Range Plan also notes the importance of the complementarity between RHIC and the Large Hadron Collider (LHC). To fully utilize information gathered from experiments at both facilities, one should measure the same observables in the same kinematic range. The increased

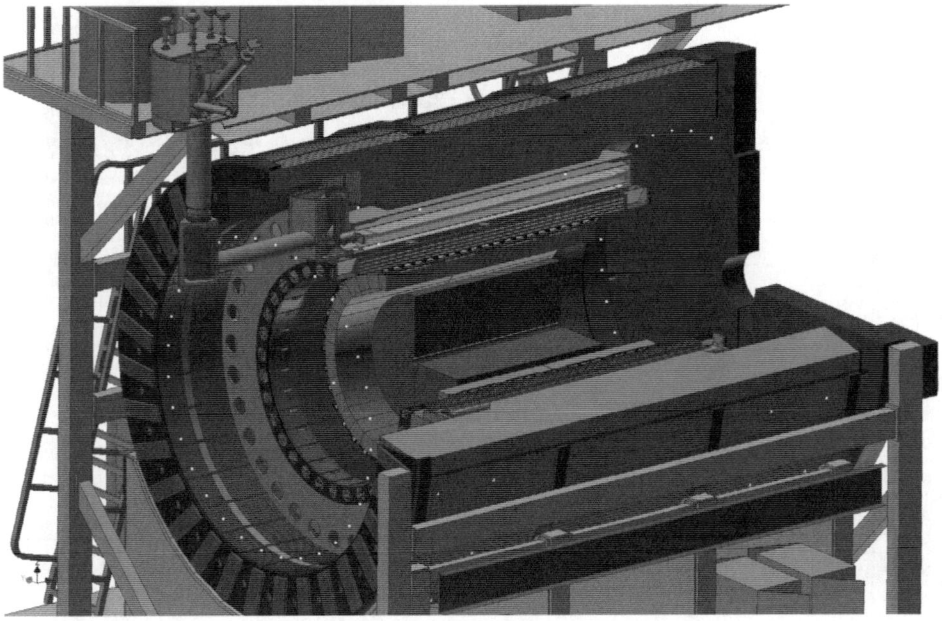

Figure 1. The sPHENIX detector design. The blue cylinder in the center shows the space for the tracking system. This is surrounded by the EMCal, which sits inside the inner HCal. The magnet and cryostat are shown in gray between the inner and outer HCal. Finally the outer HCal which encompasses the rest of the detector sits in the support structure.

kinematic range and statistics from sPHENIX will provide sufficient overlap in observables between RHIC and LHC to constrain models and improve our understanding of the properties of the QGP at different temperatures.

2. The sPHENIX Design

The sPHENIX detector [4] will be housed where PHENIX is located on the RHIC ring and will utilize the existing infrastructure. The present PHENIX detector is currently collecting its last data set during RHIC Run 16. The schedule lists sPHENIX as fully installed and ready for beam in 2022 and includes two years of data taking. It would then be available for use at an electron ion collider.

The sPHENIX detector is comprised of a tracking system surrounded by calorimetry based around a 1.5 Telsa solenoid magnet. The calorimetry includes an electromagnetic calorimeter (EMCal) and two hadronic calorimeters (HCal). The inner and outer HCals are located inside and outside the solenoid magnet respectively. The hadronic calorimeter serves as the flux return for the magnet. The solenoid magnet has already been obtained by Brookhaven National Lab (BNL) from the Babar experiment for use in sPHENIX. Recently the magnet underwent a successful round of low power cold tests. The magnet has a diameter of 2.8 m and is 3.8 m long. The sPHENIX detector will have full azimuthal coverage and span $-1.1 < \eta < 1.1$ in pseudorapidity. A schematic of the detector is shown in Figure 1.

2.1. Tracking

To complete the Υ, jet structure, and heavy flavor tagged jet programs, a high resolution tracking system is essential. The final configuration has yet to be determined however there are specific

Journal of Physics: Conference Series **736** (2016) 012027 doi:10.1088/1742-6596/736/1/012027

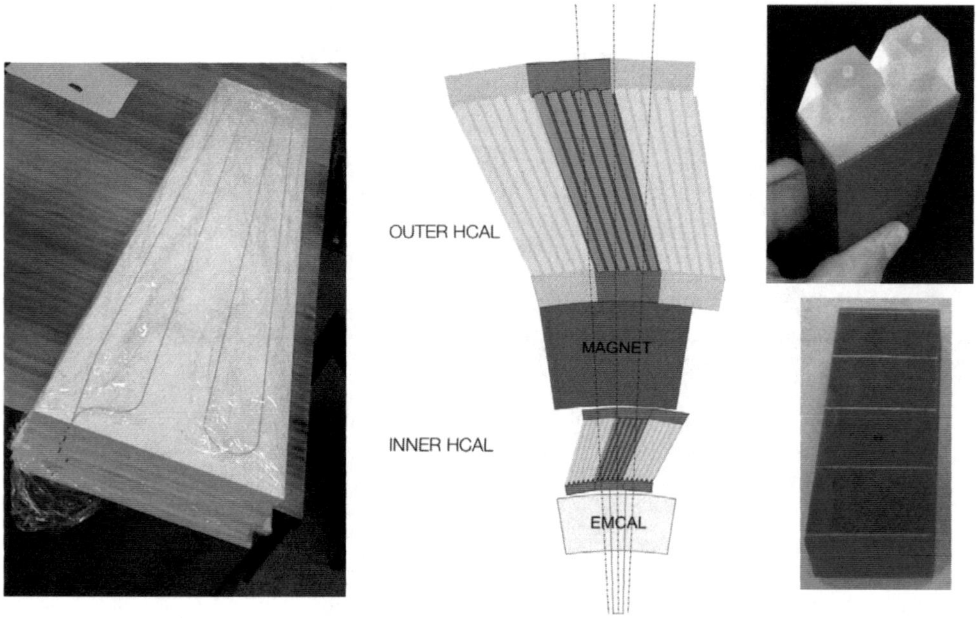

Figure 2. (Left) Fiber routed through scintillator tiles for the outer HCal. (Center) Reference design for the calorimeter system. (Right Top) EMCal tower with light guides attached. (Right Bottom) An EMCal tower with fibers embedded in a tungsten and epoxy. The mesh holding the fibers in place are visible as bands.

technologies proposed that are being fully explored before a final decision is made. The options for an inner tracking system include reusing components of the current PHENIX silicon detector, the SVX [2], or monolithic active pixel sensors (MAPS). The MAPS detector would be designed similar to the ALICE ITS upgrade [3]. Options for the outer tracking system include a new silicon tracker similar to the current PHENIX SVX or a compact Time Projection Chamber. The reference design in the sPHENIX proposal includes seven layers of silicon. However all options are being fully evaluated to determine what best satisfies the sPHENIX tracking criteria, timeline and budget. A tracking review will occur in the Fall of 2016.

2.2. Electromagnetic Calorimeter

The EMCal is designed to measure electrons and photons. The electron identification is important for measuring the different upsilon states. Direct photon measurements at high p_T will be used to identify photon-jet events, which will provide insight into partonic energy loss. The segmentation of the current EMCal design is 0.025 x 0.025 for $\Delta\eta$ x $\Delta\varphi$, which results in a total of 24,576 (96 x 256) readout channels. Each tower is assembled by threading scintillator fibers through screens that are then positioned into a mold. Tungsten powder and epoxy are poured into the mold. After the tower has hardened the mold is removed, a light guide is attached as shown in the top right picture of Figure 2. The signal is then readout by silicon photo-multipliers (SiPMs). The Moliere radius for the EMCal design is approximately 2.3 cm. 1D projective modules have been assembled and studied in prototype testing. However, 2D projective towers, which would improve the electron-pion separation are also under investigation.

Figure 3. (Top left) The steel assembly of the outer HCal. (Bottom Lef) The fully assembled inner HCal. Test beam set up with Emcal, Inner HCal, Magnet Material and Outer HCal.

2.3. Hadronic Calorimeters
The purpose of the HCals is to measure the energy from hadrons produced in the events. The energy in the HCal and EMCal is used to reconstruct the energy of jets produced in the collisions. To provide a jet energy resolution of approximately 20% the required single particle energy resolution for the HCal is $\sigma_E/E < 100\%/\sqrt{E}$.

The HCals are comprised of alternating layers of steel plates and scintillator tiles. The scintillator tiles are 7mm thick polystyrene. A 1mm wave length shifting fiber is embedded into the tiles. An unwrapped outer HCal tile is shown in the left-most photograph in Figure 2. This shows the routing pattern of the fiber used in the prototype. The ends of the fiber come together at the edge of tile and each tile illuminates an SiPM. The SiPMs of five tiles are readout by a single pre-amplifier board as a single tower. This results in 3072 (2 x 24 x 64) total readout channels. The plates are tilted such that a straight line from the center of the detector will hit four tiles in each HCal as illustrated in the middle of Figure 2. Due to the difference in size, this results in a stronger tilt angle for the inner HCal. The plates in the inner HCal are tilted in the opposite direction from the outer HCal.

2.4. Calorimeter Prototype
Prototypes of the EMCal, inner and outer HCal were constructed at BNL and tested at the test beam facility at Fermilab in April 2016. A photo of the prototype is included in Figure 3. The energy resolution of these detectors is being studied at beam energies ranging from 2 to 64 GeV. Analysis of the data collected during the beam test is ongoing.

3. Summary
The scientific case has been established for the need to build the proposed sPHENIX detector as demonstrated in the Nuclear Science Long Range Plan recommendations [1]. At present a total of 58 institutions have joined the sPHENIX collaboration, which held its first official collaboration meeting in December 2015 at Rutgers University. Open meetings for the collaboration, topical groups and detector subsystems are held regularly and participation by new members and those

Journal of Physics: Conference Series **736** (2016) 012027 doi:10.1088/1742-6596/736/1/012027

interested in joining the sPHENIX collaboration is welcomed. Early results of the April 2016 test beam study for the calorimeter prototypes are promising and should be completed soon. Another prototype test for the calorimeters is planned in early 2017. The project is on track to have sPHENIX installed by 2022 and ready to take data. The new detectors and high data rate will allow sPHENIX to make high statistics measurements over a larger kinematic range than previous RHIC experiments. This is essential for measuring important observables such as jet structure, γ-jet and heavy flavor jets as well as separating the three Υ states. sPHENIX will play a crucial role as RHIC completes its scientific mission to understand the properties of the sQGP that it first created over a decade ago.

References

[1] "Reaching For the Horizon," The 2015 Long Range Plan for Nuclear Science, `http://science.energy.gov/~/media/np/nsac/pdf/2015LRP/2015_LRPNS_091815.pdf` (DOE,NSF).

[2] Johann M. Heuser, "Vertex detector upgrade plans for the PHENIX experiment at RHIC", Nuclear Instruments and Methods, Volume 511, Issues 12, 21 September 2003, Pages 210-214, ISSN 0168-9002, http://dx.doi.org/10.1016/S0168-9002(03)01795-9. (http://www.sciencedirect.com/science/article/pii/S0168900203017959)

[3] B. Abelev *et al.* [ALICE Collaboration], "Technical Design Report for the Upgrade of the ALICE Inner Tracking System," J. Phys. G **41**, 087002 (2014). doi:10.1088/0954-3899/41/8/087002

[4] A. Adare *et al.* "sPHENIX, An Upgrade Proposal from the PHENIX Collaboration", http://arxiv.org/pdf/1501.06197v1.pdf